Lecture Notes in Computer Science 12018

More information about this series at http://www.springer.com/series/7408

Farhad Arbab · Sung-Shik Jongmans (Eds.)

Formal Aspects
of Component Software

16th International Conference, FACS 2019
Amsterdam, The Netherlands, October 23–25, 2019
Proceedings

 Springer

Editors
Farhad Arbab
CWI
University of Leiden
Leiden, The Netherlands

Sung-Shik Jongmans
Open University of the Netherlands
Heerlen, The Netherlands

ISSN 0302-9743   ISSN 1611-3349   (electronic)
Lecture Notes in Computer Science
ISBN 978-3-030-40913-5   ISBN 978-3-030-40914-2   (eBook)
https://doi.org/10.1007/978-3-030-40914-2

LNCS Sublibrary: SL2 – Programming and Software Engineering

This Springer imprint is published by the registered company Springer Nature Switzerland AG
The registered company address is: Gewerbestrasse 11, 6330 Cham, Switzerland

Preface

This volume contains the proceedings of the 16th International Conference on Formal Aspects of Component Software (FACS 2019), held during October 23–25, 2019, at the Centrum Wiskunde & Informatica (CWI), Amsterdam, the Netherlands.

Component-based software development proposes sound engineering principles and techniques to cope with the complexity of present-day software systems. However, many challenging conceptual and technological issues remain in component-based software development theory and practice. Furthermore, the advent of service-oriented and cloud computing, cyber-physical systems, and the Internet of Things has brought to the fore new dimensions, such as quality of service and robustness to withstand faults, which require revisiting established concepts and developing new ones.

FACS 2019 is concerned with how formal methods can be applied to component-based software and system development. Formal methods have provided foundations for component-based software through research on mathematical models for components, composition and adaptation, and rigorous approaches to verification, deployment, testing, and certification.

The Program Committee (PC) received 27 submissions. Each submission was reviewed by at least three reviewers. Based on the review reports, and after careful discussion, the PC decided to accept 12 papers (9 regular papers and 3 short papers) for presentation at FACS 2019 and publication in this volume. Additionally, FACS 2019 welcomed Carlo Ghezzi, Kim Larsen, and Wan Fokkink as keynote speakers and hosted Jan Friso Groote and Tim Willemse as tutorial speakers; we are delighted and grateful they accepted our invitations.

We thank the members of the PC for their time and effort to write timely and high-quality reviews. We also thank the members of the FACS Steering Committee for useful suggestions and support. Notably, we thank Eric Madelaine for managing the review process of papers with which both PC chairs had to declare conflicts. Finally, we thank Susanne van Dam and CWI for organizational support.

December 2019

Farhad Arbab
Sung-Shik Jongmans

Organization

Program Committee Chairs

Farhad Arbab   CWI and Leiden University, The Netherlands
Sung-Shik Jongmans   CWI and Open University of the Netherlands, The Netherlands

Steering Committee

Farhad Arbab	CWI and Leiden University, The Netherlands
Luis Barbosa	University of Minho, Portugal
José Luiz Fiadeiro	University of Dundee, UK
Zhiming Liu	Southwest University, China
Markus Lumpe	Swinburne University of Technology, Australia
Eric Madelaine	Inria, France
Peter Ölveczky	University of Oslo, Norway
Jose Proenca	CISTER-ISEP and HASLab-INESC TEC, Portugal

Program Committee

Kyungmin Bae	Pohang University of Science and Technology (POSTECH), South Korea
Christel Baier	TU Dresden, Germany
Luis Barbosa	University of Minho, Portugal
Simon Bliudze	Inria, France
Roberto Bruni	University of Pisa, Italy
Luís Cruz-Filipe	University of Southern Denmark, Denmark
José Luiz Fiadeiro	University of Dundee, UK
Mohamad Jaber	American University of Beirut, Lebanon
Olga Kouchnarenko	University of Franche-Comté, France
Ivan Lanese	University of Bologna, Italy
Kung-Kiu Lau	The University of Manchester, UK
Zhiming Liu	Southwest University, China
Markus Lumpe	Swinburne University of Technology, Australia
Eric Madelaine	Inria, France
Mieke Massink	CNR-ISTI, Italy
Hernan Melgratti	Universidad de Buenos Aires, Argentina
Fabrizio Montesi	University of Southern Denmark, Denmark
Catuscia Palamidessi	Inria, France
Jose Proenca	CISTER-ISEP and HASLab-INESC TEC, Portugal
Jorge A. Pérez	University of Groningen, The Netherlands
Gwen Salaün	Grenoble Alpes University, France

Francesco Santini	University of Perugia, Italy
Jacopo Soldani	University of Pisa, Italy
Anton Wijs	Eindhoven University of Technology, The Netherlands
Shoji Yuen	Nagoya University, Japan
Min Zhang	East China Normal University, China
Peter Ölveczky	University of Oslo, Norway

Additional Reviewers

Ardeshir Larijani, Ebrahim
Arellanes, Damian
Ciancia, Vincenzo
Cledou, Guillermina
Dubut, Jérémy
Dzvoti, Patrick
Feng, Hui
Krishna, Ajay
Laarman, Alfons

Osama, Muhammad
Overbeek, Roy
Qian, Chen
Quatmann, Tim
Rabinia, Amin
Wagner, Loic
Zhang, Yuanrui
Zhao, Liang

Contents

Invited Papers

Modeling Guidelines
for Component-Based Supervisory
Control Synthesis

Martijn Goorden[1], Joanna van de Mortel-Fronczak[1], Michel Reniers[1],
Wan Fokkink[1,2(✉)], and Jacobus Rooda[1]

[1] Eindhoven University of Technology, Eindhoven, The Netherlands
{m.a.goorden,j.m.v.d.mortel,m.a.reniers,j.e.rooda}@tue.nl
[2] Vrije Universiteit, Amsterdam, The Netherlands
w.j.fokkink@vu.nl

Abstract. Supervisory control theory provides means to synthesize supervisors from a model of the uncontrolled plant and a model of the control requirements. Currently, control engineers lack experience with using automata for this purpose, which results in low adaptation of supervisory control theory in practice. This paper presents three modeling guidelines based on experience of modeling and synthesizing supervisors of large-scale infrastructural systems. Both guidelines see the model of the plant as a collection of component models. The first guideline expresses that independent components should be modeled as asynchronous models. The second guideline expresses that physical relationships between component models can be easily expressed with extended finite automata. The third guideline expresses that the input-output perspective of the control hardware should be used as the abstraction level. The importance of the guidelines is demonstrated with examples from industrial cases.

Keywords: Supervisory control synthesis · Automata · Modeling

1 Introduction

The design of supervisors for cyber-physical systems has become a challenge, as these systems include more and more components to control and functions to fulfill, while at the same time market demands require verified safety, decreasing costs, and decreasing time-to-market. Model-based systems engineering methods can help in overcoming these difficulties, see [25].

For the design of supervisors, the supervisory control theory of Ramadge-Wonham [23,24] provides means to synthesize supervisors from a model of the uncontrolled plant (describing what the system *can* do) and a model of the

Supported by Rijkswaterstaat, part of the Dutch Ministry of Infrastructure and Water Management.

© Springer Nature Switzerland AG 2020
F. Arbab and S.-S. Jongmans (Eds.): FACS 2019, LNCS 12018, pp. 3–24, 2020.
https://doi.org/10.1007/978-3-030-40914-2_1

control requirements (describing what the system *may* do). Such a supervisor interacts with the plant by dynamically disabling some controllable events. Then synthesis guarantees by construction that the closed-loop behavior of the supervisor and the plant adheres to all requirements and, furthermore, is nonblocking, controllable, and maximally permissive.

The number of industrial applications of supervisory control theory reported in literature is low. In [38], two reasons are provided for this. First, it refers to the lack of tooling with sufficient computational strength to cope with the size of industrial applications. Second, it mentions the "lack of experience among control engineers with modeling and specification in the framework of automata".

Papers that do publish industrial cases often present only the final model and not the journey to arrive at this model. This makes it hard to disseminate knowledge about modeling a system for the purpose of supervisory control synthesis towards practitioners. A few exceptions exist in literature. The authors of [7,14] have indicated that modeling the system and its requirements is difficult and introduced concepts like, e.g., templates to assist the engineer in modeling correctly, i.e., such that the obtained models exhibit the behavior the engineer intended to model. The description of the case study in [4] is annotated with modeling choices, yet they are not generalized into a modeling method. In [29], a method for modeling cyber-physical systems is presented utilizing template-based modeling of [7,14]. Finally, several modeling guidelines are proposed in [35] based on experience with modeling bagage-handling systems, which is described and modeled in [34].

The purpose of this paper is to provide three modeling guidelines based on experience of modeling and synthesizing supervisors of large-scale infrastructural systems [19,28,29]. The first modeling guideline expresses that independent plant components should be modeled as asynchronous plant models, i.e., having no shared events. The second modeling guideline recommends that physical relationships between component models can be easily expressed with extended finite automata. The third modeling guideline expresses to use the abstraction level of the inputs and outputs of the control hardware for the plant models. These three modeling guidelines extend the set of modeling guidelines previously published in [10,11].

The effect of the first modeling guideline is that each individual plant model is modeled as small as possible. Besides that smaller models are easier to understand and maintain over time, having smaller plant models has a significant positive effect on the efficiency of module-based synthesis techniques, like modular synthesis of [22] and multilevel synthesis of [15], as shown in [11]. The second guideline is a natural extension to the first guideline. It may be that two independent components become dependent by their arrangement in the system, i.e., there is a physical relationship that relates the behavior of these two components together. The result of the second guideline is a set of asynchronous component models for the components and an additional automata modeling the physical relationship. The third guideline helps in determining the right abstraction level of the model. The result of following the guideline is that the first and second guideline are more often applicable.

Requirement specifications in practice often violate the aforementioned guidelines. Although the guidelines may sound somewhat obvious, it required several real-life case studies with supervisory control synthesis, see [19,28,29], to formulate them and grasp their importance.

The paper is structured as follows. Section 2 provides the preliminaries of this paper. Section 3 continues by discussing the guideline concerning the modeling of independent plant components. Section 4 discusses how to model the physical dependencies between otherwise independent plant components. In Sect. 5, the guideline concerning the input-output perspective is discussed. The paper concludes with Sect. 6.

2 Preliminaries

This section provides a brief summary of concepts related to automata and supervisory control theory relevant for this paper. These concepts are taken from [2,38]. We first explain supervisory control synthesis for automata conceptually. Sections 2.1–2.3 introduce these concepts formally.

The supervisory control theory of Ramadge-Wonham [23,24] provides means to synthesize supervisors from an automaton model of the uncontrolled plant and an automaton model of the control requirements. For industrial-size systems, the plant model and requirement model are each composed of smaller models describing a component of a system or a part of the desired behavior, respectively, where the smaller models synchronize by shared events. When a system controlled by a supervisor adheres to all specified requirements, the supervisor is called *safe*.

A supervisor interacts with the plant by dynamically disabling events. For the purpose of supervisory control synthesis, all events are classified either as controllable or as uncontrollable. Controllable events may be disabled by the supervisor, such as turning an actuator on; uncontrollable events may not be disabled by the supervisor, such as a sensor switching value. A supervisor adhering to this notion is called *controllable*.

The automata of the plant and requirement models also have marked states, which represent a final state or a safe mode-of-operation. It is desired that a controlled system should always be able to reach at least one of the marked states. A supervisor ensuring this is called *nonblocking*.

Finally, a trivial, yet undesired, supervisor is often one that disables all controllable events in order to be safe, controllable, and nonblocking. Therefore, a more desired supervisor is one that restricts the system only when it is needed to enforce safety, controllability, and nonblockingness. Such a supervisor is called *maximally permissive*. Supervisory control synthesis guarantees by construction that the supervisor is safe, nonblocking, controllable, and maximally permissive.

2.1 Finite Automata

An automaton is a 5-tuple $G = (Q, \Sigma, \delta, q_0, Q_m)$, where Q is the (finite) state set, Σ is the (finite) set of events also called the alphabet, $\delta : Q \times \Sigma \to Q$ the partial function called the transition function, $q_0 \in Q$ the initial state, and $Q_m \subseteq Q$ the set of marked states. The alphabet $\Sigma = \Sigma_c \cup \Sigma_u$ is partitioned into sets containing the controllable events (Σ_c) and the uncontrollable events (Σ_u), and Σ^* is the set of all finite strings of events in Σ, including empty string ε.

We denote with $\delta(q, \sigma)!$ that there exists a transition from state $q \in Q$ labeled with event σ, i.e., $\delta(q, \sigma)$ is defined. The transition function can be extended in the natural way to strings as $\delta(q, s\sigma) = \delta(\delta(q, s), \sigma)$ where $s \in \Sigma^*$, $\sigma \in \Sigma$, and $\delta(q, s\sigma)!$ if $\delta(q, s)! \wedge \delta(\delta(q, s), \sigma)!$. We define $\delta(q, \varepsilon) = q$ for the empty strings. The language generated by the automaton G is $\mathcal{L}(G) = \{s \in \Sigma^* \mid \delta(q_0, s)!\}$ and the language marked by the automaton is $\mathcal{L}_m(G) = \{s \in \Sigma^* \mid \delta(q_0, s) \in Q_m\}$.

A state q of an automaton is called reachable if there is a string $s \in \Sigma^*$ with $\delta(q_0, s)!$ and $\delta(q_0, s) = q$. A state q is coreachable if there is a string $s \in \Sigma^*$ with $\delta(q, s)!$ and $\delta(q, s) \in Q_m$. An automaton is called nonblocking if every reachable state is coreachable.

Two automata can be combined by synchronous composition. In a synchronous composition, transitions labeled with shared events have to be executed simultaneously.

Definition 1. Let $G_1 = (Q_1, \Sigma_1, \delta_1, q_{0,1}, Q_{m,1})$, $G_2 = (Q_2, \Sigma_2, \delta_2, q_{0,2}, Q_{m,2})$ be two automata. The synchronous composition of G_1 and G_2 is defined as

$$G_1 \parallel G_2 = (Q_1 \times Q_2, \Sigma_1 \cup \Sigma_2, \delta_{1\parallel 2}, (q_{0,1}, q_{0,2}), Q_{m,1} \times Q_{m,2})$$

where

$$\delta_{1\parallel 2}((x_1, x_2), \sigma) = \begin{cases} (\delta_1(x_1, \sigma), \delta_2(x_2, \sigma)) & \text{if } \sigma \in \Sigma_1 \cap \Sigma_2, \delta_1(x_1, \sigma)!, \\ & \text{and } \delta_2(x_2, \sigma)! \\ (\delta_1(x_1, \sigma), x_2) & \text{if } \sigma \in \Sigma_1 \setminus \Sigma_2 \text{ and } \delta_1(x_1, \sigma)! \\ (x_1, \delta_2(x_2, \sigma)) & \text{if } \sigma \in \Sigma_2 \setminus \Sigma_1 \text{ and } \delta_2(x_2, \sigma)! \\ \text{undefined} & \text{otherwise.} \end{cases}$$

Synchronous composition is associative and commutative up to reordering of the state components in the composed state set. Two automata are called asynchronous if no events are shared, i.e., they do not synchronize over any event.

A composed system $\mathcal{G}$ is a collection of automata, i.e., $\mathcal{G} = \{G_1, \ldots, G_m\}$. The synchronous composition of a composed system $\mathcal{G}$, denoted by $\parallel \mathcal{G}$, is defined as $\parallel \mathcal{G} = G_1 \parallel \cdots \parallel G_m$, and the synchronous composition of two composed systems $\mathcal{G}_1 \parallel \mathcal{G}_2$ is defined as $(\parallel \mathcal{G}_1) \parallel (\parallel \mathcal{G}_2)$. A composed system $\mathcal{G} = \{G_1, \ldots, G_m\}$ is called a product system if the alphabets of the automata are pairwise disjoint, i.e., $\Sigma_i \cap \Sigma_j = \emptyset$ for all $i, j \in [1, m], i \neq j$ [24].

Finally, let G and K be two automata with the same alphabet Σ. K is said to be controllable with respect to G if, for every string $s \in \Sigma^*$ and $u \in \Sigma_u$ such that $\delta_K(q_{0,K}, s)!$ and $\delta_G(q_{0,G}, su)!$, it holds that $\delta_K(q_{0,K}, su)!$, where the subscript G refers to elements of G and subscript K refers to elements of K.

2.2   Extended Finite Automata

In [32], extended finite automata (EFAs) are introduced for modeling systems, which are FAs augmented with bounded discrete variables. An EFA is a 7-tuple $E = (L, V, \Sigma, \rightarrow, l_0, v_0, L_m)$, where L is the (finite) location set, V the set of variables, Σ is the (finite) set of events also called the alphabet, $\rightarrow$ the extended transition relation, $l_0 \in L$ the initial location, v_0 the initial valuation, and $L_m \subseteq L$ the set of marked locations.

In an EFA, the transition relation is enhanced with guard expressions (conditions) and variable assignments (updates). Formally, the extended transition relation is $\rightarrow: L \times C \times \Sigma \times U \times L$, where C is the set of all conditions and U the set of all updates. A transition is enabled if the associated condition evaluates to true for the current variables valuation. After taking a transition, the variables valuation is updated according to the associated update.

A condition is a Boolean expression constructed from discrete variables, location variables, constants, the Boolean literals true ($\mathbf{T}$) and false ($\mathbf{F}$), and the usual arithmetical operators and logical connectives, see [20]. A location variable is a reference to a location, denoted by A.l, where A is the automaton name and l a location of automaton A. It evaluates to $\mathbf{T}$ when A is in location l.

An update consists of zero or more variable assignments of the form $v_b := c$, where ':=' denotes an assignment of the value of c to variable v_b. It is not allowed for an update to have multiple assignments for the same variable.

Two EFAs can be combined by computing the synchronous product as defined in [32]. The state of an EFA is the combination of the active location and current variables valuation. With respect of FAs, two EFAs are now called asynchronous if they do not share events, variables, or location variables.

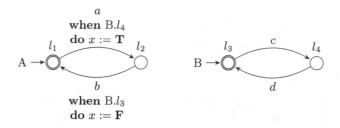

Fig. 1. An example of two EFAs.

Figure 1 shows an example of two EFAs. As shown with EFA A, keyword **when** indicates the condition of the transition and keyword **do** indicates the update. In EFA B the condition and update are omitted. An omitted condition indicates that the condition for that transition is $\mathbf{T}$. An omitted update indicates an 'I don't care' update, i.e., the value of the variables is updated by another synchronizing transition or, when no synchronizing transitions update the variable, the value remains the same.

State-based expressions are introduced in [16,17] as a modeling formalism more closely related to the textual formulation of control requirements. The state-event expression e **needs** c formulates that event e is only enabled when condition c evaluates to **T**. The EFA representation of a state-event expression is shown in Fig. 2, such that the synchronous product of two EFAs can be used to synchronize state-based expressions with EFAs or other state-based expressions.

Fig. 2. The EFA representation of state-event expression e **needs** c.

2.3   Supervisory Control Theory

The objective of supervisory control theory is to design an automaton called a supervisor which function is to dynamically disable controllable events so that the closed-loop system of the plant and the supervisor obeys some specified behavior, see [2,23,24,38]. More formally, given a plant model P and requirement model R, the goal is to synthesize supervisor S that adheres to the following control objectives.

- *Safety*: all possible behavior of the closed-loop system $P \parallel S$ should always satisfy the imposed requirements, i.e., $\mathcal{L}(P \parallel S) \subseteq \mathcal{L}(P \parallel R)$
- *Controllability*: uncontrollable events may never be disabled by the supervisor, i.e., S is controllable with respect to P.
- *Nonblockingness*: the closed-loop system should be able to reach a marked state from every reachable state, i.e., $P \parallel S$ is nonblocking.
- *Maximal permissiveness*: the supervisor does not restrict more behavior than strictly necessary to enforce safety, controllability, and nonblockingness, i.e., for all other supervisors S' satisfying safety, controllability, and nonblockingness it holds that $\mathcal{L}(P \parallel S') \subseteq \mathcal{L}(P \parallel S)$.

Given a composed system representation of the plant $\mathcal{P} = \{P_1, \ldots, P_m\}$ and a collection of requirements $\mathcal{R} = \{R_1, \ldots, R_n\}$, we define the tuple $(\mathcal{P}, \mathcal{R})$ as the *control problem* for which we want to synthesize a supervisor. Furthermore, in the context of supervisory control synthesis we call each model $P_i \in \mathcal{P}$ a *component* model, to differentiate it from the plant model $P = \parallel \mathcal{P}$.

In this paper, three different synthesis techniques are discussed: monolithic synthesis, modular synthesis, and multilevel synthesis. These synthesis techniques are introduced below.

Monolithic supervisory control synthesis results in a single supervisor S from a single plant model and a single requirement model, see [23] or, in case of EFAs, see [20]. There may exist multiple automata representations of the maximally permissive, safe, controllable, and nonblocking supervisor. When the plant model and the requirement model are given as a composed system $\mathcal{P}$ and $\mathcal{S}$, respectively, the monolithic plant model P and requirement model R are obtained by performing the synchronous composition of the models in the respective composed system.

Modular supervisory control synthesis uses the fact that the desired behavior is often specified with a collection of requirements $\mathcal{R}$ [37]. Instead of first transforming the collection of requirements into a single requirement, as monolithic synthesis does, modular synthesis calculates for each requirement a supervisor based on the plant model. In other words, given a control problem $(\mathcal{P}, \mathcal{R})$ with $\mathcal{R} = \{R_1, \ldots, R_n\}$, modular synthesis solves n control problems $(\mathcal{P}, \{R_1\}), \ldots, (\mathcal{P}, \{R_n\})$. Each control problem $(\mathcal{P}, \{R_i\})$ for $i \in [1, n]$ results in a safe, controllable, nonblocking, and maximally permissive supervisor S_i. Unfortunately, the collection of supervisors $\mathcal{S} = \{S_1, \ldots, S_n\}$ can be conflicting, i.e., $P \parallel S_1 \parallel \ldots \parallel S_n$ can be blocking. A nonconflicting check can verify whether $\mathcal{S}$ is nonconflicting, see [6,18,21]. In the case that $\mathcal{S}$ is nonconflicting, $\mathcal{S}$ is also safe, controllable, nonblocking, and maximally permissive for the original control problem $(\mathcal{P}, \mathcal{R})$ [37]. In the case that $\mathcal{S}$ is conflicting, an additional coordinator C can be synthesized such that $\mathcal{S} \cup \{C\}$ is safe, controllable, nonblocking, and maximally permissive for the original control problem $(\mathcal{P}, \mathcal{R})$, see [33].

An extension to this approach, as proposed by [22], states that instead of synthesizing each time with the complete plant $\mathcal{P}$, it suffices to only consider those automata that relate to the requirement that is considered. This extension is used in the remainder of this paper.

Multilevel supervisory control synthesis is inspired by decompositions of systems by engineers [15]. For each subsystem, a supervisor is synthesized based on requirements for only those subsystems. For synthesis, this resembles modular supervisory control in the sense that for multilevel synthesis requirements related to the same subsystem are grouped together before synthesis is performed, and a supervisor is synthesized for each such subsystem. Again, the collection of synthesized supervisors may be conflicting.

Requirements relate different component models, as events and variables mentioned in a requirement should originate from the component models. Multilevel synthesis allows to apply synthesis to a subsystem of component and requirement models, as long as all component models related to these requirement models are included in this subsystem. Therefore, it is important to formulate small requirement models, as shown in [11].

3   Modeling Independent Components

The first modeling guideline concerns the modeling of the plant. Industrial systems consist of numerous components or subsystems, of which many are clearly acting asynchronously in the uncontrolled situation. Consider for example two conveyor belts after each other, each actuated by its own motor. In the uncontrolled situation, these actuators can behave independently of each other. For the plant model, these two actuators are modeled by two asynchronous automata. Therefore, the first guideline is formulated as follows.

Model independent plant components as asynchronous component models.

Plant components that have no relationship with each other, should not be combined into a single component model. A single component model suggests a relationship, which is absent in this case. Having asynchronous models (i.e., no shared events, variables, or location variables) increases readability of the model, but also allows divide-and-conquer strategies to synthesize supervisors for smaller subsystems. We illustrate this with two examples.

3.1 Autonomous Robot

In this section, the modeling guideline will be illustrated with an industrial example. Consider an autonomous omnidirectional robot that can move on a factory floor along a grid, described by the application published in [8]. The goal of the supervisor is to ensure safe operation of the robot on the factory floor. We want to model the pose of the robot, i.e., the combination of position along the x-axis, position along the y-axis, and orientation of the front of the robot.

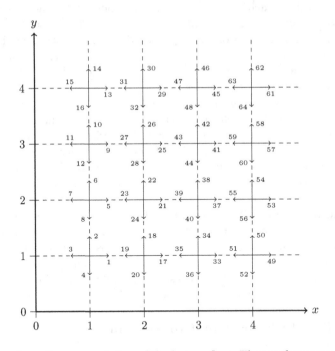

Fig. 3. The schematic representation of the factory floor. The numbers represent states, with a state being the combination of x-position, y-position, and orientation.

Figure 3 shows the schematic representation of the factory floor, where the x-axis, y-axis, and orientation are each discretized into four possible values. Each arrow indicates a pose of the robot: an x-position, y-position, and orientation.

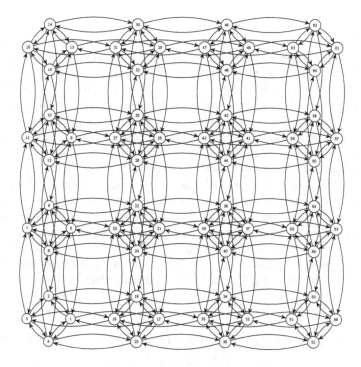

Fig. 4. The factory floor modeled as a single plant model P. Event labels are not depicted.

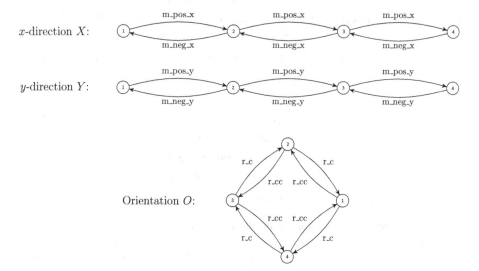

Fig. 5. The factory floor modeled as three asynchronous component models.

These poses are the states of the robot. The number indicates the state number that will be used in modeling this environment. Note that in this example the initial state is not explicitly modeled. Any state could act as the initial state.

Figure 4 shows how this factory floor could be modeled as a single component model P along the lines of [8]. This dense plant model is hard to read. The way this model is depicted unveils that there is structure in this system, which could be exploited further. Figure 5 shows three asynchronous component models X, Y, and O that in a synchronous composition model exactly the same behavior, i.e., $P = X \parallel Y \parallel O$.

For modular and multilevel synthesis, having multiple asynchronous component models instead of a single large model is an advantage. This can be demonstrated with the following requirements. Suppose that the autonomous omnidirectional robot may only move in a certain direction if it is oriented in that direction. This requirement can be formalized as follows. Requirements $R_1, \ldots, R_4$ use the single component model P, while $R'_1, \ldots, R'_4$ use the asynchronous component models X, Y, and O. Understanding and assessing the correctness of requirements $R_1, \ldots, R_4$ is more difficult than that of requirements $R'_1, \ldots, R'_4$.

$R_1:$ m_pos_x **needs** $P.1 \vee P.5 \vee P.9 \vee P.13 \vee P.17 \vee P.21 \vee P.25 \vee P.29$
$$\vee\ P.33 \vee P.37 \vee P.41 \vee P.45$$

$R_2:$ m_neg_x **needs** $P.19 \vee P.23 \vee P.27 \vee P.31 \vee P.35 \vee P.39 \vee P.43 \vee P.47$
$$\vee\ P.51 \vee P.55 \vee P.59 \vee P.63$$

$R_3:$ m_pos_y **needs** $P.2 \vee P.6 \vee P.10 \vee P.18 \vee P.22 \vee P.26 \vee P.34 \vee P.38$
$$\vee\ P.42 \vee P.50 \vee P.54 \vee P.58$$

$R_4:$ m_neg_y **needs** $P.8 \vee P.12 \vee P.16 \vee P.24 \vee P.28 \vee P.32 \vee P.40 \vee P.44$
$$\vee\ P.48 \vee P.56 \vee P.60 \vee P.64$$

$R'_1:$ m_pos_x **needs** $O.1$
$R'_2:$ m_neg_x **needs** $O.3$
$R'_3:$ m_pos_y **needs** $O.2$
$R'_4:$ m_neg_y **needs** $O.4$

In case of the single component model P, we obtain the four control problems $(P, R_i), i \in [1, 4]$. In case of the asynchronous component models X, Y, and O, we obtain the four control problems $(X \parallel O, R'_1)$, $(X \parallel O, R'_2)$, $(Y \parallel O, R'_3)$, and $(Y \parallel O, R'_3)$. Table 1 shows numerical results for synthesizing modular supervisors for the control problems mentioned before. For each supervisor, the number of states and transitions is mentioned. By modeling the subsystem as a set of asynchronous automata models, a reduction in the size of the supervisors is obtained. This reduction can be even more significant if a finer discretization

Table 1. Experimental results for synthesizing modular supervisors with the single component model and the multiple components model, with the monolithic supervisor as reference. The states and transitions are of the state space of each supervisor and $i \in \{1, 2, 3, 4\}$.

Model	Supervisor	States	Transitions
Single plant	S_i	64	284
Multiple plants	S_i'	16	47
Monolithic supervisor	S	64	176

is used. Assume that both the x and y directions are discretized in k values. Each of the four synthesized supervisors using the single component model has $4k^2$ states and $21k^2 - 13k$ transitions; each of the four synthesized supervisors using asynchronous component models has $4k$ states and $13k - 4$ transitions. So, instead of the supervisors growing quadratic in k, the overall state space can be reduced to only growing linearly in k.

3.2 Waterway Lock

In this section, the effect of the modeling guideline on the efficiency of module-based synthesis techniques is demonstrated with a large-scale industrial example. Consider a waterway lock in a river or a canal, which is an infrastructural system that maintains a difference in water levels at both sides while also allowing ships to go from one water level to the other water level. Such a system consists of actuators, such as motors to open gates, sensors, such as measuring whether a gate is open, traffic lights, to communicate with vessels, and buttons, for an operator to interact with the system.

A model of Lock III, located in Tilburg, the Netherlands, is presented in [28]. This model adheres to the proposed modeling guideline of using asynchronous component models for independent plant components. This model is adjusted such that it ignores the modeling guideline. For example, all independent components of the gate actuators on the upstream side of the lock are combined.

Table 2. Experimental results for synthesizing modular and multilevel supervisors with the models of Lock III violating or adhering to the modeling guideline.

Model	Components	Supervisor	States
Violating the guideline	35	Monolithic	$6.0 \cdot 10^{24}$
		Multilevel	$1.4 \cdot 10^{21}$
		Modular	$1.8 \cdot 10^{9}$
Adhering to the guideline	51	Monolithic	$6.0 \cdot 10^{24}$
		Multilevel	$3.5 \cdot 10^{7}$
		Modular	$6.8 \cdot 10^{5}$

Table 2 shows experimental results of synthesizing supervisors for the two different versions of the model of Lock III. By adhering to the modeling guideline, the number of component models increases from 35 automata to 51 automata. This has a significant effect on the efficiency of multilevel and modular synthesis. For the model violating the guideline, the combined size of the supervisors, which is the sum of the size of each individual supervisor, for multilevel synthesis is $1.4 \cdot 10^{21}$ states, which can be significantly reduced to $3.5 \cdot 10^7$ states if one adheres to the modeling guideline. If one adheres to the modeling guideline and deploys modular synthesis, the combined size of the supervisors can be even reduced to $6.8 \cdot 10^5$ states. These results clearly indicate the relevance of the proposed modeling guideline in practice.

4 Modeling Physical Relations

The second modeling guideline concerns the modeling of physical relations between components or subsystems. For cyber-physical systems, most actuators and sensors behave independent of each other, see the first modeling guideline in Sect. 3. Yet, some actuators and sensors are related with each other through the physical design of the component or subsystem. For example, consider a hydraulic arm, which can extend and retract, and two sensors measuring the end position, one for the fully extended position of the arm and one for the fully retracted position. If no faults occur, then these two sensors are never activated at the same time, as it is physically impossible that the hydraulic arm is fully extended and fully retracted at the same time.

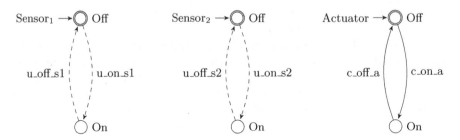

$R:$ c_off_a **needs** $Sensor_1.On \wedge Sensor_2.On$

Fig. 6. Models to illustrate the issue with omitting physical relations. Solid arrows indicate transitions labeled with controllable events, dashed transitions indicate transitions labeled with uncontrollable events.

In [39] the importance of modeling physical relationships is shown. The authors argue that models that are nonblocking, like a synthesized supervisor, may become blocking when they are implemented on actual control hardware.

This is illustrated with the models shown in Fig. 6. In this example, two sensors and an actuator are modeled without any physical relationship between them (so the modeling guideline from Sect. 3 has been applied). Requirement R expresses that the actuator may only be turned off when both sensors are on. A safe, nonblocking, controllable, and maximally permissive supervisor synthesized from these plant and requirement models disables event c_off_a when both sensors are not on at the same time and enables always all other events. When this supervisor is implemented on the system where a physical relation ensures that both sensors can never be on at the same time, the controlled system is no longer nonblocking. As event c_on_ is always enabled, the actuator can reach the state On. Subsequently, event c_off_a is permanently disabled by the physical relation between the sensors, so the actuator cannot reach a marked location from the reachable location On.

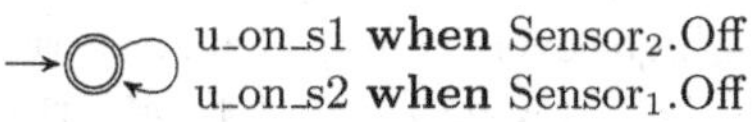

u_on_s1 **when** $Sensor_2$.Off
u_on_s2 **when** $Sensor_1$.Off

Fig. 7. The EFA model representing the physical relationship between the two sensors in the example of Fig. 6. In this drawing, the two transitions, each labeled with a different event, are visualized with only a single edge as they have the same source and target state.

EFAs are very suitable to include physical relationships into the plant model. By deploying EFAs, the actuators and sensors can be first modeled as if they do not have any physical relationship, resulting in asynchronous component models. Subsequently, a component model can be added explicitly, modeling the physical relationship. Figure 7 models the physical relationship between the two sensor models from the example in Fig. 6. This model shows clearly that $Sensor_1$ can only go on when $Sensor_2$ is off and vice versa. This example demonstrates that using EFAs for modeling physical relationships provides a clear and well maintainable model. Therefore, the second guideline is formulated as follows.

Model physical relationships between components with EFAs.

The proposed method of first modeling sensors and actuators with asynchronous component models and subsequently modeling the physical relationship with EFAs maintains the component-based modeling approach. Three other modeling approaches used in literature do not adhere to the proposed modeling guideline. The first method is to model the physically related components directly as a single component model, see, for example, the several sensors in the model of Lock III [28]. The second method is to model the physical relationship with an additional FA model, which is essentially the first method yet now keeping the original component models, see, for example, the interaction between actuators and sensors in the model of an MRI scanner [36]. The third method is to model the physical relationship directly in one of the related components, see, for example, the relationship between sensors and actuators in the model of Lock III [28].

The proposed modeling guideline has no impact on the efficiency of module-based synthesis algorithms like modular and multilevel synthesis. When synthesis is performed for a particular (set of) requirement(s), not only the directly related component models are selected, but also those indirectly related. Therefore, the models obtained by following the guideline or the other three methods mentioned in the paragraph above all have the same state-space representation in their synchronous product of the component models. The advantage of using the proposed modeling guideline is primarily in ease of modeling, understanding the model, and adjusting the model.

5   Modeling with the Input-Output Perspective

The third modeling guideline concerns the abstraction level of the model. Choosing the 'right' abstraction level for the model is often not straightforward. Often systems are modeled with high-level events, such as starting a machine, handing over a product to a buffer, or moving a robot to a certain location. In [1], the implementation of the supervisor on control hardware is considered, leading to a so-called input-output perspective modeling approach. With this perspective, events relate to the change of signal value sent to actuators or received from sensors. Furthermore, all events related to actuators are controllable and all events related to sensors are uncontrollable. It turns out that this input-output perspective has several advantages, which is explained next. Therefore, the third guideline is formulated as follows.

Use the abstraction level of the inputs and outputs of the control hardware for the plant model.

For the case studies with infrastructural systems, the goal was to eventually deploy the synthesized supervisor on hardware. Choosing the abstraction level of the inputs and outputs of the control hardware allows for the generation of control code, see [3].

Furthermore, this abstraction level leads to many small and loosely coupled models of the sensors and actuators, based on just a few templates, as introduced in [14]. Supervisory control synthesis benefits from having (almost) a product system, see Sect. 3 and the work of [5,9,12,26,35]. In software engineering, this modeling method is called component-based modeling, see [13].

Using the input-output perspective for modeling can ultimately result in skipping synthesis completely, as shown in [9]. By using the input-output perspective, textual control requirements formulated by engineers can be more easily translated into models, as the states of actuators and sensors are directly available in the plant model. This turns out to be beneficial for supervisory control synthesis, as the plant models and requirement models together already form a safe, controllable, nonblocking, and maximally permissive supervisor and no synthesis is needed.

Another modeling method called product-based modeling should be avoided when possible. An example of a model with this abstraction level is the wafer scanner logistics model of [31]. It was not possible to synthesize a monolithic supervisor for this model. In the PhD thesis [30], the wafer scanner is modeled on the action level without products (towards the input-output perspective, yet not fully there). For this adapted model, a monolithic supervisor has been synthesized. In Sect. 5.1 this example will be discussed in more detail.

5.1   Industrial Examples

In this section, modeling with the input-output perspective is demonstrated. For this purpose, three different case studies are discussed.

Production Line Buffer

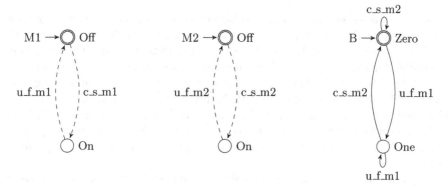

Fig. 8. A model of two machines M1 and M2 and a connecting buffer B. The letter c is an abbreviation for controllable, u for uncontrollable, s for start, and f for finish.

In the first example in Fig. 8, we model two machines M1 and M2 and a buffer B connecting the two machines, adapted from [38]. The two machines display similar behavior. The two states indicate that the machine is either on or off. A controllable event can start the machine, and an uncontrollable event indicates that the machine has finished. The start and finish events are also used in the model of the buffer. A product is placed in the buffer when the first machine is finished, and a product is taken from the buffer when the second machine starts. As can be seen, a high abstraction level is chosen to model the system.

The FESTO production line, as described and modeled in [27], also contains a buffer to temporarily store products between two work stations. This example shows which actuators and sensors are connected to the inputs and outputs of the control hardware. There is an actuator A1 present to move a product from the previous work station to the buffer, an actuator A2 to move a product from the buffer to the next work station, a sensor S1 located at the entrance of the buffer to measure whether the buffer is full, and a sensor S2 located at the exit of the buffer to measure whether the buffer is empty. The component models are shown in Fig. 9.

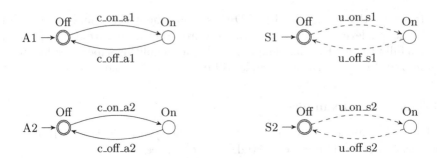

Fig. 9. An alternative model of the buffer.

Several differences can be observed between the component models in Figs. 8 and 9. First, in the high-level perspective model the machines in the proximity of the buffer are modeled, while in the input-output perspective actuators responsible for the movement of products are modeled. Secondly, the events in the high-level perspective have a complex meaning, like u_f_m1 representing that machine 1 has finished production and the product is placed in the buffer. And third, maybe the most important observation is that all component models in the high-level perspective are connected by shared events, while all component models in the input-output perspective are asynchronous and form a product system.

Requirements are formulated that express that the buffer may not overflow or underflow. Requirements R_1 and R_2 below are formulated for the high-level perspective model and requirements R_3 and R_4 for the input-output perspective model.

$$R_1 : \quad \text{c\_f\_m1 } \textbf{needs} \text{ B.Zero}$$
$$R_2 : \quad \text{c\_s\_m2 } \textbf{needs} \text{ B.One}$$

$$R_3 : \quad \text{c\_on\_a1 } \textbf{needs} \text{ S1.Off}$$
$$R_4 : \quad \text{c\_on\_a2 } \textbf{needs} \text{ S2.On}$$

While these requirements are very similar in form, the input-output perspective model and its requirements satisfy the Controllable and Nonblocking Modular Supervisors Properties as presented in [9]. Therefore, no synthesis is needed for this model and the component and requirement models are together already modular supervisors.

Waterway Traffic Light

The second example is a traffic light from a waterway lock, inspired by [28]. Such a traffic light is used to communicate with vessels whether they are allowed to enter the lock. The traffic light consists of three lamps, see Fig. 10: a red one, a green one, and another red one. Four aspects, i.e., combinations of lamps turned on, have the following legal meaning in the communication with vessels.

- *Double red aspect.* This aspect is formed by having both red lamps on and the green lamp off. It indicates that the lock is out-of-service.
- *Red aspect.* This aspect is formed by having the top red lamp on and the green and bottom red lamps off. It indicates that vessels are not allowed to enter the lock from this side of the waterway.
- *Red-green aspect.* This aspect is formed by having the top red and green lamps on and the bottom red lamp off. It indicates to vessels that they may enter the lock soon, so captains should prepare their vessels.
- *Green aspect.* This aspect is formed by having the green lamp on and both red lamps off. It indicates that vessels are allowed to enter the lock.

Fig. 10. The aspects of the lock traffic light: double red, red, red-green, and green. (Color figure online)

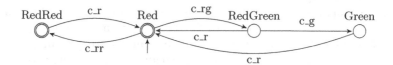

Fig. 11. The model of the traffic light as proposed in [28].

Figure 11 shows the model of the traffic light as proposed in [28]. It uses the four aspects of the traffic light as states and defines possible transitions between them. The events on the transitions do not correspond directly to a value change in one of the input or output signals of the control hardware.

Figure 12 shows the models of the traffic light when the input-output perspective is followed. Each lamp in the traffic light can be actuated separately, resulting in three asynchronous models TopRed, Green, and BottomRed. Each event now relates to a value change in the output signal of the controller hardware.

Several differences can be observed between the models in Fig. 11 and the models in Fig. 12. First, each model created with the input-output perspective is

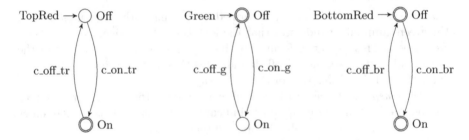

Fig. 12. The models of the traffic light using the input-output perspective. (Color figure online)

smaller than the single model with the aspect perspective, which makes understanding the model easier. Second, the three models with the input-output perspective are indeed a product system, so benefits in performance of synthesis as discussed in Sect. 3 also apply in this case. Third, the plant model with the input-output perspective describes more behavior, as it includes also illegal aspects of the traffic light. Therefore, the modeler needs explicitly exclude these illegal aspects with requirement models. These illegal aspects (and transitions to these aspects) have already been removed in the model in Fig. 11.

In this particular example, there exists a injective mapping between states of the model in Fig. 11 and states in the models in Fig. 12. For example, the state RedRed maps to state TopRed.On, Green.Off, and BottomRed.On; and the steed Green maps to RedRed.Off, Green.On, and BottomRed.Off. Mappings for states Red and RedGreen can be derived similarly.

Wafer Scanner Logistics

The third example is a model of the wafer logistics in a lithography scanner, see [31]. A lithography scanner exposes silicon wafers to manufacture integrated circuits. Besides exposing a wafer, several pre- and postprocessing steps are performed in a lithography scanner, such as conditioning, aligning and measuring. These processing steps are performed multiple times before the integrated circuits on the wafer are finished. The goal of the supervisory controller is to properly manage the wafer logistics in such a scanner.

The wafer scanner logistics model of [31] deploys a product-based modeling perspective, where the products that go through the manufacturing process are modeled, as well as all actions that are possible on the products. Figure 13 shows two of the component models of the wafer scanner logistics. The model ObsAligned_j models for each wafer $j \in J$ in the system, with J the set of all wafers, whether it is aligned or not. This component model represents a property of a product in the system. The model ReqOccupied_CH0 keeps track whether the resource CH0 is occupied by a wafer or not. As each wafer $j \in J$ may occupy this resource, this automaton needs to be able to synchronize with events from all wafers, which is in short denoted by $*$ in the model, e.g.,

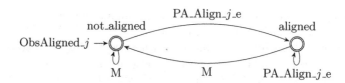

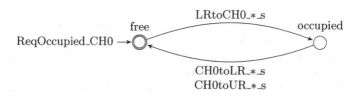

Fig. 13. The model ObsAligned_j of the alignment status of wafer j (left) and the model ReqOccupied_CH0 of the occupation of resource CH0 (right), both taken from [31].

LRtoCH0_ * _s = {LRtoCH0_j_s | $j \in J$}. This model represents a property of a resource in the system.

The product-based modeling perspective results in all component models being connected, as they need to synchronize in shared events to track the different products through the system. This is detrimental to the applicability of synthesis, as mentioned in [30], since synthesizing a monolithic supervisor was not possible. Also, due to these strongly connected component models, modular and multilevel synthesis will not ease synthesis, as for each (group of) requirement(s) all component models need to be taken into account during synthesis.

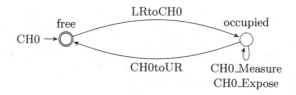

Fig. 14. The model CH0 of resource CH0, taken from [31].

In the PhD thesis [30] the explicit models of the products are removed and the remaining component models of the resources rewritten. This means that the model ObsAligned_j from Fig. 13 is no longer included in the adapted model. The model ReqOccupied_CH0 from Fig. 13 of resource CH0 is rewritten into CH0, as shown in Fig. 14. The events related to each wafer j are replaced by generalized events. Furthermore, the events CH0_Measure and CH0_Expose on the self-loop in location occupied originated from another component model, not shown in Fig. 13, which is removed in the adapted model. Now, for the adapted model a monolithic supervisor can be synthesized having 2190 states and 6969 transitions. This is a significant synthesis performance increase, as no supervisor could be synthesized for the product-based perspective model.

6   Conclusion and Future Work

This paper presents three guidelines for modeling systems for which a supervisory controller needs to be synthesized. The first one expresses that independent plant components should be modeled as asynchronous plant models. The second one recommend that physical relationships between component models can be easily expressed with extended finite automata. The third one expresses that the input-output perspective of the control hardware should be used for the plant models. Examples from practice show how the guidelines can be used and that they can result in a considerable increase in performance of supervisory control synthesis.

Acknowledgments. The authors thank Maria Angenent, Bert van der Vegt, and Han Vogel from Rijkswaterstaat for their feedback on the results.

References

1. Balemi, S.: Control of discrete event systems: theory and application. Ph.D. thesis, Swiss Federal Institue of Technology Zurich, Zurich (1992)
2. Cassandras, C.G., Lafortune, S.: Introduction to Discrete Event Systems, 2nd edn. Springer, Boston (2008). https://doi.org/10.1007/978-0-387-68612-7
3. Fabian, M., Hellgren, A.: PLC-based implementation of supervisory control for discrete event systems. In: 37th IEEE Conference on Decision and Control, vol. 3, pp. 3305–3310 (1998). https://doi.org/10.1109/CDC.1998.758209
4. Fabian, M., Fei, Z., Miremadi, S., Lennartson, B., Åkesson, K.: Supervisory control of manufacturing systems using extended finite automata. In: Campos, J., Seatzo, C., Xie, X. (eds.) Formal Methods in Manufacturing, pp. 295–314. Taylor & Francis Inc., Industrial Information Technology (2014)
5. Feng, L., Wonham, W.M.: Nonblocking coordination of discrete-event systems by control-flow nets. In: 46th IEEE Conference on Decision and Control, pp. 3375–3380. https://doi.org/10.1109/CDC.2007.4434160
6. Flordal, H., Malik, R.: Compositional verification in supervisory control. SIAM J. Control Optim. **48**(3), 1914–1938. https://doi.org/10.1137/070695526
7. Göbe, F., Ney, O., Kowalewski, S.: Reusability and modularity of safety specifications for supervisory control. In: 21st IEEE International Conference on Emerging Technologies and Factory Automation, pp. 1–8 (2016). https://doi.org/10.1109/ETFA.2016.7733498
8. Gonzalez, A.G.C., Alves, M.V.S., Viana, G.S., Carvalho, L.K., Basilio, J.C.: Supervisory control-based navigation architecture: a new framework for autonomous robots in Industry 4.0 environments. IEEE Trans. Ind. Inform. **14**(4), 1732–1743 (2018). https://doi.org/10.1109/TII.2017.2788079

9. Goorden, M.A., Fabian, M.: No synthesis needed, we are alright already. In: 15th IEEE International Conference on Automation Science and Engineering, pp. 195–202. https://doi.org/10.1109/COASE.2019.8843071

10. Goorden, M.A., van de Mortel-Fronczak, J.M., Etman, L.F.P., Rooda, J.E.: DSM-based analysis for the recognition of modeling errors in supervisory controller design. In: 21st International Dependency and Structure Modeling Conference, pp. 127–135 (2019). https://doi.org/10.35199/dsm2019.7

11. Goorden, M.A., van de Mortel-Fronczak, J.M., Reniers, M.A., Fokkink, W.J., Rooda, J.E.: The impact of requirement splitting on the efficiency of supervisory control synthesis. In: Larsen, K.G., Willemse, T. (eds.) FMICS 2019. LNCS, vol. 11687, pp. 76–92. Springer, Cham (2019). https://doi.org/10.1007/978-3-030-27008-7_5

12. Goorden, M.A., van de Mortel-Fronczak, J.M., Reniers, M.A., Fokkink, W.J., Rooda, J.E.: Structuring multilevel discrete-event systems with dependency structure matrices. IEEE Trans. Autom. Control (2019). https://doi.org/10.1109/TAC.2019.292811. Early access

13. Gössler, G., Sifakis, J.: Composition for component-based modeling. Sci. Comput. Program. **55**(1), 161–183. https://doi.org/10.1016/j.scico.2004.05.014

14. Grigorov, L., Butler, B.E., Cury, J.E.R., Rudie, K.: Conceptual design of discrete-event systems using templates. Discrete Event Dyn. Syst. **21**(2), 257–303 (2011). https://doi.org/10.1007/s10626-010-0089-0

15. Komenda, J., Masopust, T., van Schuppen, J.H.: Control of an engineering-structured multilevel discrete-event system. In: 13th International Workshop on Discrete Event Systems, pp. 103–108 (2016)

16. Ma, C., Wonham, W.: Nonblocking Supervisory Control of State Tree Structures. Lecture Notes in Control and Information Sciences, vol. 317. Springer, Heidelberg (2005). https://doi.org/10.1007/b105592

17. Markovski, J., Jacobs, K.G.M., van Beek, D.A., Somers, L.J., Rooda, J.E.: Coordination of resources using generalized state-based requirements. In: 10th International Workshop on Discrete Event Systems, pp. 300–305 (2010)

18. Mohajerani, S., Malik, R., Fabian, M.: A framework for compositional nonblocking verification of extended finite-state machines. Discrete Event Dyn. Syst. **26**(1), 33–84 (2016). https://doi.org/10.1007/s10626-015-0217-y

19. Moormann, L., Maessen, P., Goorden, M.A., van de Mortel-Fronczak, J.M., Rooda, J.E.: Design of a tunnel supervisory controller using synthesis-based engineering (2020). Accepted for ITA-AITES World Tunnel Congress

20. Ouedraogo, L., Kumar, R., Malik, R., Åkesson, K.: Nonblocking and safe control of discrete-event systems modeled as extended finite automata. IEEE Trans. Autom. Sci. Eng. **8**(3), 560–569 (2011). https://doi.org/10.1109/TASE.2011.2124457

21. Pena, P.N., Cury, J.E.R., Lafortune, S.: Verification of nonconflict of supervisors using abstractions. IEEE Trans. Autom. Control **54**(12), 2803–2815. https://doi.org/10.1109/TAC.2009.2031730

22. de Queiroz, M.H., Cury, J.E.R.: Modular supervisory control of large scale discrete event systems. In: Boel, R., Stremersch, G. (eds.) Discrete Event Systems. SECS, vol. 569, pp. 103–110. Springer, Boston (2000). https://doi.org/10.1007/978-1-4615-4493-7_10

23. Ramadge, P.J.G., Wonham, W.M.: Supervisory control of a class of discrete event processes. SIAM J. Control Optim. **25**(1), 206–230 (1987)

24. Ramadge, P.J.G., Wonham, W.M.: The control of discrete event systems. Proc. IEEE **77**(1), 81–98 (1989)

25. Ramos, A.L., Ferreira, J.V., Barceló, J.: Model-based systems engineering: an emerging approach for modern systems. IEEE Trans. Syst. Man Cybern. Part C (Appl. Rev.) **42**(1), 101–111 (2012). https://doi.org/10.1109/TSMCC.2011.2106495
26. Reijnen, F.F.H., Erens, T.R., van de Mortel-Fronczak, J.M., Rooda, J.E.: Supervisory control synthesis for safety PLCs (2020). Submitted to International Workshop on Discrete Event Systems
27. Reijnen, F.F.H., Goorden, M.A., van de Mortel-Fronczak, J.M., Reniers, M.A., Rooda, J.E.: Application of dependency structure matrices and multilevel synthesis to a production line. In: 2nd IEEE Conference on Control Technology and Applications, pp. 458–464 (2018). https://doi.org/10.1109/CCTA.2018.8511449
28. Reijnen, F.F.H., Goorden, M.A., van de Mortel-Fronczak, J.M., Rooda, J.E.: Supervisory control synthesis for a waterway lock. In: 1st IEEE Conference on Control Technology and Applications, pp. 1562–1568 (2017). https://doi.org/10.1109/CCTA.2017.8062679
29. Reijnen, F.F.H., Goorden, M.A., van de Mortel-Fronczak, J.M., Rooda, J.E.: Supervisory control synthesis for a lock-bridge combination (2019). Submitted to Discrete Event Dynamic Systems
30. van der Sanden, L.J.: Performance analysis and optimization of supervisory controllers. Ph.D. thesis, Eindhoven University of Technology (2018)
31. van der Sanden, L.J., et al.: Modular model-based supervisory controller design for wafer logistics in lithography machines. In: 18th ACM/IEEE International Conference on Model Driven Engineering Languages and Systems (2015)
32. Skoldstam, M., Åkesson, K., Fabian, M.: Modeling of discrete event systems using finite automata with variables. In: 46th IEEE Conference on Decision and Control, pp. 3387–3392 (2007). https://doi.org/10.1109/CDC.2007.4434894
33. Su, R., van Schuppen, J.H., Rooda, J.E.: Synthesize nonblocking distributed supervisors with coordinators. In: 17th Mediterranean Conference on Control and Automation, pp. 1108–1113 (2009). https://doi.org/10.1109/MED.2009.5164694
34. Swartjes, L., van Beek, D.A., Fokkink, W.J., van Eekelen, J.A.W.M.: Model-based design of supervisory controllers for baggage handling systems. Simul. Model. Pract. Theory **78**, 28–50 (2017). https://doi.org/10.1016/j.simpat.2017.08.005
35. Swartjes, L.: Model-based design of baggage handling systems. Ph.D. thesis, Eindhoven University of Technology (2018)
36. Theunissen, R.J.M., Petreczky, M., Schiffelers, R.R.H., van Beek, D.A., Rooda, J.E.: Application of supervisory control synthesis to a patient support table of a magnetic resonance imaging scanner. IEEE Trans. Autom. Sci. Eng. **11**(1), 20–32 (2013)
37. Wonham, W.M., Ramadge, P.J.G.: Modular supervisory control of discrete-event systems. Math. Control Signals Syst. **1**(1), 13–30 (1988)
38. Wonham, W.M., Cai, K.: Supervisory Control of Discrete-Event Systems, 1st edn. Springer, Heidelberg (2019). https://doi.org/10.1007/978-3-319-77452-7
39. Zaytoon, J., Carre-Meneatrier, V.: Synthesis of control implementation for discrete manufacturing systems. Int. J. Prod. Res. **39**(2), 329–345 (2001). https://doi.org/10.1080/00207540010002388

Modelling and Analysing Software in mCRL2

Jan Friso Groote, Jeroen J. A. Keiren, Bas Luttik, Erik P. de Vink,
and Tim A. C. Willemse[✉]

Faculty of Mathematics and Computer Science,
Eindhoven University of Technology,
Eindhoven, The Netherlands
{J.F.Groote,J.J.A.Keiren,S.P.Luttik,E.P.d.Vink,
T.A.C.Willemse}@tue.nl

Abstract. Model checking is an effective way to design correct software. Making behavioural models of software, formulating correctness properties using modal formulas, and verifying these using finite state analysis techniques, is a very efficient way to obtain the required insight in the software. We illustrate this on four common but tricky examples.

1 Introduction

Software consists of algorithms that manipulate data structures, and protocols that describe how software components communicate. These algorithms and protocols are expected to work correctly under all the conditions they are designed for. However, it is not easy to foresee all possible situations the software will ever encounter. Corner cases and exceptional situations are typically difficult to identify upfront. Such situations are also hard to cover using testing.

A solution is to prove the correctness of the software. A wide range of methods, such as invariants, Hoare logics, separation logics, and process algebras, have been developed that allow this. Proof checkers like Coq [3] or Isabelle [22] allow to computer check such proofs, achieving an unparalleled level of trust in the quality of such proofs. Unfortunately constructing such proofs is still a manual, labour intensive activity.

Model checking and equivalence checking are two efficient verification techniques that strike a favourable balance between the ease of automation of testing and the level of trust established by correctness proofs. Both techniques rely on models made of the software under consideration, and they are particularly effective in exploring corner cases in software. In model checking a set of high-level requirements, phrased in terms of some temporal logic, are verified against the model. In equivalence checking, a state space is abstracted into a smaller one reflecting the essential properties of the original and which is sufficiently small to be inspected. In the related technique called refinement checking, a state space of the software is generated and compared to a required state space.

© Springer Nature Switzerland AG 2020
F. Arbab and S.-S. Jongmans (Eds.): FACS 2019, LNCS 12018, pp. 25–48, 2020.
https://doi.org/10.1007/978-3-030-40914-2_2

There are a number of toolsets that support one of these two approaches. For instance, the FDR [11] toolset specialises in refinement checking, and centres around the notion of failures-divergences refinement [21,25], which facilitates a step-wise refinement software development methodology. Toolsets such as SPIN [15] and nuSMV [7] rely exclusively on model checking. CADP [10] and mCRL2 [6] are toolsets that offer both techniques.

In this paper we focus on the mCRL2 toolset and its three specification languages: the data language, which is based on the theory of abstract data types, the process specification language, which is an ACP-style process algebra, and the requirement language, which is a highly expressive extension of the modal μ-calculus. This toolset has been used successfully in a large number of case studies; see, *e.g.* [2,4,5,16,17,24], and it fulfils a role in education in various universities worldwide.

Our aim is to illustrate how to use mCRL2 to specify and analyse algorithms and data structures for which one should no longer wish to lull oneself into the belief that it does not warrant spending time on analysing their inner workings. The examples we consider, available via [1], are taken from the literature, and are presented such that they can immediately be replayed in the mCRL2 toolset.

We start by considering the well-known solution to the mutual exclusion problem, offered by Peterson's algorithm [23]. The interesting bit here is that— motivated by a non-standard exposition of the algorithm, present for some time on Wikipedia—we study the effects of different intialisations on the correctness of the algorithm. We furthermore study Knuth's dancing links [18], the concurrent data structure known as Treiber's stack [26] and Lamport's queue [20]. To facilitate a stand-alone exposition of the examples, we start with a brief primer to the languages used in mCRL2, where we assume that the reader is already familiar with similar-spirited languages.

With the current contribution, we hope to fill a void in the literature. As starting point to an uninitiated, motivated toolset user, (industrial) case studies can be discouraging due to their intrinsic complexity; on the other hand, the typical introductory problems often available in first-encounter tutorials[1] focus on language constructs and, for this reason, often lack in appeal. The aim of the current paper, therefore, is to bridge the gap between toy examples at one end, and complex case studies at the other end of the spectrum.

2   A Short Primer in mCRL2

In this section, we present a cursory overview of the most important language constructs in mCRL2. For a more elaborate introduction to the formalism, including a treatment of its real-time features, we refer to [12].

2.1   Data Types

Most (distributed) algorithms and concurrent systems revolve around data in some form or another. The mCRL2 data language is based on higher-order

---

[1] See for instance the online mCRL2 introductory tutorial on https://mcrl2.org/.

abstract data types. This allows users to define their own data types, along with operations on these. For convenience, the mCRL2 data language also includes a large number of predefined standard data types and type constructors.

User Defined Data Types. Abstract data types provide a straightforward mechanism for specifying complex data types. A user can declare new types, in this context called *sorts*, along with their (constructor) functions and their definitions, using a small number of primitives. Constructors are the atomic building blocks of a data type, allowing for an inductive definition of the type. A sort is declared using the **sort** keyword, whereas constructors are declared using the **cons** keyword; *e.g.*, for a given sort A, a declaration of an A-leaf tree could be:

```
sort Tree;
cons leaf: A -> Tree;
     node: Tree # Tree -> Tree;
```

Operators and functions that manipulate user-defined types can be declared using the **map** keyword. They are defined by a set of equations, introduced with the keyword **eqn**. These equations may refer to variables, which must be declared in a declaration block preceding the equations and announced by the keyword **var**. For most purposes, equations are interpreted as *rewrite rules*, allowing reasoning engines to manipulate and simplify (sub)expressions by matching the left-hand side of a rule and replacing these (sub)expressions by the right-hand side of a rule. For instance, assuming that Max: A # A -> A is a binary operation on sort A, a standard way of lifting that operator to a tree over A, given by function Max_Leaf, is as follows:

```
map Max_Leaf: T -> A;
var t1,t2: T; a: A;
eqn Max_Leaf(leaf(a)) = a;
    Max_Leaf(t1, t2) = Max(Max_Leaf(t1), Max_Leaf(t2));
```

Standard Data Types. For the convenience of the user, several standard data types and operations on these have been pre-defined.

One such standard data type, which, as we shall see later, is essential for the specification of conditional behaviour, is the sort *Bool*, representing the Booleans, with constructors *false* and *true*. Pre-defined operations on *Bool* include *negation* (denoted !), *conjunction* (denoted &&), *disjunction* (denoted ||), and *implication* (denoted =>). A predicate on any datatype (user-defined or standard) can be defined as a mapping from that datatype to *Bool*. On all pre-defined datatypes binary equality and inequality predicates, respectively denoted == and !=, are defined with their standard interpretation. Furthermore, the language has generic constructions for universal quantification *forall* and existential quantification *exists*, which can be applied to predicates over any datatype. Most standard binary operators can be written *infix*.

In addition to the Booleans, the positive numbers *Pos*, natural numbers *Nat*, integers *Int* and reals *Real* are available, including many of the familiar operations on these. These numbers can be written in decimal notation; *e.g.*, the expression 10 represents the number ten. There is no pre-defined limitation on the size of the elements in these data types. Whether the tools can handle a specification involving numbers depends on the available computer memory, and so computations involving numbers should be done with care.

A further useful construct is that of an *enumerated type*, called *structured sort* in mCRL2. These structured sorts are a convenient way of defining sorts with a finite set of elements, as they come with a built-in notion of equality. For instance, a data type `Colour` for the colours of a traffic light could be:

```
sort Colour = struct red | yellow | green;
```

Structured sorts are, however, more versatile than that. For instance, the A-leaf tree can alternatively be defined as follows:

```
sort Tree = struct leaf(A) | node(Tree, Tree);
```

The advantage of this definition over the one provided earlier, is that one does not need to bother defining equality as it is built-in for structured sorts.

Function Types. The mCRL2 data language also has function types. An infinite list of natural numbers is a function from $\mathbb{N}$ to $\mathbb{N}$; in mCRL2 the data type representing this set of functions is the data sort `Nat -> Nat`. Functions can be defined using lambda abstraction, or by a pointwise specification of the result of applying the function to elements of its domain (function application works as expected). There is a concise mechanism for updating a function: assuming, for instance, that the function `id: Nat -> Nat` is the identity function, the function `id[3 -> 2]` represents the identity function in which the value 3 is mapped to 2. Using lambda abstraction and function updates, operations such as removing the head of an existing list can, for instance, be modelled as follows.

```
map remove: (Nat -> Nat) -> (Nat -> Nat);
var l: Nat -> Nat;
eqn remove(l) = lambda n: Nat. l(n+1);
```

An alternative definition of the remove operation is as follows:

```
var l: Nat -> Nat; n: Nat;
eqn remove(l)(n) = l(n+1);
```

Type Constructors. The mCRL2 data language has a number of useful type constructors. Lists, sets and bags can be defined in a generic way and come with pre-defined operations. For instance, the sort `List(A)` describes the data type of finite lists containing elements of type `A`, and comes with constructor `[]: List(A)` for the empty list, and (infix) `|>: A # List(A)-> List(A)` for prefixing a list. List *concatenation* is denoted by `++`, and further operators on lists include, *e.g.*, `head`, `rhead`, `tail`, `rtail`.

Sets can be described in a way that is close to standard notation. For instance, the sort `Set(Nat)` has all sets of all natural numbers as elements. The expression `{ n:Nat | n <= 10 }` describes the (finite) set of all numbers not exceeding ten, whereas `{ n:Nat | n > 10 }` describes its (infinite) complement. Set union, set difference, *etcetera*, are defined and work as expected. In a similar fashion one can define bags: `Bag(Nat)` describes the type of bags (multi-sets) over natural numbers.

Finally, sort aliasing can be used to give more meaningful names to data sorts. An abstraction on a set of identifiers is, *e.g.*, a natural number. However, one may prefer introducing a new data sort that is syntactically distinguishable from `Nat` to better reflect its role. This can easily be achieved as follows:

```
sort Id = Nat;
```

An aliased sort such as `Id` inherits all operations of the sort that it aliases.

2.2 Processes

Arguably, the most interesting aspect of a concurrent system is its behaviour. Behaviours can be represented by *Labelled Transition Systems* (LTSs); these are essentially directed, labelled graphs, where the vertices represent a system's state and the directed edges connecting two vertices are labelled with the event that causes the state change. A process algebra such as mCRL2 allows for specifying an LTS in a compositional fashion.

Sequential Processes. A sequential process is a process that describes the possible behaviours of a system by way of actions (representing real-life or otherwise interesting events), combined sequentially, non-deterministically and using recursion. The process that cannot perform any activity (*i.e.*, is in a *deadlock*), is denoted `delta`. Somewhat different from other process algebras, the mCRL2 process algebra allows multiple actions to happen at the same time, resulting in a *multi-action*. Such a multi-action can be thought of as a multi-set of actions, all of which are assumed to happen simultaneously. The empty multi-action is denoted `tau`, whereas a multi-action of size one is often simply referred to as an action. An action may carry zero or more data arguments; this is useful for emiting relevant information of a process. Actions need to be declared explicitly using the **act** keyword:

```
act read, write: Nat;
```

Multi-actions can be constructed by listing the actions that are part of the multi-action; *e.g.*, read(0)|read(0)|write(1) denotes the multi-action consisting of two `read` actions, both with the same parameter, and one `write` action. Note that multi-action a|b|tau|c is equivalent to a|b|c.

Processes can be composed sequentially using a binary, associative sequential composition operator: process p.q denotes the process that first behaves as process p and, upon termination of p, continues to behave as process q. For instance, the process write(0).read(0) describes a process that first writes a value and subsequently reads a value.

Process p + q describes the process that chooses to behave as either process p or process q. The choice between the two processes is, in such a case, resolved by the first action that is executed: in case this action is due to process p, process p will dictate what behaviour is left, whereas when the first action is due to process q, process q will do so. In case one of the two processes cannot execute actions, the choice is automatically resolved in favour of the other process; *i.e.*, p + delta is behaviourally equivalent to p.

Note that in case the first action that is executed in p + q is offered both by p and q, the choice is resolved non-deterministically; that is, there is no guarantee which of the two processes will continue, and the choice cannot be influenced. Such non-determinism is a powerful construct for modelling unreliabilities of a (sub)system and a useful mechanism for abstracting from decisions made internally by a (sub)system. For instance, in a process write(0).read (0)+ write(0).read(-1), the execution of write(0) determines whether the same value is read as was written, or a different value; by only observing the write action, however, one cannot predict which of the two will happen.

Data can be used to influence the flow of control in a process by making process behaviour conditional: the ternary if-then-else construct b -> p <> q behaves as process p, provided that Boolean condition b holds true, and process q otherwise. For instance, the following process:

```
(reliable == true) -> write(0).read(0) <> write(0).read(-1)
```

behaves, perhaps, 'as expected' in case the Boolean reliable is true, and quirky otherwise. The binary if-then construct b -> p is short for b -> p <> *delta*.

Binary non-deterministic choice is generalised by non-deterministic choice quantification, which binds a 'local' variable. This is achieved by the construct *sum* d:D. p, in which variable d of data sort D is bound in the process expression p, and its value is chosen non-deterministically. Such a construct is useful to model, *e.g.*, a process that can write an arbitrary even value and then read it:

```
sum n: Nat. (n mod 2 == 0) -> write(n).read(n)
```

Note that, without further restrictions, this process yields an infinite state LTS; explicitly generating its transition system is therefore not feasible.

Infinite behaviours can be described using (parameterised) recursive equations, which associate behaviour to recursion variables (agents). The following process, for instance, describes the behaviour of a natural number buffer:

```
proc Buffer(b: Bool, n: Nat) =
    sum m: Nat. b -> write(m).Buffer(b = false, n = m) +
    !b -> read(n).Buffer(b = true);
```

In case parameter b is false, the value n that is currently stored in the buffer can be read through action read(n). Otherwise, an arbitrary value can be written to the buffer. The role of parameter b in the above process is to indicate whether the buffer is empty or not. Note that, in recursive calls, only the parameters that change value need to be mentioned. Thus, in the first recursive call of Buffer the parameter b is set to *false* and n gets the value m, and in the second recursive call the parameter b is set to *true* and n remains unchanged.

Parallel Processes. The parallel composition of processes p and q is denoted p || q. The action that a parallel composition p||q can execute can come either from process p, process q, or from processes p and q simultaneously. In the latter case, a multi-action consisting of an action from p and q is produced. Communication can be specified by a mapping that indicates which labels in a multi-action must synchronise; the trace of a successful communication is then a new action. Such a mapping is specified in a communication function through the keyword *comm*.

This mapping renames multi-action labels to a new action label. A successful communication is subject to matching of the parameters of the individual actions in a multi-action. For instance, a process modelling a shared variable whose value can be read through an action value_s, and another process continuously reading that value through an action value_r, may communicate to yield a new action with label value:

```
act value_s, value_r, value : Nat;
proc Variable(n: Nat) = value_s(n).Variable();
     Agent = sum m: Nat. value_r(m).Agent;
     Parallel = comm({ value_s|value_r -> value }, Variable(0) || Agent);
```

Process Parallel can execute an action value_s(0), actions value_r(0), value_r(1), ..., and all multi-actions of the form value_s(0)|value_r(v), where v is an arbitrary value not equal to 0. In addition, the process can execute action value(0). One is often only interested in the result of the communication and not in the individual actions that make up a communication. Actions can be 'filtered' using the allow operator:

```
proc Interact = allow({value}, Parallel);
```

The allow operator maps the (multi-)actions not explicitly listed to delta. There are several additional language constructs, such as block and hide; the former is, in a way, dual to the allow operator, whereas the latter maps a selected set of actions to the action tau, which is used for abstraction.

2.3 Modal Formulas

The behaviour specified using the process and data languages can be analysed in order to determine whether it satisfies certain requirements. These requirements are specified in the first-order modal μ-calculus, a fixed point language based on a first-order extension of the modal logic called *Hennessy-Milner* logic (HML) [13]. The language offers the modal operators <_>_ and [_]_, next to the familiar first-order logic constructs ||, &&, forall and exists, and predicates **val**(b), where b is an arbitrary Boolean expression from the data language. For a set of actions, represented by a formula A, the *may*-modality of the form <A>f holds true in a state whenever it allows for an action from the set A and leads to a state in which formula f holds true. The modal formula [A]f is the dual: a state satisfies this property when none of the actions from the set A lead to a state not satisfying f. For instance, process Parallel given above satisfies the properties <value(0)>true and [value(1)]false. The sets of actions used in modalities are also described using first-order logic, where, *e.g.*, && denotes intersection, || denotes union, and true denotes the set of all actions. This way, one can write [value(1)|| value(2)]false to claim that neither value 1, nor 2, can be communicated. The property, [exists n: Nat. **val**(n >= 1)&& value(n)]false, or, equivalently, forall n: Nat. [value(n)]**val**(n < 1), denotes that no value other than 0, can be communicated.

Since HML is not capable of reasoning about the behaviours of unbounded depth, recursion is needed. This enters the language through a least fixed point operator *mu* X. f and a greatest fixed point operator *nu* X. f. Informally, a set of states satisfies *mu* X. f when each state satisfies some finite unfolding of X. For instance, the formula *mu* X. (<a>X || <b>*true*) indicates that there should be some finite-depth formula <a>^{i}<b>*true* that is satisfied (although the depth i is potentially different for each state satisfying this formula). Dually, the formula *nu* X. ([a]X && [b]*false*) indicates that all possible unfoldings of the formula should hold. Essentially, this is the case when in no a-reachable state, a b-action is ever enabled.

Since fixed point formulas can be hard to understand, the mCRL2 requirement language offers regular expressions to reason about recursive processes. Using regular expressions, one can build a language from formulas describing sets of actions, sequential composition, choice and iteration. Such regular expressions can then be combined with the two modalities to yield expressions that permit to reason about processes of arbitrary or infinite depth. For example, the regular expression *true**.a represents the set of sequences consisting of zero or more arbitrary actions, followed by an action a. Consequently, formula [*true* *.a] *false* asserts that none of these sequences are possible; *i.e.*, no a-action is ever possible. In a similar vein, formula <a*.b>*true* asserts that some sequence of a-actions leads to a state that can execute a b-action. Such regular formulas can be translated to standard fixed point formulas; *e.g.*, for a regular expression R and formula f, formula [R*]f is equivalent to *nu* X. ([R]X && f).

In mCRL2, fixed points can carry parameters, which can be useful to record information about the recursions that have been taken. For instance, formula *nu* X(n:*Nat* = 0). (**val**(n < Max)&& [a]X(n)&& [b]X(n+1)) only holds in case all a-b-runs of a system contain at most Max b-actions. This parameterisation is an incredibly powerful construct.

3   Using the mCRL2 Toolset

The mCRL2 toolset has long been a collection of stand-alone tools, building on the philosophy that the user should be supported in using the transformations and solvers in as liberal a way as possible. While this philosophy has not been abandoned, the increasing popularity of the toolset has called for a more accessible way of using the toolset. For this reason, recent versions of the toolset come with a plain IDE, called mcrl2ide, which can be used to carry out basic analyses of mCRL2 specifications, without exposing the user to the overwhelming number of tools available in the toolset. The analyses for the examples we present in the next sections can be carried out from within this IDE. However, it should be noted that, for simplicity and accessibility, the IDE has opted to only implement a 'standard' workflow for all analyses, which may not always be optimal. In case non-standard, more advanced algorithms are needed for successfully carrying out an analysis, one needs to resort to the old way of working, using the file-driven environment mcrl2-gui or the command line.

4   Peterson's Mutual Exclusion Algorithm

Peterson's mutual exclusion algorithm is a well-known protocol that coordinates different processes to obtain exclusive access to a shared resource by allowing at most one process at a time to enter a critical section [23]. We focus on a setting with two processes, but the algorithm and our analysis generalise to any number of processes.

The algorithm uses three shared variables. For each of the two processes there is a shared Boolean variable *flag* which they set to *true* when they wish to enter the critical section. In addition, a variable *turn* is used to indicate which process is allowed to enter the critical section. Before entering, a process grants the other process access. If a process is granted access or the other process does not desire to enter the critical section, the critical section can be entered. In pseudo code the behaviour can be described as follows.

Global variables :	Behaviour of process i :
$flag[0]{:}\mathbb{B}$	$flag[i] := true$
$flag[1]{:}\mathbb{B}$	$turn := 1-i$
$turn{:}\mathbb{N}$	**while** $flag[1-i] = true \land turn = 1-i$ **do**
	busy wait
	critical section
	$flag[i] := false$

The initial value of the Boolean flags must be *false* for the algorithm to work correctly. However, different initialisations have appeared in online sources, cf. [27]; in such cases the behaviour is almost correct and the problem only surfaces by conducting a thorough analysis.

The mCRL2 Model. To reason about the correctness of the algorithm in mCRL2, we introduce parameterised actions wish, enter and leave to model the interesting state changes of both processes. We remark that these actions are not needed for the correct functioning of the algorithm but they help in its analysis. Action wish signals a process' desire to enter the critical section. The action enter marks the moment a process enters the critical section and leave signals the process leaving the critical section. The assignments in the algorithm itself are modelled using actions get_flag, set_flag, get_turn and set_turn through which the shared variables can be read and set. The shared variables are, as remarked in Sect. 2, typically modelled as processes, and assignment to, and checks on these variables are modelled by the communications with these processes. Peterson's algorithm, and our model by extension, uses only standard data structures. In our model, we identify each of the two processes by a number, and we use a custom mapping other to obtain the identity of the other process.

```
act wish, enter, leave: Nat;
    get_flag_r, get_flag_s, get_flag,
      set_flag_r, set_flag_s, set_flag: Nat # Bool;
    get_turn_r, get_turn_s, get_turn,
      set_turn_r, set_turn_s, set_turn: Nat;
```

```
map other: Nat -> Nat;
eqn other(0) = 1;
    other(1) = 0;
```

The processes Flag(0, *true*), Flag(1, *true*) and Turn(0) depicted below model the three shared variables. The additional argument for these processes sets the relevant initial values.

```
proc
  Flag(id: Nat, b: Bool)=
    sum b: Bool. set_flag_r(id, b).Flag(id, b) +
    get_flag_s(id, b).Flag(id, b);

  Turn(n:Nat)=
    sum n': Nat. set_turn_r(n').Turn(n') + get_turn_s(n).Turn(n);
```

A single thread of the algorithm is represented by Process. Modelling action wish and the *true*-assignment to the *flag* variable are specified to occur simultaneously, because the latter marks the wish of the process to enter the critical section. This highlights a typical use-case for multi-actions. Our model abstracts from the busy waiting loop in the algorithm by modelling the loop by a single communication. Using that actions synchronise on values, it is only necessary to check whether one of the shared variables attains a value that allows to enter the critical section.

```
Process(id: Nat) =
  wish(id)|set_flag_s(id, true).set_turn_s(other(id)).
  (get_flag_r(other(id), false) + get_turn_r(id)).enter(id).
  leave(id).set_flag_s(id, false).Process(id);
```

The initialisation of the process is shown below. Note that we explicitly hide the communications with the shared variables; this allows for focussing on the interesting state changes of both processes. As a result, the only actions, apart from *tau*, that remain in the state space of the algorithm are the enter, leave and wish actions (the latter results from the underlying theory, stating that wish(0) | *tau* is the (multi-)action wish(0)).

```
init
  hide({ get_flag, set_flag, get_turn, set_turn },
    allow({ wish|set_flag, enter, leave,
            get_flag, set_flag, get_turn, set_turn},
      comm({ get_flag_r | get_flag_s -> get_flag,
             set_flag_r | set_flag_s -> set_flag,
             get_turn_r | get_turn_s -> get_turn,
             set_turn_r | set_turn_s -> set_turn },
        Process(0) || Process(1) ||
        Flag(0,false) || Flag(1,false) || Turn(0))));
```

The Analysis. There are three fundamental requirements a mutual exclusion algorithm must meet. The first one asserts mutual exclusion: at no time is it possible to be able to do two enter actions without a leave action in between.

```
[true*.(exists id1: Nat. enter(id1)).(!exists id2 :Nat. leave(id2))*.
 (exists id3: Nat. enter(id3))]false
```

The second property says that whenever a process wishes to enter, it is allowed access within a finite number of steps. A simple formulation is the following: whenever an action wish(id) happens, an action enter(id) is guaranteed to follow within a finite number of steps. The latter requires a least fixed point.

[*true*∗] *forall* id: *Nat*. [wish(id)](*mu* Y. [!enter(id)]Y && <*true*>*true*)

The third property is called bounded overtaking: whenever one process indicates the wish to enter the critical section, the other process can at most enter the critical section twice. This is formalised by asserting that after a wish(id), at most two enter(other(id)) actions can happen without an enter(id) somewhere in between. Counting of the number of occurrences of enter(other (id)) actions is taken care of by the parameter n in the formula; each time such an action is encountered, n is increased, but in all cases, the formula asserts that it never invalidates condition **val**(n<=2).

[*true*∗] *forall* id: *Nat*. [wish(id)]
    (*nu* Y(n: *Nat* = 0). **val**(n<=2) && [!enter(id)]Y(n) && [enter(other(id))]Y(n+1))

The stronger property that if a process wishes to enter the critical section, the other process can enter the critical section only once is invalid.

The three properties above hold true for Peterson's algorithm. One may wonder whether an out-of-order execution, which is common in modern processors, affects the correctness of the algorithm. Out-of-order execution means that if a sequential process writes to unrelated memory addresses, writing can take place in any order. As the shared variables *flag* and *turn* are stored at different addresses, there is no guarantee that assignments are executed in the order as listed. It is easy to change the model and prove that it violates the mutual exclusion property, whereas the other two properties remain valid. This means that, to guarantee correctness on contemporary hardware, this algorithm requires special measures that prevent swapping certain instruction.

Remarkably, the validity of the three properties is independent of the initialisation of the algorithm. To gain some further understanding, we hide all actions, except enter and leave and apply weak trace minimisation. The corresponding labelled transition systems are depicted in Fig. 1.

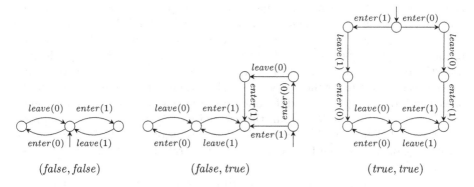

Fig. 1. The weak trace reduced behaviour of Peterson's algorithm with different initialisations.

At the left the graph with the correct initialisation (*false*, *false*) of the flags is depicted. In the middle the initialisation (*false*, *true*) and at the right (*true*, *true*) can be found. The initial states have a small incoming arrow. When the initial-isation is not correct, process 0, after it has entered the critical section once, is forced to wait until process 1 entered its critical section for the first time. For the transition system at the right, the reverse is also true.

The transition systems suggest that the second formula, expressing that a process that wishes to enter the critical section will be granted access after a finite number of steps, is insufficient, since it does not take into account that one process' capability of entering the critical section should not be dependent on the other process' desire to do so. To take this into account, we phrase the following requirement: whenever a process expresses the wish to enter the critical section, it can enter the critical section on its own accord, unless the other process already expressed a wish to enter the critical section, enters the critical section or expresses the wish to enter the critical section. We again use a counter n to record the number of processes that have expressed the wish to enter the critical section and have not yet left the critical section. In case n==0, we use a least fixed point to check that for process id wishing to enter the critical section, along all paths not involving wish or enter actions of process other(id), process id inevitably enters the critical section.

```
nu X(n: Nat = 0).
    ([exists id: Nat. wish(id)]X(n+1) && [exists id: Nat. leave(id)]X(max(0,n-1))) &&
    [!exists id: Nat. wish(id) || leave(id)]X(n) &&
    (val(n==0) => forall id:Nat. [wish(id)]
      ( mu Y. ([!enter(id) && !(wish(other(id)) || enter(other(id)))]Y &&
          <!(wish(other(id)) || enter(other(id)))> true)))
```

It turns out that this formula distinguishes between Peterson's mutual exclusion algorithm with and without proper initialisation.

5 Knuth's Dancing Links

Dancing links [14,18] is a technique to efficiently perform removal and inser-tion operations on a circular doubly linked list. It is intended to be used when elements temporarily need to be removed from the list in the course of a compu-tation, and have to be re-inserted at the same position at a later stage. Knuth used the technique in his *Algorithm X*, which solves the exact cover problem. The correctness claim for the operations is that whenever a sequence of removals of elements $x_0, \ldots, x_n$ is applied, followed by a sequence of insertions of elements $x_n, \ldots, x_0$, then the result is again the original list. A verification of this prop-erty was a challenge of the VerifyThis [9] competition in 2015. We include a verification of the dancing links technique to illustrate how mCRL2 can be used to verify the correctness of operations on data structures.

Let x be an element of a circular doubly linked list, and suppose that $L(x)$ refers to its predecessor and $R(x)$ to its successor. The operations $remove(x)$ and $insert(x)$ are defined by

$$remove(x) : L(R(x)) := L(x); \; R(L(x)) := R(x), \text{ and}$$
$$insert(x) : \;\; L(R(x)) := x; \; R(L(x)) := x.$$

The idea is that, after removal, the element x is not garbage-collected, waiting to be inserted again at its original position (*e.g.*, when the algorithm that uses the list backtracks). Also the pointers $L(x)$ and $R(x)$ remain available. The implementation should guarantee that if a sequence of removals and insertions is performed in a last-out-first-in order, then the linear ordering induced on the elements by the list is preserved.

The mCRL2 Model. The list data structure is described by a pair of functions L and R that for each element gives its left and right neighbours. For simplicity, the elements of the data structure are taken to be natural numbers from 0 up till and including some natural number MAX, set to 4 in the specification below.

The functions remove and insert are described using function updates, and they strictly follow the definition above.

```
sort D = struct pair(L: Nat -> Nat, R: Nat -> Nat);

map MAX: Nat;
eqn MAX = 4;

map remove,insert: Nat # D -> D;
var x: Nat; p: D;
eqn remove(x, p) = pair(L(p)[R(p)(x) -> L(p)(x)], R(p)[L(p)(x) -> R(p)(x)]);
    insert(x, p) = pair(L(p)[R(p)(x) -> x], R(p)[L(p)(x) -> x]);
```

For the formulation in the parameterised modal μ-calculus of the correctness claim mentioned above, we define a process UseList(d,stack), which has a doubly linked list d and a stack as parameters; the stack stores the removed elements and allows them to be reinserted in a last-out-first-in order. Initially, the list is full and the stack is empty:

```
map d_full: D;
eqn d_full = pair(lambda n:Nat.if(n == 0, MAX, max(0, n - 1)),
                  lambda n:Nat.if(n == MAX, 0, n + 1));

init UseList(d_full,[]);
```

The process UseList(d, stack) executes actions do_remove and do_insert, representing the activities of removing and inserting an element in the linked list, respectively. An element can only be removed from the list if it is in the list. Since, in our model, the full list contains the numbers 0 to MAX, whether some element x is in the list can be determined by checking whether x is between 0 and MAX and not an element of the stack. Only the top of a non-empty stack, represented by head(stack), can be re-inserted in the list. The current list structure is exposed by the actions left(x,L(d)(x)) and right(x,R(d)(x)); *e.g.*, if, for some element x of the list, the action left(x,x') is enabled in some state, then this means that x' is L(x).

```
act do_remove, do_insert: Nat;
    left, right: Nat # Nat;

proc UseList(d: D, stack: List(Nat))=
  sum x: Nat. (x > 0 && x <= MAX && !(x in stack)) ->
    do_remove(x).UseList(remove(x, d), x |> stack) +
  (stack != []) ->
    do_insert(head(stack)).UseList(insert(head(stack), d), tail(stack)) +
  sum x: Nat. (x <= MAX && !(x in stack)) ->
    (left(x, L(d)(x)) + right(x, R(d)(x))).UseList(d, stack);
```

The Analysis. It needs to be verified that, at all times, the operations of removing and inserting an element indeed have the intended removal and insertion effects and that, moreover, the linear ordering induced by the list structure on the elements currently in the list is still consistent with the ordering induced by the list structure on the initial full list. Note that, by our definition of d_full, the ordering induced by the list structure on the initial full list is the standard ordering on the natural numbers between 0 and MAX, so in the formula below we can refer to the ordering induced by the list structure on the initial full list by simply referring to the standard ordering on the natural numbers.

```
nu X(s: Set(Nat) = {n: Nat | n <= MAX}).
  (forall x: Nat. [do_insert(x)](val(!(x in s)) && X(s + {x})) &&
                  [do_remove(x)](val(x in s) && X(s - {x}))) &&
  (forall x,x': Nat. val(x in s) =>
    [right(x,x')]( X(s) && val(x' in s) &&
      forall x'': Nat.(
         val(x < x'' && x'' < x')
      || val(x' <= x && x < x'' && x'' <= MAX)
      || val(x' <= x && 0 <= x'' && x'' < x')
      ) => val(!(x'' in s)) )
  ) &&
  (forall x,x': Nat. val(x in s) =>
    [left(x,x')]( X(s) && val(x' in s) &&
      forall x'': Nat.(
         val(x' < x'' && x'' < x)
      || val(x <= x' && x' < x'' && x'' <= MAX)
      || val(x <= x' && 0 < x'' && x'' < x)
      ) => val(!(x'' in s)) )
  )
```

The formula needs to express an invariant that holds for all reachable states, so we use a greatest fixed point. The parameter of the formula is a set s, which contains the natural numbers currently in the list. The subformula under the greatest fixed point operator expresses that

- the actions do_remove(x) and do_insert(x) can only take place when the element x can be removed or inserted, respectively, and that their execution results in behaviour that is in accordance with a list structure from which x has been removed or to which x has been added, respectively;
- whenever an action right(x,x') is enabled, then x' is indeed the next element in the list: it is either the least natural number larger than x in the list, or it is less or equal to x and natural numbers larger than x or smaller than x' are not in the list; an analogous property should hold when an action left(x,x') is enabled,

The formula can be verified for reasonably large numbers of MAX. It can be investigated whether the last-out-first-in order of removals and insertions is essential by replacing the stack parameter of the UseList process by a set. Doing so, we find that for values of MAX greater than 1 the correctness requirement fails.

6 Concurrent Data Structures

6.1 Treiber's Stack

In programming it is often necessary to keep track of shared resource elements, such as chunks of memory, that are free to use. Linked lists are often used to keep track of available resources, either using a first-in-first-out or a last-in-first-out strategy. When processes are using such a list concurrently, the standard sequential release and retrieve operations do not suffice any more. It is tempting to use *compare-and-swap* operations for insertion and deletion of list elements in that case. However, R. Kent Treiber showed in [26] that this does not work either. When using compare-and-swap the so-called *ABA problem* causes a problem, cf. [8]. Treiber showed that a double compare-and-swap operation is required. The ensuing data structure is a last-in-first-out linked-list that is commonly referred to as Treiber's stack.

The erroneous compare-and-swap implementation is interesting because it is very hard to find out by means of testing that the implementation is incorrect, especially when the number of elements in the list grows. Even with a dedicated testing scheme, it may take hundreds of millions of insertions and deletions in the list before the erroneous situation is encountered. However, when the error occurs, the list structure is in total disarray, leading to elements in the list becoming inaccessible and lost for use. Probably even worse, it allows elements to be simultaneously used by different processes. In practice this means that software using the faulty implementation can run well for years, but suddenly exhibit erroneous behaviour due to inexplicably messed up data structures.

The Treiber stack is described using a shared linked-list data structure that contains available shared resources v. Each node in the list contains a pointer $v.next$ to the next element in the list. The head of the list is contained in shared variable hd. A shared resource v can be released to the stack using $release(v)$. Resources can be obtained from the stack using $retrieve$. A pseudocode description of releasing and retrieving an element is the following.

release v :	*retrieve:*
repeat	**repeat**
$\quad v.next := hd;$	$\quad v := hd;$
$\quad b := comp\_and\_swap(hd, v, v.next);$	$\quad$ **if** $v \approx 0$ **return** *nothing*;
until $b;$	$\quad b := comp\_and\_swap(hd, v, v.next);$
	until $b;$
	return $v;$

The mCRL2 Model. We model a situation where two processes p1 and p2 share a Treiber stack. The shared linked-list representing the stack is described as follows. The stack consists of N elements, that are modelled by natural numbers. One number hd represents the head of the list. Function next: `Nat -> Nat` is such that for list element v, next(v) is the next element in the list. As the data structure is global, it is modelled using a separate process `treibers_stack` that maintains hd and next. The operations on the data structure are getting and setting a next value in next using actions set_next and get_next, as well as getting the value of hd and setting it using a compare-and-swap action cmp_swp_hd(id,v1,v2,b). When this action takes place, the variable hd is set to v2 provided hd was equal to v1. The boolean b is true when the value was changed. Otherwise, it is false. Note that `id` is the identity of the process that performs the compare-and-swap.

```
sort ID = struct p1 | p2;
map N: Nat;
eqn N = 2;

act set_next_r, set_next_s, set_next,
    get_next_r, get_next_s, get_next: ID # Nat # Nat;
    cmpswp_hd_r, cmpswp_hd_s, cmpswp_hd: ID # Nat # Nat # Bool;
    get_head_r, get_head_s, get_head: ID # Nat;
    nothing: ID;
    retrieve, release: ID # Nat;

proc
  treibers_stack(hd: Nat, next: Nat -> Nat) =
     sum id: ID, a,v: Nat. set_next_r(id, a, v).treibers_stack(hd, next[a -> v]) +
     sum id: ID, a: Nat. get_next_s(id, a, next(a)).treibers_stack(hd, next) +
     sum id: ID. get_head_s(id, hd).treibers_stack(hd, next) +
     sum id: ID, v1,v2: Nat. cmpswp_hd_r(id, v1, v2, hd==v1).
        treibers_stack(if(hd==v1, v2, hd), next);
```

Process `P(id, owns)` with identifier `id` retrieves resources from the stack and releases resources to it. The resources it currently owns are stored in `owns`. The process can either retrieve an element, which is then added to `owns`, except if the list is empty in which case no element is obtained, or it can release one of the elements it owns. These procedures have been encoded in the processes `release_elmt` and `retrieve_elmt`. Note that there are two actions `release (id, v)` and `retrieve(id, v)` that are used to signal that an element `v` is released to or retrieved from the list. Action `nothing(id)` indicates that process `id` tried to retrieve an element from the empty stack.

```
proc
  release_elmnt(id: ID, v: Nat, owns: Set(Nat)) =
    sum hd: Nat. get_head_r(id, hd).
    set_next_s(id, v, hd).
    sum b: Bool. cmpswp_hd_s(id, hd, v, b).
    (b -> P(id, owns-{v})
       <> release_elmnt(id, v, owns));

  retrieve_elmnt(id: ID, owns: Set(Nat)) =
    sum v: Nat. get_head_r(id, v).
    ((v==0) -> nothing(id).P(id, owns)
            <> (sum v_next: Nat. get_next_r(id, v, v_next).
                sum b: Bool. cmpswp_hd_s(id, v, v_next, b).
                (b -> retrieve(id,v).P(id, owns+{v})
                   <> retrieve_elmnt(id, owns))));

  P(id: ID, owns: Set(Nat)) =
    retrieve_elmnt(id, owns) +
    sum v: Nat. (v in owns) -> release(id, v).release_elmnt(id, v, owns);
```

The data structure is initialised in the init section by setting `hd` to `N`, and linking each element `l` to `l-1`. The number `0` is used as a *null*-pointer, *i.e.*, an indication for the empty list.

```
init allow({ set_next, get_next, cmpswp_hd, get_head,
             nothing, retrieve, release },
       comm({ set_next_r|set_next_s -> set_next,
              get_next_r|get_next_s -> get_next,
              cmpswp_hd_r|cmpswp_hd_s -> cmpswp_hd,
              get_head_r|get_head_s -> get_head },
```

```
treibers_stack(N, lambda l: Nat. max(0, l-1)) ||
P(p1, {}) || P(p2, {})));
```

The Analysis. There are two major properties of Treiber's stack that we want to hold. The first property essentially is the following. Provided that only elements are released that are not in the list, processes can only retrieve elements that are supposed to be in the list. This is expressed in the following formula, that is structurally similar to the last property for Peterson's algorithm. Parameter free represents the set of elements supposed to be in the list, and after retrieve and release actions, it is updated accordingly. The assumption on release is characterised using the implication in **val**(!(n *in* free))=> X(free + {n}).

```
nu X(free: Set(Nat) = { n:Nat | 0 < n && n <= N }).
  (forall id: ID, n: Nat.
    [release(id, n)](val(!(n in free)) => X(free + {n})) &&
    [retrieve(id, n)](val(n in free) && X(free - {n}))) &&
  [!exists id: ID, n: Nat. (release(id, n) || retrieve(id, n))]X(free)
```

The second property states that at any moment when there are at least two elements in the list, it is always possible to retrieve an element from the list within a finite number of actions while no elements can be released. The reason that there must be at least two elements in the list is that the other process can already have put a claim on an element in the list, without actually having retrieved it. The first five lines of the property follow a structure similar to properties we have seen before. The condition **val**(*exists* n: *Nat.* (n *in* free && free - {n} != {}) encodes that free contains at least two elements (it contains n, and after removing it, the set is non-empty). As sets can contain infinitely many elements, there is no function in mCRL2 that yields the size of a set. The last two lines say that along all paths not involving release actions, a retrieve action must inevitably happen.

```
nu X(free: Set(Nat) = { n:Nat | 0 < n && n <= N }).
  (forall id: ID, n: Nat.
    [release(id, n)]X(free + {n}) &&
    [retrieve(id, n)]X(free - {n})) &&
  [!exists id: ID, n: Nat. (release(id, n) || retrieve(id, n))]X(free) &&
  (val(exists n: Nat. (n in free && free - {n} != {})) =>
    mu Y.([(!exists id: ID, n: Nat. retrieve(id, n)) &&
            (!exists id: ID, n: Nat. release(id, n))]Y &&
      <!exists id: ID, n: Nat. release(id, n)>true))
```

It turns out that both formulas are not valid for Treiber's stack when implemented using compare-and-swap, even not so if there are initially two elements only. In Fig. 2 we depict two counter examples by drawing the working of the operations on the list. These are the shortest counterexamples that exist, as we used the modal formula prover in breadth first search mode.

The sequence of pictures at the left of the figure shows that it is possible that element 1 is retrieved twice from the data structure without being returned in between. Initially, process p2 attempts to retrieve the first data element, but it stops before doing the compare-and-swap. So, v = 2 and v_next = 1. Then process p1 obtains both element 2 and 1 in a proper fashion. This brings us to the third diagram where hd is 0; the list is empty. Process p1 returns element 2

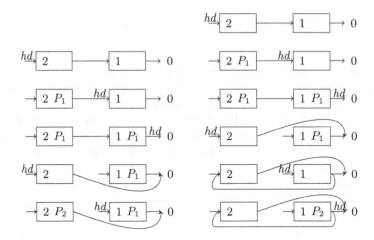

Fig. 2. Counter examples showing that Treiber's stack is incorrect

in the fourth picture. Now, process p2 carries out its compare-and-swap, setting hd to 1. The next element that will be released by the stack is 1 despite the fact that is currently owned by process p1.

The pictures at the right show how the second property is violated. Process p1 first obtains element 2. Subsequently, process p2 attempts to get element 1 setting v = 1 and v_next = 0, but again the compare-and-swap is not carried out. Process p1 obtains element 1 and returns subsequently element 2 and then 1. This yields the fifth picture from above. Process p2 carries out its compare-and-swap. As hd = 1 this is successful obtaining the situation at the list at the bottom right. If process p2 starts to release element 1, indicated by release(p2,1), but has not yet carried out the compare-and-swap, the variable free in the modal formula equals {1, 2}, so the free list contains at least two elements. However, hd = 0, so, process p1 will repeatedly fail to get an element from the list. Note that these failures can persist indefinitely, as long as p1 never completes the compare-and-swap.

Treiber's stack is correct when the compare-and-swap operation in the process retrieve_elt is replaced by a double compare-and-swap that checks both *hd* and *hd.next* have the expected value before updating them. Concretely,

> sum b: *Bool*. cmpswp_hd_s(id, v, v_next, b).

is replaced with

> sum b: *Bool*. double_cmpswp_s(id, v, v_next, v, v_next, b).

To support the double compare-and-swap operation, the treibers_stack process must be extended with the following:

```
sum id: ID, hd_old, hd_new, a, v_old: Nat.
    double_cmpswp_r(id, hd_old, hd_new, a, v_old, hd==hd_old && next(a) == v_old).
    treibers_stack(if(hd==hd_old && next(a)==v_old, hd_new, hd), next);
```

With this change Treiber's stack satisfies both correctness properties, which we could also check for slightly larger stacks.

When verifying the incorrect and correct versions of Treiber's stack one may observe that the state space of the incorrect version is much larger (approximately 300M states for N=4) than the correct version. This is a pattern that is more commonly observed in practice, and suggests that, when model checking, one should always start by analysing the smallest conceivable instance of a problems, and only increase instance sizes when these smallest instances satisfy all desired properties.

6.2 Lamport's Queue

Bounded single-producer/single-consumer queues have been studied extensively. A classical, and simple description of such a queue is given by Lamport [20]: a queue is represented by an array Q of size N, and is indexed $0 \ldots N - 1$. The queue is stored in a circular way, using indices *head* and *tail* to represent the pointers to the head and tail of the queue. Accesses to these pointers are assumed to be atomic. Reading and writing of array elements are non-atomic.

The queue supports two operations, push and pop, that are assumed to be executed from different threads. The producer repeatedly pushes elements to the queue, and the consumer repeatedly pops elements from the queue. When the queue is full, the operation push blocks until an element is removed from the queue. Likewise, when the queue is empty, the operation pop blocks until an element is added to the queue. The algorithm can be described in pseudocode as follows. Note that the algorithm is *wait-free*.

Global variables:	Behaviour push(v) :	Behaviour pop :
$Q[0 \ldots N)$: *Value*	**do**	**do**
head: $\mathbb{N}$	$\quad t := tail$	$\quad t := tail$
tail: $\mathbb{N}$	$\quad h := head$	$\quad h := head$
	while $(t + 1) \bmod N = h$	**while** $t = h$
	$Q[t] := v$	$v := Q[h]$
	$tail := (t + 1) \bmod N$	$head := (h + 1) \bmod N$

Variables t and h in both procedures are local. For correctness, the algorithm assumes so-called sequential consistency [19], *i.e.*, the operations in each of the procedures are executed in the order in which they appear in the program.

The mCRL2 Model. The indices in the array are modelled as natural numbers. Values in the queue are represented using finite sort Value; the special value garbage represents values in the array for which no concrete value is known (either the location is uninitialised, or there is an incomplete write to the location).

```
sort Value = struct garbage | d0 | d1;
```

The head pointer is modelled using the process Head(i). Parameter i stores the index to which Head points. The parameter can be set to an arbitrary index i', and the process continues recursively with this new index as its parameter. Alternatively, Head can return the current index it points to; the value of the parameter is not changed in this case. The tail pointer is modelled analogously. The mCRL2 code is as follows:

```
proc
  Head(i: Nat) = sum i': Nat. set_head_r(i').Head(i') + get_head_s(i).Head();
  Tail(i: Nat) = sum i': Nat. set_tail_r(i').Tail(i') + get_tail_s(i).Tail();
```

The queue of size N is modelled using N individual instances of the process Queue(i, v). Each of the instances represents a single position in the array, with index i, and value v that is currently stored at that position. Reads and writes are considered to be non-atomic. Therefore, writes are modelled using start_write_queue and end_write_queue. Reads are modelled using start_read_queue and end_write_queue. To accurately model non-atomicity of these operations, we keep track of threads currently reading from the position, writing to it (we store the value being written), or both. When a thread is writing, we store the value garbage in v. This enables us to verify that no value is ever read while another thread is writing.

```
map N: Pos;
eqn N=2;
proc
  Queue(i: Nat, v: Value) =
    sum v': Value. start_write_queue_r(i, v').QueueW(v = garbage, w = v') +
    start_read_queue_s(i).QueueR();
  QueueW(i: Nat, v: Value, w: Value) =
    end_write_queue_r(i).Queue(v = w) +
    start_read_queue_s(i).QueueRW();
  QueueR(i: Nat, v: Value) =
    sum v': Value. start_write_queue_r(i, v').QueueRW(v = garbage, w = v') +
    end_read_queue_s(i, v).Queue();
  QueueRW(i: Nat, v: Value, w: Value) =
    end_write_queue_r(i).QueueR(v = w) +
    end_read_queue_s(i, v).QueueW();
```

The Producer repeatedly Push-es an arbitrary (non-garbage) value to the queue. The Push process first gets the values of the tail and head pointer from the respective variables and keeps track of them locally in t and h. If the queue is full, the process blocks: it will loop and get the head and tail pointers again until space becomes available. Note that this is a less abstract way of modelling busy-waiting than that which was chosen in Peterson's algorithm. If the queue is not full, it will non-atomically store value v to the tail position t that was just obtained, and (atomically) update the tail pointer to $(t+1) \bmod$ n. The Pop process is modelled in a similar way.

```
proc
  Producer = sum v: Value. (v != garbage) -> call_push(v).Push(v).Producer;
  Push(v:Value) =
    sum t: Nat. get_tail_r(t).sum h: Nat. get_head_r(h).
    (((t+1) mod N == h) -> Push()
      <> ( start_write_queue_s(t, v).end_write_queue_s(t).
           set_tail_s((t+1) mod N).
           ret_push ));

  Consumer = call_pop.Pop.Consumer;
```

```
Pop =
  sum t: Nat. get_tail_r(t).sum h: Nat. get_head_r(h).
  ((t == h) -> Pop
      <> ( start_read_queue_r(h).
          sum v: Value. end_read_queue_r(h, v).
          set_head_s((h+1) mod N).
          ret_pop(v) ));
```

The process is initialised as follows:

```
init
  allow({ start_read_queue, end_read_queue, start_write_queue,
          end_write_queue, get_head, set_head, get_tail, set_tail,
          call_push, ret_push, call_pop, ret_pop },
    comm({ start_read_queue_s | start_read_queue_r -> start_read_queue,
           end_read_queue_s | end_read_queue_r -> end_read_queue,
           start_write_queue_r | start_write_queue_s -> start_write_queue,
           end_write_queue_r | end_write_queue_s -> end_write_queue,
           get_head_s | get_head_r -> get_head,
           set_head_r | set_head_s -> set_head,
           get_tail_s | get_tail_r -> get_tail,
           set_tail_r | set_tail_s -> set_tail },
        Queue(0,garbage) || Queue(1,garbage) || Head(0) || Tail(0) ||
        Producer || Consumer));
```

The Analysis. The main property we want to verify for Lamport's queue is that it actually behaves as a queue. The queue is *first-in-first-out*, and the capacity is never exceeded. Together, this is expressed in the following μ-calculus formula.

```
nu X(q: List(Value) = []).
  [ret_push](val(#q <= N)) &&
  forall v: Value.
    [call_push(v)](val(v != garbage) && X(v |> q)) &&
    [ret_pop(v)](val(q != [] && v == rhead(q)) && X(rtail(q))) &&
  [!(exists v': Value. call_push(v') || ret_pop(v'))] X(q)
```

This first order μ-calculus formula, similar to what we have seen before, keeps track of the contents of the queue in parameter q, which is a list of values. After every completion of a push operation it checks the current size of the queue is at most N. For every call to push with value v, it is verified that the value that is pushed is not garbage, and it recursively verifies the property for the queue extended with value v. Likewise, for every value that is returned by pop, it is verified that the queue was not empty, and the value that is returned corresponds to the oldest value in the queue. For all other values, q is checked recursively again. Since we use the greatest fixed point, we verify an invariant.

We can also verify other properties for the queue. For instance, every call to push is guaranteed to terminate. This is expressed as follows.

```
[true*.exists v: Value. call_push(v)]( mu X.[!ret_push]X && <true>true )
```

This property does not hold for the queue. A counterexample is the infinite sequence in which subsequence get_tail(0).get_head(0) is repeated indefinitely. In this case, the consumer always checks whether an element is available in the queue, but the producer never produces any value. Instead, the following property holds. It expresses that after call_push, it remains possible to do a ret_push as long as it has not been done yet.

```
[true*.exists v: Value. call_push(v).!ret_push*]<!ret_push* . ret_push>true
```

We next verify some properties for the `head` pointer; similar properties hold for `tail`. First, whenever the pointer to head is set, subsequent reads are guaranteed to return the same value, until the pointer is set again.

```
[true*] forall i: Nat. [set_head(i).(!exists j: Nat.set_head(j))*]
    forall i': Nat.[get_head(i')] val(i==i')
```

Next, the pointer is never set out of bounds.

```
[true*] forall i :Nat. [set_head(i)] val(i<N)
```

Finally, we verify that the `push` function never tries to write out of bounds. Due to the way synchronisation works, the calls `start_set_queue(i,v)` and `start_get_queue(i)` with $i \geqslant N$ will not be visible in the process as we modelled it so far. So, attempts to access these positions will silently fail. This can be circumvented by adding a process `Invalid` into the parallel composition with the following specification.

```
Invalid =
    sum i: Nat, v: Value. (i>=N) -> start_write_queue_r(i,v).Invalid +
    sum i: Nat. (i>=N) -> start_read_queue_s(i).Invalid;
```

If we incorporate this process, we can verify the absence of out of bounds writes using the following invariant.

```
[true*.exists i: Nat, v: Value. val(i>=N) && start_set_queue(i, v)] false
```

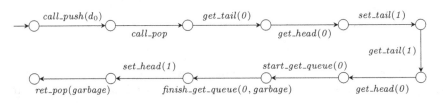

Fig. 3. Counterexample for Lamport's queue with no sequential consistency

As mentioned before, Lamport's queue assumes sequential consistency. If we drop this requirement, since there is no dependency between the assignment to $Q[t]$ and the assignment to *tail* in the *push* routine, the compiler may reorder these statements. If we change the mCRL2 model accordingly, and we verify whether the process behaves as a queue according to the first property stated above, the tool will observe the property does not hold. The counterexample that it generates is shown in Fig. 3. From this example we clearly see that the consumer reads invalid values from the queue. The underlying reason for this is that variable *head* is used for synchronisation between the producer and consumer; incrementing this variable signals to the consumer that a new value has been added to the queue. However, when allowing for reordering of operations, this is now signalled before the write has completed.

References

1. https://github.com/mCRL2org/mCRL2/tree/master/examples/software_models
2. Bartholomeus, M., Luttik, B., Willemse, T.: Modelling and analysing ERTMS hybrid level 3 with the mCRL2 toolset. In: Howar, F., Barnat, J. (eds.) FMICS 2018. LNCS, vol. 11119, pp. 98–114. Springer, Cham (2018). https://doi.org/10.1007/978-3-030-00244-2_7
3. Bertot, Y., Castéran, P.: Interactive Theorem Proving and Program Development - Coq'Art: The Calculus of Inductive Constructions. Texts in Theoretical Computer Science. An EATCS Series. Springer, Berlin (2004). https://doi.org/10.1007/978-3-662-07964-5
4. van Beusekom, R., et al.: Formalising the Dezyne modelling language in mCRL2. In: Petrucci, L., Seceleanu, C., Cavalcanti, A. (eds.) FMICS/AVoCS -2017. LNCS, vol. 10471, pp. 217–233. Springer, Cham (2017). https://doi.org/10.1007/978-3-319-67113-0_14
5. Bouwman, M., Janssen, B., Luttik, B.: Formal modelling and verification of an interlocking using mCRL2. In: Larsen, K.G., Willemse, T. (eds.) FMICS 2019. LNCS, vol. 11687, pp. 22–39. Springer, Cham (2019). https://doi.org/10.1007/978-3-030-27008-7_2
6. Bunte, O., et al.: The mCRL2 toolset for analysing concurrent systems. In: Vojnar, T., Zhang, L. (eds.) TACAS 2019. LNCS, vol. 11428, pp. 21–39. Springer, Cham (2019). https://doi.org/10.1007/978-3-030-17465-1_2
7. Cimatti, A., et al.: NuSMV 2: an opensource tool for symbolic model checking. In: Brinksma, E., Larsen, K.G. (eds.) CAV 2002. LNCS, vol. 2404, pp. 359–364. Springer, Heidelberg (2002). https://doi.org/10.1007/3-540-45657-0_29
8. Dechev, D., Pirkelbauer, P., Stroustrup, B.: Understanding and effectively preventing the ABA problem in descriptor-based lock-free designs. In: 13th IEEE International Symposium on Object/Component/Service-Oriented Real-Time Distributed Computing (ISORC 2010), Carmona, Sevilla, Spain, 5–6 May 2010, pp. 185–192. IEEE Computer Society (2010). https://doi.org/10.1109/ISORC.2010.10
9. Ernst, G., Huisman, M., Mostowski, W., Ulbrich, M.: VerifyThis – verification competition with a human factor. In: Beyer, D., Huisman, M., Kordon, F., Steffen, B. (eds.) TACAS 2019. LNCS, vol. 11429, pp. 176–195. Springer, Cham (2019). https://doi.org/10.1007/978-3-030-17502-3_12
10. Garavel, H., Lang, F., Mateescu, R., Serwe, W.: CADP 2011: a toolbox for the construction and analysis of distributed processes. STTT 15(2), 89–107 (2013)
11. Gibson-Robinson, T., Armstrong, P., Boulgakov, A., Roscoe, A.W.: FDR3—a modern refinement checker for CSP. In: Ábrahám, E., Havelund, K. (eds.) TACAS 2014. LNCS, vol. 8413, pp. 187–201. Springer, Heidelberg (2014). https://doi.org/10.1007/978-3-642-54862-8_13
12. Groote, J.F., Mousavi, M.R.: Modeling and Analysis of Communicating Systems. MIT Press, Cambridge (2014). https://mitpress.mit.edu/books/modeling-and-analysis-communicating-systems
13. Hennessy, M., Milner, R.: Algebraic laws for nondeterminism and concurrency. J. ACM 32(1), 137–161 (1985). https://doi.org/10.1145/2455.2460
14. Hitotumatu, H., Noshita, K.: A technique for implementing backtrack algorithms and its application. Inf. Process. Lett. 8(4), 174–175 (1979). https://doi.org/10.1016/0020-0190(79)90016-4
15. Holzmann, G.J.: The model checker SPIN. IEEE Trans. Softw. Eng. 23(5), 279–295 (1997). https://doi.org/10.1109/32.588521

16. Hwong, Y.L., Keiren, J.J.A., Kusters, V.J.J., Leemans, S., Willemse, T.A.C.: Formalising and analysing the control software of the Compact Muon Solenoid experiment at the large Hadron Collider. Sci. Comput. Program. **78**(12), 2435–2452 (2013). https://doi.org/10.1016/j.scico.2012.11.009

17. Keiren, J.J.A., Klabbers, M.D.: Modelling and verifying IEEE Std 11073–20601 session setup using mCRL2. Electron. Commun. EASST **53** (2013). https://doi.org/10.14279/tuj.eceasst.53.793

18. Knuth, D.E.: Dancing links (2000). arXiv:cs/0011047

19. Lamport, L.: How to make a multiprocessor computer that correctly executes multiprocess programs. IEEE Trans. Comput. **C-28**(9), 690–691 (1979). https://doi.org/10.1109/TC.1979.1675439

20. Lamport, L.: Specifying concurrent program modules. ACM Trans. Program. Lang. Syst. **5**(2), 190–222 (1983). https://doi.org/10.1145/69624.357207

21. Laveaux, M., Groote, J.F., Willemse, T.A.C.: Correct and efficient antichain algorithms for refinement checking. In: Pérez, J.A., Yoshida, N. (eds.) FORTE 2019. LNCS, vol. 11535, pp. 185–203. Springer, Cham (2019). https://doi.org/10.1007/978-3-030-21759-4_11

22. Nipkow, T., Wenzel, M., Paulson, L.C. (eds.): Isabelle/HOL - A Proof Assistant for Higher-Order Logic. LNCS, vol. 2283. Springer, Heidelberg (2002). https://doi.org/10.1007/3-540-45949-9

23. Peterson, G.L.: Myths about the mutual exclusion problem. Inf. Process. Lett. **12**(3), 115–116 (1981)

24. Remenska, D., Willemse, T.A.C., Verstoep, K., Templon, J., Bal, H.: Using model checking to analyze the system behavior of the LHC production grid. Future Gener. Comput. Syst. **29**(8), 2239–2251 (2013). https://doi.org/10.1016/j.future.2013.06.004

25. Roscoe, A.W.: Understanding Concurrent Systems. Texts in Computer Science. Springer, London (2010). https://doi.org/10.1007/978-1-84882-258-0

26. Treiber, R.K.: Systems programming: coping with parallelism. Technical Report RJ 5118 (53162). International Business Machines Incorporated, Thomas J. Watson Research Center, San Jose, California (1986)

27. Wikipedia. http://en.wikipedia.org/wiki/peterson's_algorithm (2015). Accessed 17 May 2015

Regular Papers

A Formally Verified Model of
Web Components

Achim D. Brucker[1]([✉])[ID] and Michael Herzberg[2][ID]

[1] Department of Computer Science, University of Exeter, Exeter, UK
a.brucker@exeter.ac.uk
[2] Department of Computer Science, The University of Sheffield, Sheffield, UK
msherzberg1@sheffield.ac.uk

Abstract. The trend towards ever more complex client-side web applications is unstoppable. Compared to traditional software development, client-side web development lacks a well-established component model, i. e., a method for easily and safely reusing already developed functionality. To address this issue, the web community started to adopt *shadow trees* as part of the Document Object Model (DOM). Shadow trees allow developers to "partition" a DOM instance into parts that should be safely separated, e. g., code modifying one part should not unintentionally affect other parts of the DOM.

While shadow trees provide the technical basis for defining web components, the DOM standard neither defines the concept of web components nor specifies the safety properties that web components should guarantee. Consequently, the standard also does not discuss how or even if the methods for modifying the DOM respect component boundaries.

In this paper, we present a formally verified model of web components and define safety properties which ensure that different web components can only interact with each other using well-defined interfaces. Moreover, our verification of the application programming interface (API) of the DOM revealed numerous invariants that implementations of the DOM API need to preserve to ensure the integrity of components.

Keywords: Web component · Shadow tree · DOM · Isabelle/HOL

1 Introduction

The trend towards ever more complex client-side web applications is unstoppable. Compared to traditional software development, client-side web development lacks a well-established component model which allows easily and safely reusing implementations. The Document Object Model (DOM) essentially defines a tree-like data structure (the *node tree*) for representing documents in general and HTML documents in particular.

Shadow trees are a recent addition to the DOM standard [24] to enable web developers to partition the node tree into "sub-trees." The vision of shadow trees

© Springer Nature Switzerland AG 2020
F. Arbab and S.-S. Jongmans (Eds.): FACS 2019, LNCS 12018, pp. 51–71, 2020.
https://doi.org/10.1007/978-3-030-40914-2_3

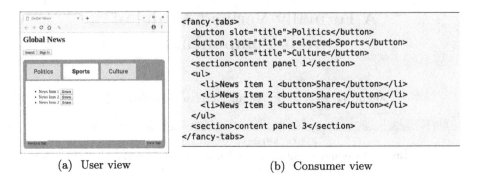

(a) User view   (b) Consumer view

Fig. 1. Our running example: a fancy tab component.

is to enable web developers to provide a library of re-usable and customizable widgets. For example, let us consider a multi-tab view called *Fancy Tab*, which is a simplified version of [3].

The left-hand side of Fig. 1 shows the rendered output of the widget in use while the right-hand side shows the HTML source code snippet (we will discuss the implementation of Fancy Tab in Sect. 3). It provides a custom HTML tag <fancy-tabs> using an HTML template that developers can use to include the widget. Its children will be rendered inside the widget, more precisely, inside its *slots* (elements of type slot). It has a slot called "title" and a default slot, which receives all children that do not specify a "slot" attribute.

It is important to understand that slotting does *not change* the structure of the DOM (i.e., the underlying pointer graph): instead, slotting is implemented using special element attributes such as "slot," which control the final rendering. The DOM standard specifies methods that inspect the effect of these attributes such as assigned_slot, but the majority of DOM methods do not consider the semantics of these attributes and therefore do not traverse into shadow trees.

This provides an important boundary for client-side code. For example, a JavaScript program coming from the widget developer that changes the style attributes of the "Previous Tab" and "Next Tab" buttons in the lower corners of the widget will not affect buttons belonging to other parts coming from outside, i.e., the application of the widget consumer. Similarly, a JavaScript program that changes the styles of buttons outside of Fancy Tab, such as the navigation buttons, will not have any effect on them, even in the case of duplicate identifiers.

Sadly, the DOM standard neither defines the concept of web components nor specifies the safety properties that they should guarantee, not even informally. Consequently, the standard also does not discuss how or even if the methods for modifying the node tree respect component boundaries. Thus, shadow roots are only the very first step in defining a safe web component model.

Earlier [7], we presented a formalization of the "flat" DOM (called Core DOM) without any support for shadow trees or components. In this paper, we extend the Core DOM with a formal model of shadow trees and slots which we use for defining a *formally verified model of web components* in general and, in particular, the notion of *weak* and *strong component safety*. For all methods that query, modify, or transform the DOM, we formally analyze their level of component safety. In more detail, the contribution of this paper is four-fold:

1. We provide a formal model of web components and their safety guarantees to web developers, enabling a compositional development of web applications,
2. for each method, we formally verify that it is either weakly or strongly component safe, or we provide a proof showing that it is not component safe,
3. we fill the gaps in the standard by explicitly formalizing invariants that are left out in the standard. These invariants are required to ensure that methods in the standard preserve a valid node tree. Finally,
4. we present a formal model of the DOM with shadow roots including the methods for querying, modifying, and transforming DOM instances with shadow roots.

Overall, our work gives web developers the guarantee that their code will respect the component boundaries as long as they abstain from or are careful when using certain DOM methods such as **appendChild** or **ownerDocument**.

On the Relationship Between the "DOM Standard" and the "Shadow DOM Standard." The development of shadow trees started in the context of the dedicated "Shadow DOM" standard [23]. This standard has been considered obsolete since at least March 2018, as shadow trees have been integrated into the DOM standard [24] itself. Hence, we will in the following only refer to the "DOM standard." Our formalization is faithful with respect to the standard in the sense that our formalization passes all relevant compliance test cases provided by the standard authors (the detailed discussion of this matter is out of scope of this paper).

2   Background

In this section, we will introduce the formal background of our work and the formalization of the Core DOM [6,7], without support for shadow roots.

2.1   Isabelle and Higher-Order Logic

Isabelle/HOL [17] is a generic theorem prover supporting Higher-order Logic (HOL). It supports conservativity checks of definitions, datatypes, primitive and well-founded recursion, and powerful generic proof engines.

HOL [2,9] is a classical logic with equality enriched with total polymorphic higher-order functions. HOL is strongly typed, i. e., each expression **e** has a type **'a**, written **e::'a**. In Isabelle, we denote type variables with a prime (e. g., **'a**) instead of Greek letters (e. g., α) that are usually used in textbooks. The type constructor for the function space is written infix: **'a $\Rightarrow$ 'b**. HOL is centered around the extensional logical equality $\_ = \_$ with type **'a $\Rightarrow$ 'a $\Rightarrow$ bool**, where **bool** is the fundamental logical type. The type discipline rules out paradoxes such as Russel's paradox in untyped set theory. Sets of type **'a set** can be defined isomorphic to functions of type **'a $\Rightarrow$ bool**; the element-of-relation

_ ∈ _ has the type 'a ⇒'a set ⇒ bool and corresponds basically to the function application; in contrast, the set comprehension {_ . _} (usually written {_ | _} in textbooks) has type 'a set ⇒ ('a ⇒ bool) ⇒'a set and corresponds to the λ-abstraction. Isabelle/HOL allows defining abstract datatypes. For example, the following statement introduces the option type:

datatype 'a option = None | Some 'a

Besides the *constructors* None and Some, there is the match-operation case x of None ⇒ F | Some a ⇒ G a, which acts as a case distinction on x::'a. The option type allows us to represent *partial functions* (often called *maps*) as total functions of type 'a ⇒'b option. For this type, we introduce the shorthand 'a ⇀'b. We define dom f, called the *domain* of a partial function f by the set of all arguments of f that do not yield None. By restricting the domain of a map to be finite, we can define a type that represents finite maps:

typedef ('a, 'b) fmap = {m. finite (dom m)}::('a ⇀ 'b) set

In this paper, we will use a short-hand that hides type variables when they are identical to the declaration of a datatype or a type definition. For example, we will write (_) fmap instead of ('a,'b) fmap. This short-hand notation is provided by [6] as an Isabelle theory, i.e., it is fully supported by Isabelle.

We represent DOM methods directly as HOL functions (*shallow embedding*) using a monad-like syntax that looks similar to Haskell programs. Additionally, we introduce the syntax h ⊢ f x →$_r$ y, which is a predicate that evaluates to true if and only if the method f invoked on heap h and with argument x returns the value y. Similarly, h ⊢ f x →$_h$ h', refers to the (potentially modified) new heap h'. In addition, we introduce syntax for binding the result directly to a variable y: y = |h ⊢ f x|$_r$.

2.2   The Core DOM

Our work is based on the formalization of the Core DOM presented in [6,7]. The Core DOM describes a tree-like data structure *without* shadow roots. Figure 2 shows the most important interfaces of the Core DOM using Web IDL [22], a formal notation used in the standard [24]. For each class (expressed as an interface in Web IDL), the Core DOM formalization introduces a HOL datatype representing a typed pointer. For example, the pointer for Node is modeled as:

datatype 'node_ptr node_ptr = Ext 'node_ptr

Here, object-oriented sub-typing (inheritance) is modeled using the type variable 'node_ptr. Classes are HOL records, e. g.,

record Node = Object + nothing :: unit

where we see that Node inherits from Object.

The Core DOM formalization provides an object-oriented data model of a tree-like data structure, which is called *node tree* in the DOM standard, where 1. the root of the tree is an instance of Document, 2. instances of the class Element can be internal nodes or leaves, and 3. instances of the class CharacterData can

```
interface Document {                 interface CharacterData : Node {
 readonly attribute                    attribute DOMString data;
         DocumentType? doctype;      };
 readonly attribute                  interface Element : Node {
         Element? documentElement;    readonly attribute
};                                            DOMString tagName;
                                      readonly attribute
interface Node {                              NodeList childNodes;
 readonly attribute                   readonly attribute
         Document? ownerDocument;             NamedNodeMap attributes;
 readonly attribute                   readonly attribute
         Node? parentNode;                    ShadowRoot? shadowRoot;
};                                   };
```

Fig. 2. The interface specification of the Core DOM.

only appear as leaves. Moreover, the Core DOM formalization defines a heap for "storing documents," i.e., instances of the DOM data model. A DOM heap is a finite map from object pointers to objects:

```
datatype ('object_ptr, 'Object) heap
   = Heap (the_heap: ('object_ptr object_ptr, 'Object Object) fmap)
```

On top of this data model, the Core DOM formalization also defines methods for creating, querying, and modifying DOM heaps:

- get_attribute returns the attribute (e. g., id or class) of an element,
- set_attribute sets the attribute of an element,
- get_tag_name returns the tag type (e. g., div) of an element,
- set_tag_name sets the tag type of an element,
- get_child_nodes returns the children of an element or the document element of a document,
- get_ancestors returns a list of ancestor nodes, with the first node being the argument itself, the second one being the parent, and so on,
- get_parent returns the parent.
- get_root_node returns the root node, the node that is obtained after repeatedly calling get_parent,
- insert_before inserts the given node into the children of the argument, possibly removing it first from its former parent,
- get_element_by_id traverses the tree in depth-first pre-order (called tree-order in the standard) and returns the first element matching the given id,
- get_owner_document returns the owner document.

The formal signatures of these methods in Isabelle/HOL are: Fig. 3 provides an overview of the formal signatures in Isabelle/HOL: All methods return a program of type (_, 'result) dom_prog, where 'result can be interpreted as the "real" return type of the methods. A dom_prog takes a heap and returns either an error or a result along with a new heap.

```
get_attribute :: (_) element_ptr ⇒ attr_key
                       ⇒ (_, attr_value option) dom_prog
set_attribute :: (_) element_ptr ⇒ attr_key ⇒ attr_value option
                       ⇒ (_, unit) dom_prog
get_tag_name :: (_) element_ptr ⇒ (_, tag_type) dom_prog
set_tag_name :: (_) element_ptr ⇒ tag_type ⇒ (_, unit) dom_prog
get_child_nodes :: (_) object_ptr ⇒ (_, (_) node_ptr list) dom_prog
get_ancestors :: (_::linorder) object_ptr
                       ⇒ (_, (_) object_ptr list) dom_prog
get_parent :: (_) node_ptr ⇒ (_, (_) object_ptr option) dom_prog
get_root_node :: (_) object_ptr ⇒ (_, (_) object_ptr) dom_prog
insert_before :: (_) object_ptr ⇒ (_) node_ptr ⇒ (_, unit) dom_prog
get_element_by_id :: (_) object_ptr ⇒ attr_value
                       ⇒ (_, (_) element_ptr option) dom_prog
get_owner_document :: (_) object_ptr ⇒ (_, (_) document_ptr) dom_prog
```

Fig. 3. The methods for creating, querying, and modifying the DOM. All functions return a program of type (_, 'result) dom_prog, where 'result can be interpreted as the "real" return type of the function. A dom_prog takes a heap and returns either an error or a result and a new heap.

Not all objects in a heap are necessarily a regular node of a DOM instance. The DOM also introduces the concept of *disconnected* nodes (e. g., for freshly created objects or local variables), which are unreachable from the main document until they are inserted into the node tree by, e. g., using insert_before.

3   Motivating Example: Fancy Tab

In this section, we discuss our running example Fancy Tab from Fig. 1. Figure 4a focuses on the HTML part of defining Fancy Tab. As the DOM standard does not allow for creating shadow roots statically (i. e., using pure HTML), the definition of shadow roots requires JavaScript to create them at run-time. In our example, we assign the actual definition to innerHTML of an already created shadow root.

Figure 4b shows an attempt to provide the functionality of Fancy Tab *without* using shadow roots. While this alternative definition provides—at first glance—a similar "look and feel," it does not provide any form of run-time separation.

For both variants, we assume that a web developer consumes Fancy Tab without being familiar with its implementation and would like to style their navigation buttons by changing the label text to upper case. Let us assume that the web developer uses the JavaScript snippet from Fig. 5c to change the button texts, which traverses the document's node tree starting from its root node, looking for buttons, and changing their innerText attribute.

Now, let us observe the results: Fig. 5b shows the version without shadow roots; here, all buttons, including the navigation buttons on the bottom (which belong to the widget and should be abstracted away), turned upper case, as they are part of the same scope. We consider this undesired behavior, because the developer inadvertently modified the internal representation of Fancy Tab: we would like the Fancy Tab developer to be protected from these kinds of effects.

```
shadowRoot.innerHTML = '
  <style>...</style>
  <div id="tabs">
    <slot id="tabsSlot" name="title"></slot>




  </div>
  <div id="panels">
    <slot id="panelsSlot"></slot>








  </div>
  <button id="left">
    Previous Tab</button>
  <button id="right">
    Next Tab</button>
';
```

```
<div>
  <style>...</style>
  <div id="tabs">
    <button slot="title">
      Politics</button>
    <button slot="title" selected>
      Sports</button>
    <button slot="title">
      Culture</button>
  </div>
  <div id="panels">
    <section>content panel 1</section>
    <ul><li>News Item 1
        <button>Share</button></li>
      <li>News Item 2
        <button>Share</button></li>
      <li>News Item 3
        <button>Share</button></li></ul>
    <section>
      content panel 3</section>
  </div>
  <button id="left">
    Previous Tab</button>
  <button id="right">
    Next Tab</button>
</div>
```

(a) Excerpt of the source of the Fancy Tab widget. We assign the HTML definition to the innerHTML of an already created shadow root.

(b) Defining Fancy Tab without shadow roots would require mixing the code of Fancy Tab and the consuming app, *losing any kind of separation properties.*

Fig. 4. The source code of Fancy Tab with shadow roots (left) and without (right).

Figure 5a shows the version with shadow roots, where we can see that only the buttons in the top navigation bar turned upper case—the navigation buttons on the bottom remain unaffected, because they are not part of the same scope (i. e., they are not part of the document). To understand the difference, we need to look at the DOM representation with shadow roots as shown in Fig. 6; by calling document.getElementsByTagName("button"), we enumerate all buttons, starting from the root document, and thus traverse the tree along the solid arrows. The method getElementsByTagName traverses the tree in depth-first pre-order (called *tree order* in the DOM standard), which does not descend along the dotted line.

4   Formalizing Shadow Trees

In this section, we describe how we formalize the addition of shadow trees, i. e., sub-trees of a DOM whose root node is a shadow root, to the DOM. In particular, we 1. extend the data model of the Core DOM to support shadow roots, 2. elicit and formalize invariants that are not made explicit in the DOM standard, and 3. formalize methods for querying and modifying shadow roots.

 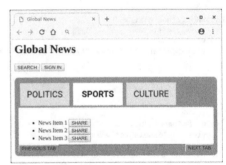

(a) Styling Fancy Tab with shadow root only affects buttons outside of Fancy Tab, but not inside.

(b) Styling fancy tabs without shadow root affects additionally buttons inside of Fancy Tab.

```
for (let btn of document.getElementsByTagName("button")) {
    btn.innerText = btn.innerText.toUpperCase();
}
```

(c) A simple JavaScript snippet that converts all button labels to upper case.

Fig. 5. Difference of modifying a website with shadow roots and one without.

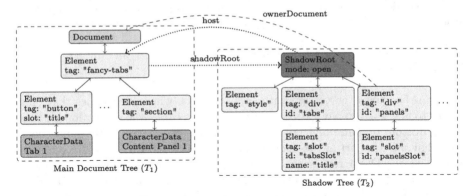

Fig. 6. Representation of the internal DOM structure of our running example Fig. 1.

4.1 Data Model and Basic Accessors

To represent shadow trees, we introduce a new type: **ShadowRoot**. Using Web IDL, the interface of a shadow root is given as:

```
interface ShadowRoot {
    readonly attribute ShadowRootMode mode;
    readonly attribute Element host;
    readonly attribute NodeList childNodes;
}
```

Shadow roots can only be contained in an `Element`, where they behave as a special kind of child (successor) node. Instead of using `get_child_nodes` or `get_parent`, we will introduce the methods `get_shadow_root` and `get_host`, respectively, that act similarly. We formalize these signatures as follows:

get_shadow_root ::
    (_) element_ptr ⇒ (_, (_) shadow_root_ptr option) dom_prog
get_mode :: (_) shadow_root_ptr ⇒ (_, shadow_root_mode) dom_prog
get_host :: (_) shadow_root_ptr ⇒ (_, (_) element_ptr) dom_prog

Shadow roots manage a list of children, like an `Element` does, but also a flag `mode` that indicates whether the sub-tree shall be accessible from the outside. For this purpose, this flag affects methods such as `get_shadow_root`, i.e., they will not return any nodes if the shadow tree is closed. Figure 6 illustrates how the new node fits into the concept of a tree structure as defined by the Core DOM.

4.2 Tree Order and DOM Invariants

The data model given in the DOM standard describes a directed object graph, but not necessarily a tree-like data structure. The fact that a valid DOM needs to be a tree-like data structure is only given implicitly: The standard informally defines the concept of a *tree order* as "pre-order depth-first search." The tree order is defined in two variants: *(shadow-excluding) tree order* and *shadow-including tree order*. The former ignores shadow roots while the latter traverses (open) shadow roots prior to traversing the "regular" child nodes. We formalize the shadow-including tree order as follows:

```
partial_function (dom_prog) to_tree_order_si
  :: (_) object_ptr ⇒ (_, (_) object_ptr list) dom_prog where
  to_tree_order_si ptr = do {
    children ← get_child_nodes ptr;
    shadow_root_part ← (case cast ptr of
      Some element_ptr ⇒ do {
        shadow_root_opt ← get_shadow_root element_ptr;
        (case shadow_root_opt of
          Some shadow_root_ptr ⇒ return [cast shadow_root_ptr] |
          None ⇒ return [])
      } |
      None ⇒ return []);
    treeorders ← map_M to_tree_order_si
      ((map cast children) @ shadow_root_part);
    return (ptr # concat treeorders)
  }
```

While not explicitly stated by the standard, it implicitly assumes that the algorithm computing a shadow-including tree order terminates for all "valid" instances of the DOM. In our formalization, this is expressed as the property that "for all well-formed heaps, the (partial) function `to_tree_order_si` does not produce an error." In the following, we will discuss the requirements that are necessary to formally prove this property.

A well-formed heap needs to fulfill several properties that can either be modeled as typing constraints or predicates. An example for the former is the property "an element has at most one attached shadow root," which is, both in the DOM standard and our formalization, enforced by the type system. As an example for the latter, we model the requirement that "shadow roots in a node tree are always attached to a valid host" by using a predicate:

```
definition shadow_root_valid :: (_) heap ⇒ bool where
  shadow_root_valid h = (∀shadow_ptr |∈| fset (shadow_root_ptr_kinds h).
    (∃host. host |∈| element_ptr_kinds h ∧
      |h ⊢ get_tag_name host|r ∈ safe_shadow_root_element_types ∧
      |h ⊢ get_shadow_root host|r = Some shadow_ptr))
```

The constraint shadow_root_is_valid is described in the standard informally as requirement that a DOM does not contain "disconnected" shadow roots.

Moreover, the standard required that shadow roots cannot belong to more than one host:

```
definition distinct_lists :: (_) heap ⇒ bool where
  distinct_lists h = distinct (concat (
    map (λelement_ptr. (case
      |h ⊢ get_shadow_root element_ptr|r of
        Some shadow_root_ptr ⇒ [shadow_root_ptr] |
        None ⇒ []))
    |h ⊢ element_ptr_kinds_M|r
))
```

To ensure that the DOM with shadow roots is a tree-like data structure, we need to ensure that the underlying object-graph is acyclic. In HOL, we model this in two steps. First, we define a relation between hosts and shadow roots:

```
definition host_shadow_root_rel :: (_) heap ⇒
    ((_) object_ptr × (_) object_ptr) set where
  host_shadow_root_rel h = (λ(x, y).
    (cast x, cast y)) ' {(host, shadow_root).
      host |∈| element_ptr_kinds h
      ∧ |h ⊢ get_shadow_root host|r = Some shadow_root}
```

This relation captures the requirement that the "link" between shadow roots and hosts is a reversible relation. Second, we make use of the pre-defined acyclic predicate of HOL for arbitrary relations to postulate that this relation is acyclic.

Now, we can formally capture the concept of a well-defined heap:

```
definition heap_wf :: (_) heap ⇒ bool where
  heap_wf h ⟷
  Core_DOM.heap_wf h ∧
    acyclic (Core_DOM.parent_child_rel h ∪ host_shadow_root_rel h)
    ∧ all_ptrs_in_heap h ∧ distinct_lists h ∧ shadow_root_valid h
```

More precisely, a well-defined heap requires that the regular parent-child relation (defined in the Core DOM [6]) *together* with the shadow root-host relation is acyclic, so we combine them and need `parent_child_rel h ∪ host_shadow_root_rel h` to be acyclic. Also, we require that all pointers in a DOM instance are pointing to instances that are members of the well-defined heap (this is captured by `all_ptrs_in_heap h`).

Additionally, we introduce a short-hand predicate `valid_heap` (which we will use in lemmas throughout this paper) that captures `heap_wf`, but also ensures that the heap only contains pointers and objects whose types correspond (`type_wf`) and pointers that are "known" (`known_ptrs`), which is a property related to the extensibility (see [5,8] for details on how to encode extensible object-oriented data models in HOL) of the formal model:

definition valid_heap :: (_) heap ⇒ bool **where**
  valid_heap h = heap_wf h ∧ type_wf h ∧ known_ptrs h

We can now formally prove, in Isabelle/HOL, that `to_tree_order_si` will always terminate for well-defined heaps, meaning its execution is error-free (captured by the predicate `ok`):

lemma to_tree_order_si_ok:
  assumes valid_heap h
  assumes ptr |∈| object_ptr_kinds h
  shows h ⊢ ok (to_tree_order_si ptr)

This ensures termination, since `to_tree_order_si` is a partial functions in Isabelle/HOL that maps the case of non-termination to a value of our error type.

5 Web Components

We will now focus on the semantics of web components. While the DOM standard introduces the API for working with shadow trees, it neither defines the concept of a component nor specifies the safety guarantees that should be provided to authors or consumers of components.

5.1 A Formal Definition of Web Components

Many DOM methods, e.g., `get_element_by_id`, traverse the node tree top-down exclusively along the `childNodes` relation (i.e., in shadow-excluding tree order). These methods will not traverse the DOM along the `shadowRoot` relation. Similarly, methods, such as `get_root_node` only traverse the tree bottom-up using the `parent` relation; they will not continue along the `host` relation.

Intuitively, the `shadowRoot` relation acts as a "component boundary" that can only be crossed by explicitly calling a method that is defined to traverse the `shadowRoot`-`host` relation.

The standard informally introduces a *(shadow-excluding) tree order* computation (in the standard this is an abstract concept, i.e., not a method available directly to web developers) that returns, in depth-first pre-order, all nodes reachable from a given node by traversing the `childNodes` relation. We use its formalization `to_tree_order` to provide a formal definition of web components:

Definition 1 (Web Component). *A* (Web) Component *of an object o is defined as the list of all objects in* tree order *that are reachable from the root node of o. Formally, we define:*

```
definition
  get_component :: (_) object_ptr ⇒ (_, (_) object_ptr list) dom_prog
where
  get_component ptr = get_root_node ptr ≫= to_tree_order
```

Informally, an object *o* belongs to a component *c* if and only if *o* is in the list of nodes that are reachable from the root of *c* via the `childNodes` relation. In our running example (see Fig. 6) the set of all objects is divided into two components: T_1 and T_2.

Our component definition naturally allows distinguishing three different types of components, based on the type of their root node.

Definition 2 (Document Component). *A* Document Component *is a web component where the root node is of type* **Document**. *Formally, we define:*

```
definition is_document_component :: (_) object_ptr list ⇒ bool
  where is_document_component c = is_document_ptr_kind (hd c)
```

Since an object of type **Document** can only occur as the root node of a node tree, a document component can be considered the main part of a node tree.

Definition 3 (Shadow Root Component). *A* Shadow Root Component *is a web component where the root node is of type* **ShadowRoot**. *Formally, we define:*

```
definition is_shadow_root_component :: (_) object_ptr list ⇒ bool
  where is_shadow_root_component c = is_shadow_root_ptr_kind (hd c)
```

A shadow root component might be considered the "canonical component." It encapsulates its contained nodes from outside components and uses slots to interact with the outer component.

Finally, we define a disconnected component as a component only containing *disconnected nodes* (recall Sect. 2), i.e., nodes that are not reachable by traversing the DOM (not even in shadow-including tree order) from its `ownerDocument`.

Definition 4 (Disconnected Component). *A* Disconnected Component *is a web component where the root node is of type* **Node**. *Formally, we define:*

```
definition is_disconnected_component :: (_) object_ptr list ⇒ bool
  where is_disconnected_component c = is_node_ptr_kind (hd c)
```

Disconnected components will not take part in the rendering of the final node tree. Usually, disconnected components will contain freshly crated object graphs that will become a part of a "regular" DOM instance by passing them as argument to methods such as append_child.

5.2 Component Safety

Web components should provide a certain form of safety guarantee to both component developers and consumers of components. Informally speaking, neither should a DOM method unintentionally modify the consuming web application when called in the context of the component, nor the other way round. This is particularly important for web components that are developed in JavaScript, a language without (static) typing and with concepts that support the run-time extension of classes using prototype inheritance.

To address this issue, we introduce the notion of component safety for DOM methods that captures which part of a DOM can be modified by a method. We distinguish three types of methods; ones that

1. only operate within the components of their arguments, as one could argue that it is expected that most methods only operate within their proximity. We will call these methods *strongly component-safe*.
2. only operate within the components of their arguments and any newly created components. While these methods operate outside their perceived boundaries, they at least leave other, existing components untouched. We will call these methods *weakly component-safe*.
3. operate on arbitrary parts of a DOM instance. We will call these methods *unsafe*.

In the following, we will introduce our different levels of component safety. In our formalization, we analyze the level of component safety for all methods of the DOM standard. Due to the limitations of using a shallow embedding in HOL, we cannot provide a single HOL predicate that captures the level of component safety. Instead, we will provide two predicates (one for strong safety, one for weak safety) that capture the essence of the safety definitions, along with a proof pattern that we apply to each DOM method.

We start by defining strong component safety, followed by discussing the formal proofs for showing strong component safety.

Definition 5 (Strong Component Safety). *A DOM method is* strongly component safe *if and only if it does not create, delete, return, or modify any objects outside of the components given by its arguments.*

In HOL, we define this property as follows:

```
definition is_strongly_component_safe :: (_) object_ptr set ⇒
    (_) object_ptr set ⇒ (_) heap ⇒ (_) heap ⇒ bool where
  is_strongly_component_safe S_arg S_result h h' =
    let outside_ptrs = fset (oeject_ptr_kinds h)  -
        (⋃ptr ∈ S_arg. set |h ⊢ get_component ptr|_r) in
    let outside_ptrs' = fset (object_ptr_kinds h') -
        (⋃ptr ∈ S_arg. set |h' ⊢ get_component ptr|_r) in
    outside_ptrs = outside_ptrs' ∧
    S_result ∩ outside_ptrs = {} ∧
    (∀outside_ptr ∈ outside_ptrs. preserved (get_M outside_ptr id) h h')
```

The predicate takes the set of pointers that are arguments of the method invocation, the set of pointers that is returned by the method invocation, and the state of the heap before and after the method invocation. It then builds the set of pointers that lie outside of the arguments' components before and after the method invocation, and then makes three assertions: both sets of pointers must be equal, i.e., no pointers have been created or deleted; the intersection of the result pointer set and the set of pointers of the arguments' components must be empty; and all pointers outside of these components must remain unmodified (`get_M outside_ptr id` returns the whole object, which is compared to other objects by comparing all their fields).

We will show how this predicate is used to show the strong component safety of a DOM method. For this purpose, we will look at the proofs of component safety for `get_child_nodes` and `get_element_by_id`, which are both *strongly* component safe. The complete formal proofs for these and all other supported DOM methods are included in our Isabelle formalization. In order to show that `get_child_nodes` is strongly component safe, we prove the following lemma:

```
lemma get_child_nodes_strongly_component_safe:
  assumes valid_heap h
  assumes h ⊢ get_child_nodes ptr →_r children
  assumes h ⊢ get_child_nodes ptr →_h h'
  shows is_strongly_component_safe {ptr} (cast ' set children) h h'
```

The first argument of `is_strongly_dom_component_safe` is S_{arg}, which is the set of all pointers that are arguments to the DOM method call in question—in the case of `get_child_nodes`, that is only `ptr`. The second argument is S_{result}, the set of all pointers that are returned by the method, in this case the returned `children` after they have been appropriately cast to object pointers. The last two arguments, h and h' refer to the heap states before and after the method call. For `get_child_nodes` they will both be the same, so all that is to show is that none of the `children` are outside of the component of `ptr`. Since the component is constructed by iteratively invoking `get_child_nodes` (`to_tree_order` in Definition 1), it follows that the `children` are indeed inside that component and therefore not outside of it.

Methods that iterate in shadow-excluding tree order are also strongly component safe, as in the case of `get_element_by_id`:

```
lemma get_element_by_id_is_strongly_component_safe:
  assumes valid_heap h
  assumes h ⊢ get_element_by_id ptr id →ᵣ Some result
  assumes h ⊢ get_element_by_id ptr id →ₕ h'
  shows is_strongly_component_safe {ptr} {cast result} h h'
```

This is the variant for the case that such an element pointer, `result`, is indeed found. The case that no such element is found is a separate lemma and trivial to prove. The proof idea for this kind of lemma is similar to proof idea for `get_child_nodes`; any object has the same root as its parent, and any node found by `get_element_by_id` has the same root as the anchored object—from the definition of `get_component` it then follows that they also have the same component. As these methods do not modify the heap, this is all we need to show for strong component safety.

We continue by defining weak component safety:

Definition 6 (Weak Component Safety). *A DOM method is* weakly component safe *if and only if it does not delete, return, or modify any objects outside of the components given by its arguments.*

The only difference between weak component safety and strong component safety (Definition 5) is that weak safety allows for the creation of new objects outside of the given components. Clearly, strong component safety implies weak component safety, i.e., any strongly component safe method is also weakly component safe.

In general, methods that create new objects are weakly component safe, because most of them return a new object which is not part of any component yet, thus effectively creating a new one.

Examples of weakly component safe methods are **create_element** (creating a new disconnected component), **create_character_data** (creating a new disconnected component), **create_document** (creating a new document component), and **attach_shadow_root** (creating a new shadow root component). For example, for **create_element** we prove the following lemma:

```
lemma create_element_is_weakly_component_safe:
  assumes valid_heap
  assumes h ⊢ create_element document_ptr tag →ᵣ result →ₕ h'
  shows is_weakly_component_safe {cast document_ptr} {cast result} h h'
```

The proof idea is that the only object (that existed in h) that is modified is referenced by **document_ptr**, which has the newly created element added to its list of disconnected nodes. The new element pointer forms its own, new component, as it does not belong to the one of **document_ptr** or any other one. This is allowed by the definition of weak component safety.

Now we know that **create_element** is weakly component safe, but we would also like to show that it is indeed *not* strongly component safe. In order to do so, we will again leverage our `is_strongly_component_safe` predicate, but invert our proof pattern:

lemma `create_element_not_strongly_component_safe`:
  obtains h **and** h' **and** document_ptr **and** new_element_ptr **and** tag
where
  valid_heap h **and**
  h ⊢ create_element document_ptr tag →$_r$ new_element_ptr →$_h$ h' **and**
  ¬ is_strongly_component_safe
      {cast document_ptr} {cast new_element_ptr} h h'

The structure of this kind of lemma is different from the lemmas showing safety; we use the Isar **obtains** concept to show that there exists at least one heap for which the method is not strongly component safe. It therefore suffices to construct a counter-example, which in the case of `create_element` can be as small as a heap containing nothing but a single `document_ptr`. Since our model is completely executable, we can use the symbolic execution engine of Isabelle to show that this heap indeed fulfills the lemma.

The next class of methods includes ones that concern shadow roots, which are expectedly generally unsafe. In case of a closed shadow tree (`mode` is set to `Closed`), methods trying to look inside Shadow Root Components (e.g., `get_shadow_root` and `assigned_slot`) will return an error, making these methods strongly component-safe in this case—methods trying to break out (e.g., `get_host`, `get_composed_root_node`, and `get_assigned_nodes`) are not affected by the mode and thus remain unsafe. In the case of `assigned_slot`, this looks as follows:

lemma `assigned_slot_not_weakly_dom_component_safe`:
  obtains h **and** node_ptr **and** slot_opt **and** h'
where
  valid_heap h **and**
  h ⊢ assigned_slot node_ptr →$_r$ slot_opt →$_h$ h' **and**
  ¬ is_weakly_component_safe
      {cast node_ptr} (cast ' set_option slot_opt) h h'

We use the same construction as we did for showing that `create_element` was not strongly safe, but now use `is_weakly_component_safe`. We use the function `set_option` to convert the return value of `assigned_slot` into a pointer set, so we can allow both possible outcomes of the method call, regardless of whether a slot has been found. Recall that we are constructing a counter example here, so we only need to find one valid instantiation of variables. The proof follows the usual schema, though the counter-example is more complex than before. We need to create a heap that contains an element, a shadow root, slots and slotables, but otherwise does not require any special configuration.

5.3   Component Safety of the DOM Methods

In the following, we will discuss to what extent the methods defined in the DOM standard are component safe. By doing this we effectively evaluate how suitable shadow roots are for providing separation. We will see that some methods, in particular **append_child**, break our separation in unexpected ways.

Table 1 summarizes our classification. All shown lemmas with their proofs can be found in our formalization (in the formalization, the lemma names start with the method name, followed by _is or _not, followed either by the suffix _strongly_component_safe, _weakly_component_safe, or, if the method is not safe at all, _component_unsafe). The two methods **get_shadow_root** and **assigned_slot** are only safe if the DOM only contains closed shadow roots.

Table 1. Classification of the DOM methods into whether they are strongly or weakly component safe, or not at all. The last column (closed) classifies the methods for the special case that the DOM instance only contains closed shadow roots.

Method	Component safety	
	Open	Closed
get_child_nodes	Strong	Strong
get_parent	Strong	Strong
get_root_node	Strong	Strong
get_element_by_id	Strong	Strong
get_elements_by_class_name	Strong	Strong
get_elements_by_tag_name	Strong	Strong
create_element	Weak	Weak
create_character_data	Weak	Weak
create_document	Weak	Weak
attach_shadow_root	Weak	Weak
get_shadow_root	Unsafe	Safe
get_host	Unsafe	Unsafe
get_composed_root_node	Unsafe	Unsafe
get_assigned_nodes	Unsafe	Unsafe
assigned_slot	Unsafe	Safe
adopt_node	Unsafe	Unsafe
remove_child	Unsafe	Unsafe
insert_before	Unsafe	Unsafe
append_child	Unsafe	Unsafe
get_owner_document	Unsafe	Unsafe

From the view of a web developer, the fact that **get_owner_document** is unsafe is particularly worrisome: if the root node of a given pointer is not a document, then this method will return a document that is outside of the current component. Thus, if a library developer uses this method for setting up their component, they might inadvertently break out and change arbitrary objects outside of their component.

Surprisingly and unfortunately, many of the heap-modifying methods such as `adopt_node`, `remove_child`, `insert_before`, and `append_child` are unsafe, too, because they all access and modify the list of disconnected nodes of owner documents. For example, if an element is removed by using `remove_child`, it gets added to the list of disconnected nodes of the argument's owner document, which is outside of the component of the removed child.

We have seen that there are a number of DOM methods which we could prove to be unsafe with regard to components. This is undesirable, as this means that these methods break the expectations that a developer might have when working with shadow root components.

5.4   Recommendations

Web components based on shadow trees are an important step forward for a component-based web development approach. They allow web developers to define components with well-defined interfaces (called slots) for interacting with the embedding application or other components (components can be nested arbitrarily). However, our formal analysis shows that there are subtle ways to accidentally break the component boundaries: most prominently, the enclosing owner document is easily accessible from inside a shadow root component by using the `ownerDocument()` method on any node of that component, which corresponds to the ubiquitous `document` reference in any (Web) JavaScript context. We suggest changing this behavior and instead providing a reference to the root of the current component, thus strengthening the component separation against accidental interference with other components. This would, on the one hand, remove the most unexpected way of breaking up the component boundaries and, on the other hand, simplify the overall definition of web components. This change would also simplify the notion of component safety by removing boundary cases for disconnected nodes.

A second point of concern is that methods such as `remove_child` can have unexpected effects outside of shadow trees, even if all arguments lie within that shadow tree. Such removed nodes get added to the context of the surrounding document, from which they might added again onto other, unrelated components of the same document (DOM instance).

6   Related Work

To the best of our knowledge, we are the first to formalize the concept of shadow roots. The most closely related works are our own formalization of the DOM [7] *without* shadow roots that we use as basis of our component model, and the works of Gardner et al. [12], Raad et al. [19], Smith [20]. In the latter ones, the authors present a non-executable, non-extensible, and non-mechanized operational semantics of a minimal DOM and show how this semantics can be used for Hoare-style reasoning for analysis heaps of DOMs. The authors focus on providing a formal foundation for reasoning over client-side JavaScript programs that modify the DOM. Neither of these works defines formally the concept of web components nor do they define component safety or formally analyze the behavior of DOM methods in the context of shadow trees.

Our work shares a common goal with ownership type systems [10]. For example, Poetzsch-Heffter et al. [18] use type annotations to give objects in object-oriented programs a notion of ownership. This enables them to allow certain components only read-access to an object, while the owner might have full read and write-access. This line of work is orthogonal to ours; it is certainly possible to create an access-control layer on top of our web components, but we are more concerned with components inside a tree-like structure and how a given set of methods behave regarding the boundary induced by shadow roots.

A more informal model of the DOM that focuses on the needs of building a static analysis tool for client-side JavaScript programs is presented by Jensen et al. [15]. This model does not focus on the DOM as such, instead the authors focus on the representation of HTML documents on top of the DOM.

There are also very few formalizations of data structures for manipulating XML-like document structures available. The most closely related one is presented by Sternagel and Thiemann [21]. The authors present an "XML library" for Isabelle/HOL. The purpose of this library is to provide XML parsing and pretty printing facilities for Isabelle. As such, it is not a formalization of XML or XML-like data structures in Isabelle/HOL.

Shadow roots seem to achieve a very similar goal as the <iframe>-tag of the HTML standard. Still, the motivation for both differ significantly: while iframes were introduced to allow the *secure* integration of content from different websites, shadow roots were introduced to allow component-based web development similar to, for example, using components in the .net framework. The limitations of shadow roots to ensure the privacy of data processed by web applications have already been discussed by Légaré et al. [16] and Freyberger et al. [11].

Finally, there are several works, e. g., [1,4,13,14] on formalizing parts of web browsers for analyzing their security. These works use high-level specifications of web browsers and do not contain a formalization of the DOM itself.

7   Conclusion

We present a formal model of web components and component safety, and we formally verify the level of component safety for the DOM API as defined in the DOM standard [24].

Our formalization of the DOM with shadow roots and its API is an important step towards providing formal guarantees for a modular development approach to web applications as well as increasing the security and safety of large web applications. While the current proposal clearly has weaknesses, our analysis also shows that moderate changes to the concept of shadow roots can make the web components a much more powerful and stronger concept that, hopefully, also can make developing secure applications easier.

On a technical level, our formalization is based on a shallow embedding of the DOM with shadow roots into Isabelle/HOL. We use only conservative extensions of HOL (i. e., we do not introduce any axioms). Hence, our formalization is consistent by construction. Overall, it consists of more than 10 000 lines of Isabelle code, including conservative definitions and proofs. To ensure the compliance of our formalization to the official DOM standard, we aimed for an executable formalization: an executable specification allows for symbolically evaluating the official compliance test suite in Isabelle/HOL. Thus, as our formalization passes the test suite, our formalization adheres to the same compliance standards as the widely used web browser engines.

Future Work. While web components based on shadow roots provide some form of isolation for JavaScript developers, they have weaknesses and, clearly, cannot provide the isolation necessary for enforcing security guarantees similar to iframes. While iframes are a rather old concept that is defined on top of the DOM in the HTML standard [25], shadow roots are a very recent concept that is integrated into the DOM [24]. On the first glance, the two concepts do not have much in common. Having a closer look reveals that the concepts are closely related: on the one hand, it seems desirable to introduce security concepts to shadow roots and, on the other hand, iframes would clearly benefit from interfaces allowing web developers to adapt certain aspects of an included iframe. Thus, the question emerges whether shadow roots can, in the long term, replace iframes. To answer this question, we plan to formalize the core of the HTML standard on top of our DOM formalization. This allows us to compare both concepts formally and also formally investigate the impact of adding security features to shadow roots.

References

1. Akhawe, D., Barth, A., Lam, P.E., Mitchell, J., Song, D.: Towards a formal foundation of web security. In: IEEE Computer Security Foundations Symposium (CSF), pp. 290–304. IEEE Computer Society (2010). https://doi.org/10.1109/CSF.2010.27
2. Andrews, P.B.: Introduction to Mathematical Logic and Type Theory: To Truth through Proof, 2nd edn. Kluwer Academic Publishers, Dordrecht (2002)
3. Bidelman, E.: Shadow DOM v1: self-contained web components (2017). https://developers.google.com/web/fundamentals/getting-started/primers/shadowdom
4. Bohannon, A., Pierce, B.C.: Featherweight firefox: formalizing the core of a web browser. In: Usenix Web Application Development (WebApps) (2010)
5. Brucker, A.D.: An interactive proof environment for object-oriented specifications. Ph.D. thesis, ETH Zurich (2007). ETH Dissertation No. 17097
6. Brucker, A.D., Herzberg, M.: The core DOM. Archive of formal proofs (2018). http://www.isa-afp.org/entries/Core_DOM.html. Formal proof development
7. Brucker, A.D., Herzberg, M.: A formal semantics of the Core DOM in Isabelle/HOL. In: Champin, P., Gandon, F.L., Lalmas, M., Ipeirotis, P.G. (eds.) The 2018 Web Conference Companion (WWW), pp. 741–749. ACM Press (2018). https://doi.org/10.1145/3184558.3185980
8. Brucker, A.D., Wolff, B.: An extensible encoding of object-oriented data models in HOL. J. Autom. Reasoning **41**, 219–249 (2008). https://doi.org/10.1007/s10817-008-9108-3

9. Church, A.: A formulation of the simple theory of types. J. Symbolic Logic **5**(2), 56–68 (1940)
10. Clarke, D., Östlund, J., Sergey, I., Wrigstad, T.: Ownership types: a survey. In: Clarke, D., Noble, J., Wrigstad, T. (eds.) Aliasing in Object-Oriented Programming. Types, Analysis and Verification. LNCS, vol. 7850, pp. 15–58. Springer, Heidelberg (2013). https://doi.org/10.1007/978-3-642-36946-9_3
11. Freyberger, M., He, W., Akhawe, D., Mazurek, M.L., Mittal, P.: Cracking shadowcrypt: exploring the limitations of secure I/O systems in internet browsers. In: PoPETs, vol. 2018, no. 2, pp. 47–63 (2018). https://doi.org/10.1515/popets-2018-0012
12. Gardner, P., Smith, G., Wheelhouse, M.J., Zarfaty, U.: DOM: towards a formal specification. In: Programming Language Technologies for XML (PLAN-X). ACM (2008)
13. Guha, A., Fredrikson, M., Livshits, B., Swam, N.: Verified security for browser extensions. In: IEEE Symposium on Security and Privacy, pp. 115–130 (2011). https://doi.org/10.1109/SP.2011.36
14. Jang, D., Tatlock, Z., Lerner, S.: Establishing browser security guarantees through formal shim verification. In: Kohno, T. (ed.) USENIX, pp. 113–128. USENIX (2012)
15. Jensen, S.H., Madsen, M., Møller, A.: Modeling the HTML DOM and browser API in static analysis of JavaScript web applications. In: ESEC/FSE, pp. 59–69. ACM (2011). https://doi.org/10.1145/2025113.2025125
16. Légaré, J., Sumi, R., Aiello, W.: Beeswax: a platform for private web apps. In: PoPETs, vol. 2016, no. 3, pp. 24–40 (2016)
17. Nipkow, T., Wenzel, M., Paulson, L.C. (eds.): Isabelle/HOL. LNCS, vol. 2283. Springer, Heidelberg (2002). https://doi.org/10.1007/3-540-45949-9
18. Poetzsch-Heffter, A., Geilmann, K., Schäfer, J.: Infering ownership types for encapsulated object-oriented program components. In: Reps, T., Sagiv, M., Bauer, J. (eds.) Program Analysis and Compilation, Theory and Practice. LNCS, vol. 4444, pp. 120–144. Springer, Heidelberg (2007). https://doi.org/10.1007/978-3-540-71322-7_6
19. Raad, A., Santos, J.F., Gardner, P.: DOM: specification and client reasoning. In: Igarashi, A. (ed.) APLAS 2016. LNCS, vol. 10017, pp. 401–422. Springer, Cham (2016). https://doi.org/10.1007/978-3-319-47958-3_21
20. Smith, G.D.: Local reasoning about web programs. Ph.D. thesis, Imperial College London, London, UK (2011)
21. Sternagel, C., Thiemann, R.: XML. Archive of formal proofs (2014). http://isa-afp.org/entries/XML.shtml. Formal proof development
22. W3C: Web IDL (2017). https://heycam.github.io/webidl/
23. W3C: Shadow DOM (2018). https://www.w3.org/TR/2018/NOTE-shadow-dom-20180301/. Last Updated 1 March 2018
24. WHATWG: DOM - living standard (2019). https://dom.spec.whatwg.org/commit-snapshots/7fa83673430f767d329406d0aed901f296332216/. Last Updated 11 February 2019
25. WHATWG: HTML - living standard (2019). https://html.spec.whatwg.org/commit-snapshots/b8c084e9d5461b858180e7f80ad6ca19c7963723/. Last Updated 19 February 2019

Minimizing Characterizing Sets

Kadir Bulut[1], Guy Vincent Jourdan[2], and Uraz Cengiz Türker[1]

[1] Gebze Teknik Üniversitesi, 41400 Gebze, Kocaeli, Turkey
kdrblt93@gmail.com, urazc@gtu.edu.tr
[2] University of Ottawa, Ottawa, ON, Canada
gjourdan@uottawa.ca

Abstract. A characterizing set (CS) for a given finite state machine (FSM) defines a set of input sequences such that for any pair of states of FSM, there exists an input sequence in a CS that can separate these states. There are techniques that generate test sequences with guaranteed fault detection power using CSs. The number of inputs and input sequences in a CS directly impacts the cost of the test: the higher the number of elements, the longer it takes to generate the test. Despite the direct benefits of using CSs with fewer sequences, there has been no work focused on generating minimum sized characterizing sets. In this paper, we show that constructing CS with fewer elements is a **PSPACE-Hard** problem and that the corresponding decision problem is **PSPACE-Complete**. We then introduce a heuristic to construct CSs with fewer input sequences. We evaluate the proposed algorithm using randomly generated FSMs as well as some benchmark FSMs. The results are promising, and the proposed method reduces the number of test sequences by 37.3% and decreases the total length of the tests by 34.6% on the average.

Keywords: Model-based testing · Characterization set · Complexity

1 Introduction

Testing is an indispensable aspect of a development-cycle for any kind of system. One promising approach for automating testing is Model Based Testing (MBT). MBT techniques and tools use behavioural models and usually operate on either finite state machines (FSMs), extended finite state machines (EFSMs) or labelled transition systems (LTSs) that define the semantics of the underlying model. There has been significant interest in automating testing based on an FSM model in areas such as sequential circuits [11], lexical analysis [2], software design [7], communication protocols [6–8, 26, 28, 29, 34, 35], object-oriented systems [4], and web services [3, 18, 33, 39]. Such techniques have also been shown to be effective when used in large industrial projects [16].

The literature contains many formal methods to automatically generate test sequences from FSM models of systems [1, 13, 15, 19, 32, 34, 37, 41]. These methods are based on *fault detection experiments* [24]. Formal methods for generating fault detection experiments are based on particular types of sequences that

© Springer Nature Switzerland AG 2020
F. Arbab and S.-S. Jongmans (Eds.): FACS 2019, LNCS 12018, pp. 72–86, 2020.
https://doi.org/10.1007/978-3-030-40914-2_4

are derived from the specification M. Different methods use different types of sequence. Among such special sequences, *Unique Input Output sequences* (UIO) which are used to verify the current state of the implementation; *Characterization Sets* (W-set), *Preset Distinguishing Sequences* (PDS), and *Adaptive Distinguishing Sequences* (ADS), which are used to identify the current state of the implementation. A W-set is a set of input sequences such that for any pair (s,s') of states of M there exists an input sequence in the W-set such that s and s' are separated [14].

It has been reported that every minimal complete and initially connected FSM has a W-set and every deterministic (not minimal and partial) FSM can be converted into deterministic complete and minimal FSM [22,38]. Therefore, there exist a substantial number of formal methods for constructing fault detection experiments relying on W-sets [7,10,12,13,20,21,23,25,30,31,34,40]. A survey of such methods can be found in [9].

1.1   Motivation and Problem Statement

Characterizing sets are widely used in test generation. For example, a well known fault detection experiment is the so-called *W-method* (given in [7,40]) consists of two steps: state recognition and transition verification. In the W-method, to recognize a state (s), each element $\bar{x}$ of the W-set is applied after M is reset and is brought to s. To verify the transitions out of a state (s), for each element (x) of the input alphabet and for each element $(\bar{x})$ of the W-set, the input $x\bar{x}$ is applied after M is reset and is brought to s.

Moreover, the well known HSI-method [30] requires *harmonized state identifiers* (HSIs) to construct test sequences for a given FSM M. Construction of HSIs requires one to harmonize the elements of a W-set of M. That is, in order to construct HSI for M, one has to construct a W-set [30].

It is thus clear that the overall cost of these methods is directly impacted by the number of elements in the W-set. Of course, this cost is increasing when the reset operation is time consuming. Therefore there has been an interest in constructing fault detection experiments with the minimum number of resets. The work in this line of research tries to reduce the number of test cases [7,10,12,13,20,21,23,25,30,31,34,40] and [9].

This research revisits this long-standing problem and aims at reducing the number of test cases by minimizing the number of elements in W-sets. To the best of our knowledge, this is the first attempt at addressing this question. In this paper, we focus on the following problem, its complexity and the algorithmic approaches that solve it:

Definition 1. *Let M be a minimal deterministic completely specified finite state machine, and let K be a positive integer. In the K-W-set problem we are asked to decide if there exists a W-set $\mathcal{W}$ for M such that $|\mathcal{W}| \leq K$.*

1.2 Results

We show that the K-W-set problem is PSPACE complete. We introduce a heuristic that relies on breadth first search strategy to find a W-set of minimal size. We perform and present a set of experiments, which indicate that we can reduce the number of tests by 37.3% and can decrease the total length of test by 36.4% on the average.

1.3 Practical Implications of Our Results and Future Directions

W-sets are used in many formal methods that generate fault detection experiments, therefore the methods that use these sequences will directly benefit from the proposed work. In order to assess the effect of minimizing W-sets more accurately, as future work we will extend the empirical evaluation to other fault detection experiment generation algorithms.

1.4 Summary of the Paper

This paper is structured as follows. In the next section, we provide the terminology used throughout the paper. In Sect. 3, we provide the hardness result of the K-W-set problem. This section is then followed by a section in which we introduce the heuristic approach. In Sect. 5, we provide the results of our experiments. We conclude the paper by providing some future directions in Sect. 6.

2   Preliminaries

An FSM M is defined by a tuple $(S, s_0, X, Y, \delta, \lambda)$ where $S = \{s_1, s_2, \ldots, s_n\}$ is a finite set of states, $s_0 \in S$ is the initial state, $X = \{x_1, x_2, \ldots, x_p\}$ and $Y = \{y_1, y_2, \ldots, y_q\}$ are finite sets of inputs and outputs, $\delta : S \times X \to S$ is the transition function, and $\lambda : S \times X \to Y$ is the output function. If x is applied when M is in state s, M moves to state $s' = \delta(s, x)$ and produces output $y = \lambda(s, x)$. This defines the *transition* $\tau = (s, x/y, s')$ and we say that x/y is the *label* of τ, s is the *start state* of τ, and s' is the *end state* of τ. If there exists a sequence of transitions from the initial state to any other state then M is called *initially connected*. M is *completely specified* if δ and λ are total functions, i.e., for every state s and every input x, $\delta(s, x)$ and $\lambda(s, x)$ are both defined.

We use juxtaposition to denote concatenation, and so, for example, if x_1, x_2, and x_3 are input symbols then $x_1 x_2 x_3$ is an input sequence. The transition and output functions can be extended to input sequences. By abusing of notation, we will still use δ and λ for the extended functions. These extensions are defined as follows: let ϵ be the empty sequence, let $x \in X$ and $\bar{x} \in X^*$. $x\bar{x}$ is defined at s: $\delta(s, \epsilon) = s$, $\delta(x\bar{x}) = \delta(\delta(s, x), \bar{x})$; $\lambda(s, \epsilon) = \epsilon$ and $\lambda(s, x\bar{x}) = \lambda(s, x)\lambda(\delta(s, x), \bar{x})$. States $s, s' \in S$ are *equivalent* if for all $\bar{x} \in X^*$, $\bar{x}$ is defined in state s iff it is defined in s' and then $\lambda(s, \bar{x}) = \lambda(s', \bar{x})$. If there exists $\bar{x} \in X^*$ defined in s and s' such that $\lambda(s, \bar{x}) \neq \lambda(s', \bar{x})$, then $\bar{x}$ *separates* s and s', and $\bar{x}$ is a *separating*

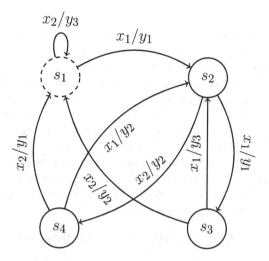

Fig. 1. FSM M_1 and a W-set for M_1 is $W = \{x_1, x_2\}$. Note that the initial state is highlighted with a dashed line.

sequence. FSM M is *minimal* if all pair of states of M are separable. In this paper, we assume that the FSM is completely specified, minimal, deterministic, initially connected and have a reliable reset functionality (Fig. 1). We now define *W-Sets.*

Definition 2. *Given FSM M, $W = \{\bar{x} | \bar{x} \in X^*\}$ is a W-Set for M if for any pair of distinct states of M there exists a separating sequence $\bar{x} \in W$ such that $\bar{x}$ separates the pair, i.e., $\forall s, s' \in S$, such that $s \neq s'$, $\exists \bar{x} \in W$ we have that $\lambda(s, \bar{x}) \neq \lambda(s', \bar{x})$.*

3   Hardness of K-W-set Problem

We now show that K-W-set problem is in **PSPACE**. First note that every deterministic minimal FSM has a W-set with at most $n(n-1)/2$ elements, therefore we implicitly consider that K is a value in the range $1 \leq K < n(n-1)/2$. Second, as the length of a separating sequence $\bar{x}$ for a given pair of states for an FSM M with n states is at most $n - 1$ [24]. Therefore we keep the bound of such sequence as $n-1$; hence we can reason about the space a non-deterministic Turing Machine might use when applying an algorithm that generates K-W-set.

Proposition 1. *It is possible for a non-deterministic Turing Machine to decide whether a state set S of completely specified deterministic FSM M has a K-W-set in $O(n^2 log(n))$ space.*

Proof. We will show how a non-deterministic Turing Machine can solve the K-W-set problem within this space. This Turing Machine will guess one input at a time. It will keep tuples of the form (s, s', c) where $s, s' \in S$ and $c \in \{0, 1\}$. Besides it will guess K input sequences $\bar{x}_0, \bar{x}_1, \ldots, \bar{x}_{K-1}$ each with length at most $n - 1$. As there are $n(n-1)/2$ pairs, and encoding each state id and each input require $log(n)$ space the Turing Machine requires $O(n^2 log(n)) + O(Knlog(n))$ space. At each iteration, K next inputs are guessed non-deterministically and the pairs are updated in the natural way (i.e. $(s, s', c_b) \rightarrow (\delta(s, x), \delta(s', x), c)$ where $c = 1$ if $c_b = 1$ or $\lambda(s, x) \neq \lambda(s', x)$, and $c = 0$ otherwise). If for every tuple we have $c = 1$ then the process terminates with success, while if not all of the c are 1 then a K-W-set has not been found. Finally we add a counter ctr, which is increased in each iteration, in order to ensure termination. Thus, the Turing Machine also terminates with failure if the counter exceeds the bound $n - 1$. This counter thus takes $O(log(n))$ and thus, the overall space requirement is of $O(log(n) + n^2 log(n) + Knlog(n))$ space.

We now show that K-W-set problem is `PSPACE-Complete`.

Theorem 1. *K-W-set problem for completely specified deterministic FSMs is* `PSPACE-complete`*.*

Proof. First, by Proposition 1 we know that a non-deterministic Turing Machine can solve the problem in $O(n^2 log(n))$ space. Since K is bounded above by a polynomial in terms of n, this gives a polynomial upper bound on the space requirements and so the problem is in `PSPACE`.

We now show that the problem is `PSPACE-hard`. Consider the case where $K = 1$. Then there is a K-W-set for M if and only if there is a PDS for M. Thus, any algorithm that can decide whether an FSM has a K-W-set can also be used to decide whether an FSM has a PDS. Since the problem of deciding whether an FSM has a PDS is `PSPACE-hard` [27], we have that the problem of deciding whether an FSM has a K-W-set is also `PSPACE-hard`. The result thus follows.

4   Algorithm to Construct Minimum W-Sets

The proposed algorithm tries to generate a W-set with fewer elements without specifying a bound (K) on the number of elements. The proposed algorithm relies on breadth first search (BFS). The algorithm receives an FSM M and starts to form a breadth-first search tree exhaustively either one of the termination conditions hold. In our case they are (i) a W-set is formed and (ii) an upper bound for the depth of the tree is reached. We set this bound as $n - 1$ as for a machine with n states, as the length of a separating sequence is $n - 1$ [14].

Upon receiving its input, the algorithm initiates a *pair set* Δ (Line 1 of Algorithm 1). The pair set keeps the set of all possible 2-combinations (i.e., the set defined by $\binom{S}{2}$) and will be used to check which pairs have been separated so far. After this, the algorithm constructs BFS tree iteratively. A BFS tree is defined by a set of vertices V such that a vertex $v \in V$ keeps four pieces of information, a set of *current states* v_c, a set of *initial states* v_I, an input sequence $v(\bar{x})$, and finally an output sequence $v(\bar{y})$ such that $\delta(v_I, v(\bar{x})) = v_C$ and $\lambda(v_I, v(\bar{x})) = v(\bar{y})$.

BFS tree is constructed as follows. At each level, algorithm process the set V. Initially, V has a single element v such that $v_C = S$, $v_I = S$, and $v(\bar{x}) = v(\bar{y}) = \varepsilon$ (Line 2 of Algorithm 1). While processing the vertices in V, algorithm receives the next vertex v and *applies* all the inputs from set X to v_C (Lines 8–14 of Algorithm 1).

The application of an input x to v_C is carried out as follows, for every state s of v_C, we bring M to s and apply x to M. By considering output produced by M as a response to x, v is *partitioned*, i.e., $\forall y \in \cup_{s \in v_C} \lambda(s, x)$, we introduce a new vertex v^y such that $v_C^y = \cup_{s \in v_C, \lambda(s,x)=y} \delta(s, x)$, $v^y(\bar{x}) = v(\bar{x})x$, $v^y(\bar{y}) = v(\bar{y})y$, and we form v_I^y by simply inheriting the corresponding initial state of s. We denote the set of vertices that are created by applying x on set v_C as $P(v_C, x)$.

Afterwards, the algorithm pushes new vertices to another set of vertices ($P(v_C, x)$) called *next level vertex set*. All the new partitions generated during this level are stored in V^* (Line 14 of Algorithm 1). When all the vertices of current level are processed, algorithm copies V^* to V (Line 15 of Algorithm 1), increments the level variable (ℓ) by one (Line 16 of Algorithm 1) and initiates a set of sequences Seq (Line 17 of Algorithm 1).

This is then followed by analyzing the outcome of BFS step. In order to do this, algorithm first gathers all the distinct input sequences from V^* and form set $\bar{X}$. Then for each input sequence $\bar{x} \in \bar{X}$, it counts the number of pairs that can be separated from Δ and keeps this value as χ (Lines 18–20 of Algorithm 1). Note that the algorithm does not remove a pair at this step, it only counts the number of possible pairs that can be removed by the input sequence under consideration. Then algorithm maps this value with the input sequence, i.e., $(\bar{x}, \chi)$ and stores this in set Seq, afterwards algorithm sorts Seq according to the χ values (Line 21 of Algorithm 1).

After this, algorithm moves to an iterative step at which it drops pairs (through a heuristic step) from Δ. At every iteration algorithm performs a heuristic step; it selects an input sequence ($\bar{x}$) that is associated with one of the biggest χ value. That is it selects an input sequence that separates as much state as possible. If the algorithm can eliminate a pair from Δ with $\bar{x}$, the algorithm adds $\bar{x}$ to $\mathcal{W}$. This process continues until all pairs are separated (or it runs out of input sequences) (Lines 22–25 of Algorithm 1). Note that at a given (jth) iteration selected input sequence $\bar{x}$ may not remove χ number of elements from Δ. This stems from the fact that it is possible that previously processed input sequence (i.e., input sequences considered before jth iteration) may remove the pairs that are also separated by $\bar{x}$.

After analyzing the result of BFS, algorithm decides what to do next. If Δ is empty, the algorithm returns $\mathcal{W}$ otherwise algorithm continues to execute (Lines 26–27 of Algorithm 1). Please see Algorithm 1 for details.

We now show that if Algorithm 1 terminates with success, $\mathcal{W}$ defines a W-set for M.

Algorithm 1. Minimum W-set for M

Input: FSM M
Output: A W-set for M
begin

```
1      Δ ← {(sᵢ, sⱼ)|sᵢ, sⱼ ∈ S and i < j}
2      v ← (S, S, ε, ε), push(v, V)
3      ℓ ← 0.
4      while l ≤ n − 1 do
5          V* ← ∅
6          while V ≠ ∅ do
7              v ← pop(V)
8              foreach input symbols x ∈ X do
9                  Retrieve P(v_C, x).
10                 foreach vʸ ∈ P(v_C, x) do
11                     v_C^y = ∪_{s∈v_C,λ(s,x)=y} δ(s, x)
12                     v_I^y = ∪_{s∈v_I,λ(s,v(x̄)x)=v(ȳ)y∧δ(s,v(x̄)x)∈v_C^y}
13                     v^x(x̄) = v(x̄)x, vʸ(ȳ) = v(ȳ)y
14                     push(vʸ, V*)

15             V ← V*
16             ℓ ← ℓ + 1
17             Seq ← ∅
18             foreach v ∈ V do
19                 χ ← number of pairs removed from set Δ by v(x̄)
20                 Seq ← Seq ∪ {χ}

21             Sort(Seq)
22             foreach χ ∈ Seq do
23                 Remove pairs separated by x̄* from Δ
24                 if A pair has been separated then
25                     W ← W ∪ {x̄*}

26             if Δ = ∅ then
27                 Return W
```

Proposition 2. *Let $v \in V$ be a vertex such that $\exists (s, s') \in \Delta$ and $\lambda(s, v(\bar{x})) \neq \lambda(s', v(\bar{x}))$ then $v(\bar{x})$ is a separating sequence for s and s'.*

Proof. The result follows from the definition of a separating sequence.

Therefore for each pair of Δ the algorithm can compute a separating sequence. Since algorithm terminates when all the pairs are dropped from Δ, we reach to the following result.

Theorem 2. *Let M be an initially connected completely specified minimal FSM, then when Algorithm 1 receives M and returns non-empty set W, then W is a W-set for M.*

Finally, since the proposed algorithm constructs a BFS tree by applying every input until a terminating condition is met. The time complexity of the algorithm is exponential. As we show in the following section, the proposed algorithm reduces the number of tests and the number of inputs of tests 37.3% and 36.4% on the average respectively in a timely manner.

5   Empirical Evaluation

In this section we present the result of our experiments. We used an Intel I7 CPU with 32 GB RAM to carry out these experiments. We implemented the W-set construction algorithm as given in [14], the W-method as given in [7] and the proposed method using C++ language on Microsoft Visual studio .Net 2013.

5.1   FSMs Used in the Experiments, Experiment Settings and Evaluation

In order to compare the proposed method and the existing W-set generation method, we generated different classes of FSMs and we also used a series of FSMs that are available as benchmark sets. In this section, we overview the FSMs we used throughout the experiments and how we evaluate the results of our experiments.

FSMs in Class I. The FSMs in first class (C1) were generated as follows. First, for each input x and state s we randomly assigned the values of $\delta(s, x)$ and $\lambda(s, x)$. After an FSM M was generated we checked its suitability as follows. We checked whether M was minimal, and had a W-set. If the FSM passes all these tests, we included it into C1, otherwise we omitted this FSM and produced another one. Consequently, all generated FSMs were initially connected, minimal, and had W-sets.

Following this procedure, we constructed six sets of 1000 FSMs with n states, where $n \in \{50, 60, \ldots, 150\}$. The number of the input and the output symbols were 3. In total, we constructed 11000 FSMs for the first class of FSMs.

FSMs in Class II. Note that for FSMs in C1, the next state of each transition is randomly selected, hence the in-degree of the states are close to one another. In contrast, while generating the FSMs in the second class (C2), we provided a nonuniform in-degree distribution. To create such a distribution, we first randomly select a subset $\bar{S}$ of states which will have higher in-degree values than the states in $S \setminus \bar{S}$. To create a higher in-degree values for the states in $\bar{S}$, we randomly select a subset of transitions Γ (where each element of Γ is a pair (s, x) denoting the transition of state s for input symbol x). We then force the transitions in Γ to end in states in $\bar{S}$.

The key point of constructing weighted FSMs was choosing the cardinalities of $\bar{S}$ and Γ. If $|\Gamma|$ was too large and $|\bar{S}|$ was too small then one might not able to construct a connected FSM, or might not be able to construct an FSM with a W-set. On the other hand, if $|\Gamma|$ was too small and $|\bar{S}|$ was too large then the in-degrees of states became similar.

In these experiments we chose $|\bar{S}|$ to be 10% of the states and we set $|\Gamma|$ to be 30% of the transitions. We observed that if the percentage for $|\Gamma|$ is increased further, it takes too much time to construct an FSM with a W-set.

As in the case of generation of FSMs in C1, after an FSM M was generated we checked its suitability.

We constructed 11000 weighted FSMs, with number of states $n \in \{50, 60, \ldots, 150\}$, where for each n there were 1000 FSMs. The cardinalities of the input and the output alphabets were 3.

Benchmark FSMs. In addition to the randomly generated FSMs, we also used some FSM specifications retrieved from the ACM/SIGDA benchmarks, a set of test suites (FSMs) used in workshops between 1989–1993 [5].

The benchmark suite has 59 FSM specifications ranging from simple circuits to advanced circuits obtained from industry. The FSM specifications were available in the *kiss2* format. In order to process FSMs, we converted the *kiss2* file format to our FSM specification format. We only used FSMs from the benchmark that were minimal, deterministic, had W-set and had fewer than 10 input bits[1]. 19% of the FSMs had more than 10 input bits, 38% were not minimal, 15% of the FSMs had W-set, 48% of the FSMs were nondeterministic.

Consequently, 8.5% of the FSM specifications passed all of the tests. They are *DVRAM*, *Ex4*[2], *Log*, *Rie*, and *Shift Register*. In Table 1, we present the number of states and the number of transitions of these FSMs.

Table 1. Benchmark FSMs and their sizes

Name	No of states	No of transitions
Shift register	8	16
Ex4	14	896
Log	17	8704
DVRAM	35	8960
Rie	29	14848

5.2   Evaluation

In order to evaluate the relative performance of different approaches, for each FSM M, we computed W-sets using the proposed algorithm (P) and the existing algorithm (EA) separately.

In order to compare the algorithms, we use three measures. The first measure $(M1)$ gives the ratio of the number of test cases computed by the W-Method

---

[1] Since the circuits receive inputs in bits, and since b bits correspond to 2^b inputs, we do not consider FSMs with $b \geq 10$ bits.

[2] FSM specification Ex4 is partially specified. We complete the missing transitions by adding self looping transitions with a special output symbol, and do not use these inputs for W-set construction.

that is fed with the W-sets computed by P and EA methods respectively. For an FSM M, let $P(M)$ (and $EA(M)$) be the set of tests computed for M by P (and EA) method, and $T(.)$ be the number of test cases given by the parameter '.'. Then $M1$ is computed as follows:

$$M1 = T(P(M))/T(EA(M))$$

The second measure $(M2)$ is the ratio of the number of elements in W-sets constructed by the proposed algorithm and the existing algorithm i.e.,

$$M2 = |\mathcal{W}_P|/|\mathcal{W}_{EA}|$$

Final measure $(M3)$ is the ratio of total number of inputs in of test cases computed by the W-Method that is fed with the W-sets computed by P and EA methods respectively. $M3$ is computed as follows:

$$M3 = L(P(M))/L(EA(M))$$

where $L(.)$ returns the sum of the inputs of test cases given in parameter '.'. Results conducted on C1 and C2 are given in Fig. 2. The results of the experiments are as expected: the educated selection of the separating sequence can reduce the number of elements of the W-set and number of test cases. The results also show that we can construct smaller test cases when the proposed algorithm is used. The results suggest that as the number of states increases, the proposed method constructs W-sets with fewer elements and the W-method that uses the W-set constructed by the proposed method can have fewer test cases. The average reduction of the number of test cases is 37.3% and the average reduction of the number of elements in the W-set is 44.67%. Moreover, we observe that the effect of non-uniform transition distribution (C2) is negligible[3].

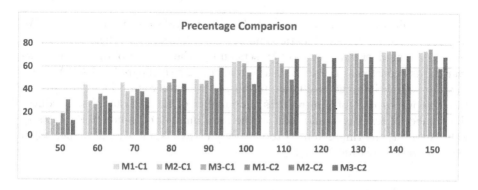

Fig. 2. Results of experiments conducted on C1 and C2. X axis labels the number of states, and Y axis increases with the percentage (%).

---

[3] One way ANOVA test accepts null hypothesis [36].

Moreover, we observe that the average reduction on the total number of inputs is 36.4% on the average, when the proposed algorithm is used to construct W-method. Moreover, we observe that the reduction increases with the number of states.

We provided the time required to construct the W-sets by EA and P, respectively. Results are given in Fig. 3. Results indicate that the proposed algorithm is slower (2.218 times on the average), yet it can easily handle relatively large FSMs. Note that, for a given FSM M, we run the proposed algorithm once, that is, it is a one time computation and the shorter test suite that you generate is expected to be run probably thousands of times, so it makes sense to "invest" in the generation of a better test suite.

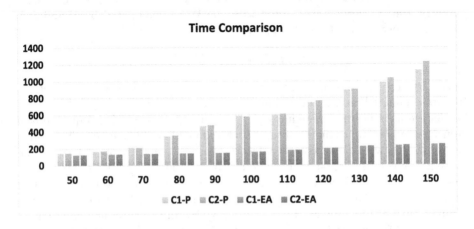

Fig. 3. Time comparison of the W-set generation algorithms. X axis labels the number of states, and Y axis labels the time (seconds).

The results of experiments conducted on benchmarks are given in Table 2. We see that except for the specifications named DVRAM and Rie, there is no difference on the number of elements of W-sets and on the number of test cases. However this is expected as these FSMs are smaller in size compared to DVRAM and Rie. For large FSMs, the results are promising: we observe that for DVRAM and Rie, the proposed algorithm reduces the number of elements in W-set, reduces the number of test cases, and reduces the total number of inputs of the tests as well. However, the time required to generate W-sets is, in general, higher when the proposed algorithm is used.

5.3   Threats to Validity

We try to identify some threats to the validity of experimental results in this section.

First, we analyse the performance of our algorithm by using randomly generated FSMs. It is possible that the performance of our proposed algorithm differs

Table 2. Results of experiments conducted on benchmark FSMs

Name	M1	M2	M3	P (msec.)	EA (msec.)
Shift Register	1	1	1	25	24
Ex4	1	1	1	270	195
Log	1	1	1	260	234
DVRAM	40%	32%	29%	567	267
Rie	34%	41%	31%	689	304

for FSMs used in real-life situations. Although using random FSMs is a usual approach for the works in this field, in order to evaluate the generalization of the proposed algorithm, we also test our solution using case studies obtained from benchmark FSM specifications as explained in Sect. 5.1. We see that the experiment results obtained from random FSMs coincided with the results obtained from the benchmark FSMs when the size of FSM is large enough. However, for small sized FSMs, we did not observe any difference between our method and the existing method.

Another threat to validity could be our incorrect implementation of proposed, existing W-set generation algorithm and the W-method. To eliminate this threat, we also used two existing tools that are used for research [17, 21]. The first tool checks if a given set of sequences is a W-set for an FSM. The second tool checks whether or not a given set of input sequences defines a checking experiment for M.

6   Conclusion

Software testing is typically performed manually and is an expensive, error prone process. This has led to interest in automated test generation, including significant interest in model based testing (MBT). Most MBT techniques generate tests from either finite state machines (FSMs) or labelled transition systems.

In this work, we studied this problem and investigate the problem of computing a minimum W-set for a given deterministic, minimal and complete FSM which has a reliable reset. We introduced the minimization problem and we showed that the problem of finding a minimum W-set is PSPACE Complete. Our initial motivation for minimizing W-set is the use of W-sets in the context of fault detection experiment generation. Generating a minimum W-set is important in such a context, since state recognitions and state verifications are performed by using a W-set, and a considerable part of a fault detection experiment consists of such state recognitions/verifications. Therefore the size of a fault detection sequence generated by using a W-set correlates with the number of elements of the W-set. Due to the hardness of W-set minimization, heuristic algorithms can be used to generate compact W-sets. In order to validate our initial motivation, we conducted experiments. The result suggests that when the fault detection sequences are generated by the W-method using W-sets with fewer elements, the number of test sequences reduces by 37.3%, and the total number of inputs reduces by 36.4% on the average.

There are several lines of future work. First, it would be interesting to explore realistic conditions under which the decision and optimisation problems can be solved in polynomial time. Such conditions might lead to new notions of testability.

Although the results of the experiments suggest that the use of compact W-sets produce test suites that require fewer resets, it would be interesting to extend the experiments and possibly also to consider the Wp, HIS and SPY algorithms [13,30,34]. Finally, it would be interesting to extend this work to non-deterministic FSMs.

Acknowledgements. This work is supported by the scientific and technological council of Turkey (TUBITAK) under the grant 117E987.

References

1. Aho, A.V., Dahbura, A.T., Lee, D., Uyar, M.U.: An optimization technique for protocol conformance test generation based on UIO sequences and rural Chinese postman tours. In: Protocol Specification, Testing, and Verification, North-Holland, Atlantic City, vol. VIII, pp. 75–86. Elsevier (1988)
2. Aho, A., Sethi, R., Ullman, J.: Compilers, Principles, Techniques, and Tools. Addison-Wesley Series in Computer Science. Addison-Wesley Publishing Company (1986)
3. Betin-Can, A., Bultan, T.: Verifiable concurrent programming using concurrency controllers. In: Proceedings of the 19th IEEE International Conference on Automated Software Engineering, pp. 248–257. IEEE Computer Society (2004)
4. Binder, R.V.: Testing Object-Oriented Systems: Models, Patterns, and Tools. Addison-Wesley (1999)
5. Brglez, F.: ACM/SIGMOD benchmark dataset. http://www.cbl.ncsu.edu/benchmarks/Benchmarks-upto-1996.html (1996). Accessed 12 Feb 2014
6. Brinksma, E.: A theory for the derivation of tests. In: Proceedings of Protocol Specification, Testing, and Verification, North-Holland, Atlantic City, vol. VIII, pp. 63–74 (1988)
7. Chow, T.S.: Testing software design modelled by finite state machines. IEEE Trans. Softw. Eng. **4**, 178–187 (1978)
8. Dahbura, A., Sabnani, K., Uyar, M.: Formal methods for generating protocol conformance test sequences. Proc. IEEE **78**(8), 1317–1326 (1990). https://doi.org/10.1109/5.58319
9. Dorofeeva, R., El-Fakih, K., Maag, S., Cavalli, A.R., Yevtushenko, N.: FSM-based conformance testing methods: a survey annotated with experimental evaluation. Inf. Softw. Technol. **52**(12), 1286–1297 (2010)
10. Dorofeeva, R., El-Fakih, K., Yevtushenko, N.: An improved conformance testing method. In: Wang, F. (ed.) FORTE 2005. LNCS, vol. 3731, pp. 204–218. Springer, Heidelberg (2005). https://doi.org/10.1007/11562436_16
11. Friedman, A., Menon, P.: Fault Detection in Digital Circuits. Computer Applications in Electrical Engineering Series. Prentice-Hall (1971)

12. Friedman, A.D., Menon, P.R. (eds.): Fault Detection in Digital Circuits. Prentice-Hall Englewood Cliffs, N.J (1971)
13. Fujiwara, S., Bochmann, G.V., Khendek, F., Amalou, M., Ghedamsi, A.: Test selection based on finite state models. IEEE Trans. Softw. Eng. **17**(6), 591–603 (1991)
14. Gill, A.: Introduction to the Theory of Finite State Machines. McGraw-Hill, New York (1962)
15. Gonenc, G.: A method for the design of fault detection experiments. IEEE Trans. Comput. **19**, 551–558 (1970)
16. Grieskamp, W., Kicillof, N., Stobie, K., Braberman, V.A.: Model-based quality assurance of protocol documentation: tools and methodology. Softw. Test. Verif. Reliab. **21**(1), 55–71 (2011). https://doi.org/10.1002/stvr.427
17. Güniçen, C., Türker, U.C., Ural, H., Yenigün, H.: Generating preset distinguishing sequences using SAT. In: Gelenbe, E., Lent, R., Sakellari, G. (eds.) Computer and Information Sciences II, pp. 487–493. Springer, London (2011). https://doi.org/10.1007/978-1-4471-2155-8_62
18. Haydar, M., Petrenko, A., Sahraoui, H.: Formal verification of web applications modeled by communicating automata. In: de Frutos-Escrig, D., Núñez, M. (eds.) FORTE 2004. LNCS, vol. 3235, pp. 115–132. Springer, Heidelberg (2004). https://doi.org/10.1007/978-3-540-30232-2_8
19. Hennie, F.C.: Fault-detecting experiments for sequential circuits. In: Proceedings of Fifth Annual Symposium on Switching Circuit Theory and Logical Design, pp. 95–110. Princeton, New Jersey, November 1964
20. Hierons, R.M.: Minimizing the number of resets when testing from a finite state machine. Inf. Process. Lett. **90**(6), 287–292 (2004)
21. Hierons, R.M., Türker, U.C.: Parallel algorithms for generating harmonised state identifiers and characterising sets. IEEE Trans. Comput. **65**(11), 3370–3383 (2016). https://doi.org/10.1109/TC.2016.2532869
22. Hopcroft, J.E.: An n log n algorithm for minimizing the states in a finite automaton. In: Kohavi, Z. (ed.) The theory of Machines and Computation, pp. 189–196. Academic Press (1971)
23. Hsieh, E.P.: Checking experiments for sequential machines. IEEE Trans. Comput. **20**, 1152–1166 (1971)
24. Kohavi, Z.: Switching and Finite State Automata Theory. McGraw-Hill, New York (1978)
25. Koufareva, I., Dorofeeva, M.: A novel modification of w-method. Joint Bull. Novosibirsk Comput. 69–81 (2002)
26. Lee, D., Sabnani, K., Kristol, D., Paul, S.: Conformance testing of protocols specified as communicating finite state machines-a guided random walk based approach. IEEE Trans. Commun. **44**(5), 631–640 (1996). https://doi.org/10.1109/26.494307
27. Lee, D., Yannakakis, M.: Testing finite-state machines: state identification and verification. IEEE Trans. Comput. **43**(3), 306–320 (1994)
28. Lee, D., Yannakakis, M.: Principles and methods of testing finite-state machines - a survey. Proc. IEEE **84**(8), 1089–1123 (1996)
29. Low, S.: Probabilistic conformance testing of protocols with unobservable transitions. In: 1993 International Conference on Network Protocols, pp. 368–375 (Oct). https://doi.org/10.1109/ICNP.1993.340890
30. Luo, G., Petrenko, A., v. Bochmann, G.: Selecting test sequences for partially-specified nondeterministic finite state machines. In: Mizuno, T., Higashino, T., Shiratori, N. (eds.) Protocol Test Systems. ITIFIP, pp. 95–110. Springer, Boston, MA (1995). https://doi.org/10.1007/978-0-387-34883-4_6

31. Petrenko, A., Bochmann, G.V., Dssouli, R.: Conformance relations and test derivation. In: Proceedings of Protocol Test Systems VI (C-19), pp. 157–178 (1993)
32. Petrenko, A., Yevtushenko, N.: Testing from partial deterministic FSM specifications. IEEE Trans. Comput. **54**(9), 1154–1165 (2005)
33. Pomeranz, I., Reddy, S.M.: Test generation for multiple state-table faults in finite-state machines. IEEE Trans. Comput. **46**(7), 783–794 (1997)
34. Sabnani, K., Dahbura, A.: A protocol test generation procedure. Comput. Netw. **15**(4), 285–297 (1988)
35. Sidhu, D.P., Leung, T.K.: Formal methods for protocol testing: a detailed study. IEEE Trans. Software Eng. **15**(4), 413–426 (1989)
36. Teetor, P.: R Cookbook, 1st edn. O'Reilly (2011). http://oreilly.com/catalog/9780596809157
37. Ural, H., Zhu, K.: Optimal length test sequence generation using distinguishing sequences. IEEE/ACM Trans. Netw. **1**(3), 358–371 (1993)
38. Ural, H.: Formal methods for test sequence generation. Comput. Commun. **15**(5), 311–325 (1992). https://doi.org/10.1016/0140-3664(92)90092-S
39. Utting, M., Pretschner, A., Legeard, B.: A taxonomy of model-based testing approaches. Softw. Test. Verif. Reliab. **22**(5), 297–312 (2012)
40. Vasilevskii, M.P.: Failure diagnosis of automata. Cybernetics **4**, 653–665 (1973)
41. Vuong, S.T., Chan, W.W.L., Ito, M.R.: The UIOv-method for protocol test sequence generation. In: The 2nd International Workshop on Protocol Test Systems, Berlin (1989)

A Bond-Graph Metamodel:
Physics-Based Interconnection of Software Components

Reynaldo Cobos Méndez[1]($\boxtimes$)(iD), Julio de Oliveira Filho[2]($\boxtimes$)(iD),
Douwe Dresscher[1]($\boxtimes$)(iD), and Jan Broenink[1]($\boxtimes$)(iD)

[1] Robotics and Mechatronics Group, University of Twente,
Enschede, The Netherlands
{r.cobosmendez,d.dresscher,j.f.broenink}@utwente.nl
[2] TNO, The Hague, The Netherlands
julio.deoliveirafilho@tno.nl

Abstract. Composability and modularity in relation to physics are useful properties in the development of cyber-physical systems that interact with their environment. The bond-graph modeling language offers these properties. When systems structures conform to the bond-graph notation, all interfaces are defined as physical "power ports" which are guaranteed to exchange power. Having a single type of interface is a key feature when aiming for modular, composable systems. Furthermore, the facility to monitor energy flows in the system through power ports allows the definition of system-wide properties based on component properties. In this paper we present a metamodel of the bond-graph language aimed to facilitate the description and deployment of software components for cyber-physical systems. This effort provides a formalized description of standardized interfaces that enable physics-conformal interconnections. We present a use-case showing that the metamodel enables composability, reusability, extensibility, replaceability and independence of control software components.

Keywords: Bond-graph · Metamodeling · Power port · Component software · Cyber-physical systems

1   Introduction

On developing cyber-physical systems that exchange power with the environment (e.g., mechatronic and robotic applications), their relation to the physics domain brings additional concerns to software developers. These include exchange, transformation and conservation of power; change of state; and geometrical constraints. Additionally, the interaction with physical systems leads to tight

---

This research has received funding from the RobMoSys project (EU project No. 732410) under the subproject EG-IPC. https://robmosys.eu/eg-ipc/.

F. Arbab and S.-S. Jongmans (Eds.): FACS 2019, LNCS 12018, pp. 87–105, 2020.
https://doi.org/10.1007/978-3-030-40914-2_5

requirements on safety and reliability [26]. Examples are the teleoperation applications with force feedback, which control architectures require a reliable bilateral interconnection of force and velocity [8,27]. The interaction of controllers and other software components with the physical world determines stability, performance and safety properties. In other words, the interaction of the system follows physical laws [38].

Conformance to physics and preservation of composition-related properties (e.g., composability, reusability, replaceability and independance) of software components facilitate the development of cyber-physical systems. Such conformance not only allows software models to seamlessly interact with the physical world but also to behave as physical elements. This leads to the need of a description method to connect component-based software for cyber-physical systems with physical laws.

Energy-based modeling languages describe physical systems using *power* as the universal interaction currency or *lingua franca* between elements [21,40]. These methods rely on the principle of conservation of energy and changes of states to determine state variables, which are necessary and sufficient to unambiguously describe a system [5]. The automatic conservation of energy, the possibility to incorporate geometrical constraints, and its graphical representation make the bond-graph notation a straightforward form of object-oriented modeling language for physical systems [12,13]. A bond-graph model is a labelled and directed graph in which the edges represent an ideal energy connection between its vertices representing (real-world) physical elements [9]. When system structures conform to the bond-graph language, the format of all interfaces are guaranteed to exchange power [21], facilitating their design and later implementation.

In this paper we integrate physics description to cyber-physical system models using bond-graph notation. We present a metamodel of the bond-graph language that formalises the features of energy-based modeling. The metamodel describes standard interfaces that are guaranteed to exchange power and provide hierarchy, inheritance and encapsulation. This effort aims to facilitate the description and deployment of power-exchanging cyber-physical systems by closing relationship gaps between software components and physical models - e.g. impedance controllers in haptic devices. It is worth stressing that incorporating the metamodel in tooling is beyond the scope of this work. However, the metamodel serves as strict guideline to provide of physics interpretation to software components by interconnecting them in a power-consistent way.

This paper is structured as follows: Sect. 2 is a review of related work associating the bond-graph language to software engineering paradigms, metamodeling and the component-based approach. Section 3 is a discussion of the abstraction level of the bond-graph language and its higher-order relations. Section 4 presents the analysis of the separation and classification of the bond-graph constructs into metamodeling entities. Section 5 contains the formal definition of the bond-graph entities, their properties and constraints. Section 6 is a discussion of the developed metamodel. A use-case example is presented in Sect. 7, followed by the conclusions of this work in Sect. 8.

2 Related Work on Bond-Graph Language and Tooling

The concept of bond-graph was originated in the early 60's by Paynter [29] and further developed by Karnopp and Rosenberg [24]. Later, Breedveld [4,6] provided an insight to describe multi-dimensional physical systems using bond-graph notation. In this section, we mention work on mapping the bond-graph language to paradigms of software-engineering. Then, we make a brief review on effort relating the bond-graph language with the component-based approach for systems modeling.

2.1 Relation to Software-Modeling Languages

Broenink [12] defined the bond-graph notation as a form of object-oriented modeling language for physical systems. The elements and properties described by Object Oriented Modeling (OOM) (i.e., *objects*, *hierarchies*, *inheritances*, and *encapsulation*) are usually illustrated by graphical representations such as Unified Modeling Language (UML) [20]. Some works explore describing physical systems using UML [34] and System Modeling Language (SysML) [16,18]. Other works aim on integrating software models with physical models using architectural description languages (ADLs) [3,19]. The mentioned approaches describe physical models for cyber-physical systems from a software-modeling perspective using standard representations for software engineering. In contrast, we integrate physical laws to software architectures from a physical-modeling perspective using the bond-graph notation. The later is particularly relevant as a straightforward enforcement of physical constraints.

On the search of a bond-graph metamodel, [35] presents a UML Class Diagram in bond-graph notation that specify the relations and data properties of the elements composing a certain physical model. [39] contributes on integrating the bond-graph language to SysML. This is done by mapping the bond-graph entities to SysML constructs. Another example is the work of [31], which propose a framework for component modeling using bond-graph-based metamodeling techniques. These efforts identify non-trivial relations of the bond-graph language to other metamodels and languages; however, details on bond-graph primitive constructs, constraints and relations to higher-order-knowledge are still missing. Our work tackles this gap by capturing the relation to physics and mathematics of the bond-graph notation into the metamodel.

2.2 Bond-Graph Language in Tooling

Various modeling tools supporting bond-graph notation are available. These tools use their own language to represent bond-graph models. General application platforms, like Simulink[1], use their native block diagram representation [2,22]. Other approaches explore implementing the bond-graph language in object-oriented environments using Modelica [12,15,17]. More dedicated tooling, like 20-SIM[2], use a bond-graph-based language for modeling, simulation and generation of implementable code [1,7,11].

Relating the component-based approach to bond-graph modeling language is non-trivial. An example is given in [10], which discusses how 20-SIM uses libraries of models and sub-models, exploiting the encapsulation provided by the bond-graph paradigm. [28] proposes (semi-)automating the generation of simulation models in bond-graph notation by using *off-the-shelf* component implementations. As a metamodel of the bond-graph language is missing (or at least inaccessible), the development and integration between tools is problematic. As a first step on the metamodel formulation, the following section positions the bond-graph language among the abstraction levels and addresses its relation to higher-order-knowledge.

3   Methods on Metamodeling

From the Object-oriented Modeling perspective, any bond-graph (sub-)model can be seen as an *object* containing the mathematical description of a physical system [9]. Mathematically speaking, the three models in Fig. 1 represent exactly the same system. The bond-graph model in Fig. 1 is a more abstract representation of the physical system compared to the iconic diagrams. In this section, the abstraction levels of physical system modeling is explored, as well as the relation of the bond-graph language to other known metamodels.

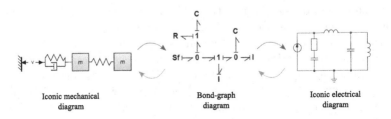

Iconic mechanical diagram   Bond-graph diagram   Iconic electrical diagram

Fig. 1. Different representations of the same dynamic system.

3.1   Levels of Abstraction

In Fig. 1 there is a bond-graph model and two domain models represented by iconic diagrams - one mechanical and one electrical. The bond-graph model

---

[1] Available Simulink library for bond-graph: https://nl.mathworks.com/matlabcentral/fileexchange/11092-bond-graph-add-on-block-library-bg-v-2-1.

[2] More details about 20-Sim: https://www.20sim.com/.

represents both domain models as the same dynamic system. This identifies the bond-graph model as a Multi-Domain-Specific Language (MDSL) with respect to the iconic diagrams. The different levels of abstraction are represented in Fig. 2, where the bond-graph metamodel is located at a higher level (M2) with respect to the bond-graph model and iconic diagrams (M1). More details about levels of abstraction are presented in [14].

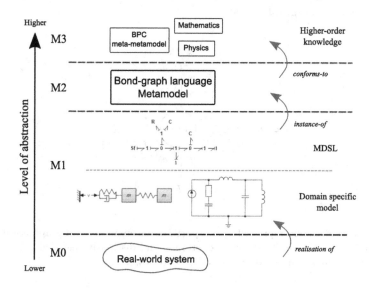

Fig. 2. Different levels of abstraction of a physical system.

As illustrated in Fig. 2, the bond-graph language conforms to higher-order knowledge (M3) - the Block-Port-Connector (BPC) metamodel, physics and mathematics. The bond-graph model in Fig. 1 would not be capable of describing physical interactions without its relation to other meta-metamodels, physics and mathematics. Therefore, it can be said that the bond-graph language adheres to other known metamodels.

3.2 Conforming to Higher-Order Knowledge

Mathematical abstractions of the bond-graph notation are port-Hamiltonian systems and Dirac structures [23]. The diagram in Fig. 3 is a port-Hamiltonian representation of an ideal physical system in bond-graph notation [37]. A port-Hamiltonian system can describe network models of physical systems that exchange power through ports [33]. The physical interaction among elements is done by the allocation of effort e and flow f variables on such ports and bonds. The mathematical relation between e and f characterizing the behavior of each (sub-)system is known as *constitutive relation*.

The energy of the system in Fig. 3 is characterized by a Hamiltonian equation, $H(e, f)$, and a Dirac structure, $D(e, f)$, representing power-conserving interconnections. Energy (defined as integral of power over time) is a conserved quantity, meaning that, in a closed system, it is at most transferred, converted or dissipated to the environment as free energy[3]. The energy conservation property is described by the power-conserving composition of $D(e, f)$ [32]. Thus, the port-Hamiltonian theory and Dirac structures serve as higher-order mathematical formulations of bond-graph models.

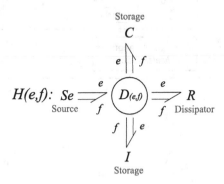

Fig. 3. A physical system represented as a port-Hamiltonian in bond-graph notation. A Dirac structure interconnects the source, storage and dissipative elements. The bonds denote power exchange as a product of effort and flow variables.

3.3   Conforming to Known Meta-Metamodels

When a model conforms-to a metamodel, the elements that are being used in the model satisfy the constraints on the relations that are made explicit in the metamodel [14]. This is the case of the Block-Port-Connector (BPC) meta-metamodel, whose modeling primitives are present in the bond-graph language known as *elements, junctions, power port* and *power bond*. In Table 1, a relation of the bond-graph constructs with the BPC primitives is shown, along with their higher-order knowledge link.

However, there are properties and constraints in the bond-graph language that cannot be sufficiently described by BPC to represent real physical interactions. For instance, physical interaction among systems require a bi-directional property in the connectors, as the energy is exchanged in both directions [6,29,33]. Given this, the bond-graph metamodel has to make such constraints explicit to avoid ambiguity when conforming to other meta-metamodels and paradigms.

The properties of OOM (i.e. *encapsulation, inheritance* and *hierarchy*) are essential when modeling physical systems [12]. The encapsulation into submodels

---

[3] Also known as the first law of thermodynamics.

and the inheritance property allow maintaining libraries of basic bond-graph elements and exchanging classes of components within a model. The hierarchy property allows complex bond-graph models to be embedded within other systems through power ports [21]. By conforming-to other metamodels (e.g, BPC), we can bring those entities, properties, attributes and relations required to describe the bond-graph language.

The realization of the metamodel is based on the separation of the elements of the modeling language into metamodeling concepts. In the following section, we describe the entities, relations and constraints of the bond-graph modeling language by identifying the properties and attributes of its elements.

Table 1. Relation between Block-Port-Connector primitives to bond-graph constructs.

BPC primitive	Bond-graph constructs	Physics/mathematics link
Block	- Dissipative elements - Storage elements - Source elements	*Hamiltonian theory*
	- Power junction - Transformer - Gyrator	*Dirac structures*
Port	- Power port	*port-Hamiltonian theory*
Connector	- Power bond	

4 Analysis of Bond-Graph Entities

As mentioned earlier, a bond-graph model is a graph which edges and vertices represent energetic interactions. Breedveld [6] provided a classification of the bond-graph vertices based on their energetic behavior. Such classification is represented in Table 2. The diagram in Fig. 4 is a representation of a bond-graph model whose elements are classified in four classes: *elements, junction structures, power port* and *power bond*. These classes have specific purposes in the bond-graph language and can be allocated and described in a higher level of abstraction.

Song [36] defines an *entity* as a primary thing that exists as itself and can be identified. Following this definition, the bond-graph *elements, junction structures, power ports* and *power bonds* can be classified as entities as they are independent things that can be clearly identified. Table 3 contains the properties and attributes of the bond-graph entities, which are essential to later formalize the modeling language.

Table 2. Classification of the bond-graph vertices based on their energetic behavior.

	Classification		Bond-graph constructs
Block	*Elements*	*Energic*	- Storage
		Entropic	- Dissipator
		Boundary	- Source
	Junction structures		- Power junction - Transformer - Gyrator

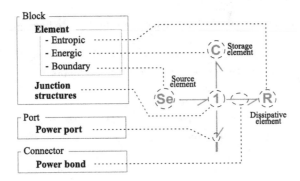

Fig. 4. Classification of bond-graph entities into 'classes'.

4.1 Classification of Bond-Graph Vertices

The UML diagram in Fig. 5 illustrates the classification of the bond-graph vertices based on their energetic properties and presence of parameters. The storages, dissipators and sources have at least one parameter - i.e., capacitance/compliance, inductance/mass, resistance/friction. We label these parametric elements as *BondElements*. The rest of the vertices are labeled as *Junction-Structures* as they interconnect *BondElements* (and other *JunctionStructures*) in a power continuous way.

As shown in Table 3, transformers and gyrators could also be classified as *BondElements* as they have parameters - namely transformation/gyration ratio. For the bond-graph metamodel, we propose classifying transformers and gyrators as parametric-*JunctionStructures* to distinguish them from the 0-/1-junctions. Having identified the entities and properties of the bond-graph language, we move forward on formally defining them into a metamodel.

5   Formalization of the Bond-Graph Metamodel

This section addresses the formal definition of the bond-graph language. Here, we capture the properties and constraints of the entities identified and classified in previous sections into a metamodel.

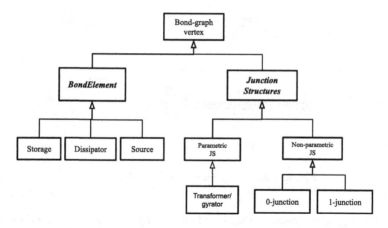

Fig. 5. Identification of bond-graph vertices based on their energetic behavior and presence of parameters

Table 3. Identification of properties and attributes of the bond-graph entities. Note: Transformers and gyrators are both Elements and Junction structures.

Bond-graph entity		Property	Attribute
Element	Storage	- Has constitutive relation	- Name
			- Symbol
			- Number of ports
			- Size of ports
	Dissipator		- Size of elements
	Source		- Has parameters
Element & Junction structure	Transformer/ gyrator	- Has power continuity - Has constitutive relation	- Name - Symbol - Number of ports - Has parameters
Junction structure	0-junction/ 1-junction	- Has power continuity - Has constitutive relation	- Name - Symbol - Number of ports
Power port		- Has 1 to 1 relation with bonds - Has 1 port has 2 variables in the constitutive relation	- Name
Power bond		- Has bidirectionality - Has power continuity - Connects 2 ports	- Name

5.1   Formal Definition of Entities, Properties and Constraints

The UML class diagram in Fig. 6a represents the association of entities of Table 3, based on the vertex classification of Fig. 5. The relation between *PowerPort* and *PowerBond* classes is represented in Fig. 6b. The definitions are enforced using *Description Logic* (DL) as UML is insufficient to describe the correct application of the bond-graph language. The formal semantics provided by DL let humans and computer systems exchange the same language, avoiding ambiguity [25].

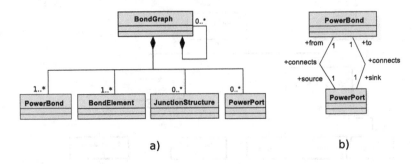

Fig. 6. UML diagram of bond-graph classes.

The bond-graph classes in Fig. 6 and their properties are expressed in detail in Table 4. The chosen language is the Web Ontology Language 2 (OWL2), which is based on DL. The axioms in Table 4 map the bond-graph classes to Block-Port-Connector (BPC) primitives (as in Table 1), facilitating the description of the bond-graph language.

Despite the bond-graph language conforms-to BPC, it is not allowed for a power bond to connect more than two power ports. Given this, it is required to constrain the relation between *PowerPort* and *PowerBond* with two '1-to-1 connects' shown in Fig. 6b. In the *PowerPort-PowerBond-PowerPort* relation, one *PowerPort* is declared as source and the other as sink. However, the energy exchange in both directions of the *PowerBond* is guaranteed by the *bidirectionality* property. Just like the previous example, the essential constraints of the bond-graph classes are formally expressed in Table 5 using DL[4].

5.2   Formal Definition of Power Variables

The bond-graph formalism indicates that the only variable to be 'transferred' among vertices through *PowerPorts* and *PowerBonds* is 'power' as a product of colocated effort and flow. To constrain this, it is proposed to define effort and flow as quantities containing a given value and a given unit. In the bond-graph metamodel ontology, we use the Ontology of Units of Measure described in [30]. The complete definition of the power variables is represented in Table 6.

5.3   A Note on Causality

The power-bond determines the effort and flow variable as a bilateral signal flow [5,29]. Meaning that, when one variable is given as an input for one port, its conjugated variable is automatically the output for that particular port. The inverse is true for the port at the other side. Causality is the policy that assigns the order of computations by determining whether the effort or the flow is either input or output. Causal analysis is addressed in [11] for computer-aided modeling and simulation purposes. In practice, causality is required to generate easy

---

[4] See the Appendix for more details about the symbols used in the formal definitions.

Table 4. Formal definitions of the bond-graph classes.

Class	OW2 axiom	Related properties
BondGraph	Declaration (Class(:BondGraph))	- hasPowerBond - hasJunctionStructure - hasBondElement
BondElement	Declaration (Class(:BondElement))	- hasParameter - hasConstitutiveRelation - hasPowerPort - DisjointWith Junction-
	SubClassOf (:BondElement :Block)	Structure - Contains max 1 BondGraph
PowerPort	Declaration (Class(:PowerPort))	- isPortOf exactly 1 (BondElement or Junction-
	SubClassOf (:PowerPort :Port)	Structure)
PowerBond	Declaration (Class(:PowerBond))	- Connects exactly 2 PowerPort - hasBidirectionality
	SubClassOf (:PowerBond:Connector)	
JunctionStructure	Declaration (Class(:JunctionStructure))	- hasPowerPort - hasConstitutiveRelation - DisjointWith BondElement
	SubClassOf (:JunctionStructure :Block)	

simulatable cyber-physical models. However, a model with the same structure but different causality is in fact the same model; therefore, this policy is not considered part of the bond-graph metamodel itself.

This section has addressed the structural allocation of the entities of the bond-graph language along with their formal definition and constraints. The following section is a discussion of the properties of the developed metamodel.

6   Metamodel Discussion

The bond-graph metamodel has to contain the properties and elements of other known meta-metamodels, in addition to the intrinsic characteristics of the bond-graph notation, as described in Sects. 3 and 4. This section is a discussion of the properties of the bond-graph metamodel as well as the resulting ambiguities as consequence of the limitations of the formalization language.

6.1   Properties of the Bond-Graph Metamodel

As mentioned earlier, the bond-graph language has *encapsulation, hierarchy* and *inheritance* properties. In the metamodel, encapsulation is represented in the

Table 5. Essential constraints of bond-graph classes

DL syntax	Description
Class: BondElement	
$BondElement \sqcap JunctionStructure = \bot$	No bond-graph vertex can be at the same time a BondComponent and a JunctionStructure
$BondElement \sqsubseteq \forall\, hasPowerPort.Port$	BondElements have any number of power ports
$BondElement \sqsubseteq\, \leq 1$ $contains.BondGraph$	BondElements may contain at most 1 BondGraph
Class: PowerPort	
$PowerPort \sqsubseteq\, 1$ $isPortOf.(BondElement \sqcup JunctionStructure)$	Any PowerPort is a port of exactly one BondElement or one JunctionStructure
$PowerPort \sqsubseteq\, \leq 1\; connects.PowerBond$	PowerPorts can connect to at most 1 PowerBond
Class: PowerBond	
$PowerBond \sqsubseteq 2\; connects.PowerPort$	A PowerBond connects exactly two PowerPort. Equivalently, a PowerBond cannot connect to more than two power ports, nor have one of its connection points loose
Class: JunctionStructure	
$JunctionStructure \sqcap BondElement = \bot$	No bond graph vertex can be at the same time a BondElement and a JunctionStructure
$JunctionStructure \sqsubseteq \forall\, hasPowerPort.PowerPort$	JunctionStructure have any number of power ports
$JunctionStructure \sqsubseteq 0\; contains.\top$	JunctionStructure is a block that contains nothing
Class: BondGraph	
$BondGraph \sqsubseteq \forall has$ $BondComponent.BondComponent$	BongGraphs have BondElements
$BondGraph \sqsubseteq \forall has$ $PowerBond.PowerBond$	BondGraphs have PowerBonds
$BongGraph \sqsubseteq \forall has$ $JunctionStructure.JunctionStructure$	BondGraphs have JunctionStructures

relation between the *BondGraph* class and the rest of the classes in Fig. 6. As expressed in Table 5, an instance of a *BondGraph* is a collection that can contain elements, junction structures and power bonds. In a similar way, a *BondGraph* instance can contain other *BondGraph* models itself.

The encapsulation of bond-graph models is illustrated in Fig. 7 as follows: Model 4 is a *BondGraph* model composed of *BondElements*, *JunctionStructures* and other *BondGraph* models interconnected by *PowerBonds* (represented by half arrows) and *PowerPorts* (represented by black squares). Model 1, Model 2 and Model 3 are *BondGraphs* models containing at least one *BondElement* each. These constructions are allowed by the metamodel, providing *hierarchy* to the system.

Given the guaranteed interconnection of vertices and the encapsulation property, the construction of complex models can be simplified by using generic submodels and only modifying parameters and attributes. For instance, Model 2

Table 6. Definition of power variables and their constraints

Data property	DL definition	Description
effort	$\exists$ *effort.* $\top \sqsubseteq PowerBond$	Domain of effort is always a PowerBond
	Quantity $\sqsubseteq \forall$ *effort.PowerBond*	Range of 'effort' is always a 'Quantity'
	Quantity $\sqsubseteq \leq 1$ *effort.PowerBond*	'effort' is functional, and additionally for each individual moment in time
flow	$\exists$ *flow.* $\top \sqsubseteq PowerBond$	Domain of flow is always a PowerBond
	Quantity $\sqsubseteq \forall$ *flow.PowerBond*	Range of 'flow' is always a 'Quantity'
	Quantity $\sqsubseteq \leq 1$ *flow.PowerBond*	'flow' is functional, and additionally for each individual moment in time

and Model 3 in Fig. 7 can be described as storage elements with different *ConstitutiveRelation*, parameter value and symbol. Therefore, *inheritance* is supported by the metamodel.

6.2   Note on Completeness of the Bond-Graph Metamodel

The implementation of the bond-graph metamodel into tooling is out of the scope of this paper. This represent a limitation on the assessment of the completeness of the metamodel to describe physical systems. However, it is possible to determine whether or not the formal definitions adhere to the the bond-graph notation to construct models.

The interaction between vertices has to be done only through the association between *PowerPort* and *PowerBond* classes. The UML diagram in Fig. 6 and the DL definitions in Table 5 enforce this constraint. Still, there is no formal definition that prevents a *PowerBond* to be connected at both sides to the same *BondElement* or *BondGraph* model (through different ports) as shown in the system in Fig. 8a. Nevertheless, such an unusual connection does not compromise the correct application of the bond-graph language to describe the physical behavior of the system. For instance, Fig. 8b is a representation of the same system shown in Fig. 8a. On the other hand, the structure in Fig. 9a is not allowed due to lack of information required to describe the power exchange between *BondElements*. Changing the splitter to a *JunctionStructure* block - either 0-junction or 1-junction[5] - in Fig. 9b is essential to allow a physics-conformal interconnection.

---

[5] Depending on the model, the *JunctionStructure* could be either a 1-junction or a 0-junction as they denote different Diract structures.

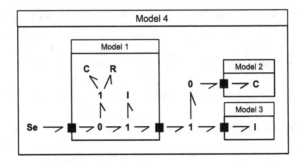

Fig. 7. A bond-graph model encapsulating other bond-graph (sub-)models.

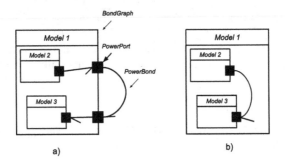

Fig. 8. Two equal systems (a) Unusual connection to same element. (b) Encapsulation of the PowerBond.

7   Interfacing Software Components Using Bond-Graph Entities

By incorporating the bond-graph metamodel to the component-based approach, software models like in Fig. 10 can be realized. The software components have instances of *PowerPort* class connected by instances of *PowerBond* class. The components themselves are *BondElement* instances which *ConstitutiveRelation* (see Table 4) is the implementation. Since power exchange (product of effort and flow variables) is enforced on the interfaces, the components can be exchanged or replaced depending on the application. Thus, the component implementation can be extended without compromising its independence from the interfaces. In this section we present a use case example where the bond-graph metamodel is applied.

7.1   Use-Case Example: Haptic Telemanipulation

We applied the entities described by the bond-graph metamodel on the teleoperation with force feedback use case depicted in Fig. 11. The system model is in bond-graph notation, which means that power is exchanged between the elements. A human operator telemanipulates a robot ('Slave') using another robot

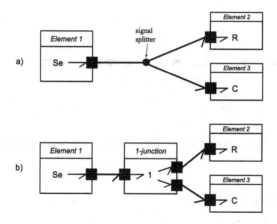

Fig. 9. Interaction between BondElements: (a) Ill-connection through a signal splitter not allowed by the metamodel. (b) Correct application of the bond-graph language. Note: the 1-junction in (b) could be a 0-junction.

('Master') as haptic device. The impedance controller component is provided of power ports and its implementation is bond-graph conformal - that is, dealing with effort and flow signals and other constraints as formalized in the metamodel. This is the same for the geometric Jacobian components.

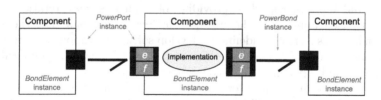

Fig. 10. Software components interacting via power ports and power bonds. Given the interfaces, it can be assumed that *power* is exchanged among them.

The *PowerPort* and *PowerBond* instances provide a power-consistent interconnection between the physical and virtual environments in Fig. 11. In other words, the impedance controller has a 'physical' link with the target environment and the operator. Thus, a (force) feedback loop is enforced. The interfaces allow the use-cased system to be composable and its components replaceable. The impedance controller and geometric Jacobians can be replaced by other components as long as they have the required power ports. The implementation of each component can also be extended. If the application changes - for instance, adding another 'slave' robot - the controller and Jacobian can be reused. Additionally, system developers can get the flexibility and independence offered by the component software approach by having *application-specific* and *tool-specific* components clearly separated.

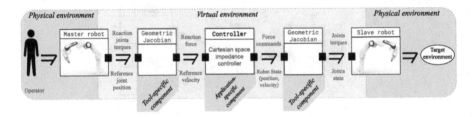

Fig. 11. Control software model of the teleoperation application in bond-graph notation.

8 Conclusions

We presented a metamodel that formalizes the bond-graph modeling language. This effort aims to ease the development of software components for cyber-physical systems that interact with their environment. This is done by enforcing the correct application of the bond-graph notation to describe physical interconnections among components. The elements, properties and constraints of the modeling language were identified and characterized using Description Logics. The result is a set of definitions of bond-graph classes, their object properties and relations that conform to physics, mathematics and Block-Port-Connector meta-metamodel.

As discussed in Sect. 6, the bond-graph metamodel supports encapsulation, hierarchy and inheritance, while providing of a physics interpretation to the component-based approach. The use case in Sect. 7 suggests that the metamodel can serve as a strict guideline to develop software components interfacing with physical systems. It was shown that composability, reusability, extensibility, replaceability and independence of components are present.

Further work can be done on integrating the bond-graph metamodel to software-modeling tooling. This means merging physical laws to *off-the-shelf* software components. Such incorporation can help reducing development times and improve cost-effectiveness by facilitating the task of control-software developers.

Moreover, further work can be on the formal description of causality. Thus the presented metamodel becomes more rich. Tools using this extended bond-graph metamodel can be enhanced with automated causal analysis functions, exploiting transformation from energy relations to signal input-output relations - as done by dedicated tooling for bond-graph modeling and simulation, e.g., 20-SIM.

Despite the tooling limitation, the presented metamodel is useful to provide the required descriptions to make already existent control components abide physics through guaranteed power exchange. The facility to monitor energy flows in the system through power ports can allow the definition of system-wide properties based on component properties, most notably passivity. Therefore, the bond-graph metamodel can contribute on reliability and safety of component-based cyber-physical systems.

Appendix

See Table 7.

Table 7. Symbols used in DL definitions

Symbol	Legend
$\sqcap$	Meet semilattice
$\perp$	Minimum element or bottom
$\sqsubseteq$	Partial order relation
$\forall$	For all
$\exists$	There exist
$\top$	Maximum element or top
$\sqcup$	Disjoint union

References

1. Automated modelling. In: Borutzky, W. (ed.) Bond Graph Methodology: Development and Analysis of Multidisciplinary Dynamic System Models, pp. 469–560. Springer, London (2010). https://doi.org/10.1007/978-1-84882-882-7_11
2. Antic, D., Vidojkovic, B.: Obtaining system block diagrams based on bond graph models and application of bondsim tools. Int. J. Model. Simul. **21**(4), 257–262 (2001). https://doi.org/10.1080/02286203.2001.11442210
3. Bhave, A.Y., Garlan, D., Krogh, B., Rajhans, A., Schmerl, B.: Augmenting software architectures with physical components. In: Embedded Real Time Software and Systems Conference (2010)
4. Breedveld, P.C.: Multibond graph elements in physical systems theory. J. Franklin Inst. **319**(1), 1–36 (1985). https://doi.org/10.1016/0016-0032(85)90062-6
5. Breedveld, P.: Integrated modeling of physical systems - dynamic systems, vol. 1. University of Twente, Enschede, The Netherlands (2014)
6. Breedveld, P.C.: Physical Systems Theory in Terms of Bond Graphs. Twente University of Technology, Department of Electrical Engineering, Enschede (1984). oCLC: 852801415
7. Breunese, A.P.J., Broenink, J.F.: Modeling mechatronic systems using the SIDOPS+ language. Simul. Ser. **29**(1), 301 (1997). oCLC: 106228295
8. Brodskiy, Y.: Robust autonomy for interactive robots (2014). https://doi.org/10.3990/1.9789036536202
9. Broenink, J.: Introduction to physical systems modelling with bond graphs (1999)

10. Broenink, J.F.: 20-sim software for hierarchical bond-graph/block-diagram models. Simul. Pract. Theory **7**(5), 481–492 (1999). https://doi.org/10.1016/S0928-4869(99)00018-X

11. Broenink, J.F.: Computer-aided physical-systems modeling and simulation: a bond graph approach, March 1990

12. Broenink, J.F.: Object-oriented modeling with bond graphs and Modelica. In: Proceedings of the 1999 International Conference on Bond Graph Modeling and Simulation, pp. 163–168, February 1999

13. Brown, F.T.: Engineering System Dynamics: A Unified Graph-Centered Approach, 2nd edn. CRC Press (2006). https://doi.org/10.1201/b18080

14. Bruyninckx, H., Scioni, E., Hubel, N., Reniers, F.: Composable control stacks in component-based cyber-physical system platforms, April 2018

15. de la Calle, A., Cellier, F.E., Yebra, L.J., Dormido, S.: Improvements in BondLib, the Modelica bond graph library. In: 2013 8th EUROSIM Congress on Modelling and Simulation, pp. 282–287, September 2013. https://doi.org/10.1109/EUROSIM.2013.58

16. Cao, Y., Liu, Y., Fan, H., Fan, B.: SysML-based uniform behavior modeling and automated mapping of design and simulation model for complex mechatronics. Comput. Aided Des. **45**(3), 764–776 (2013). https://doi.org/10.1016/j.cad.2012.05.001

17. Cellier, F.E., Nebot, À.: The Modelica Bond Graph Library, p. 10 (2005)

18. Chen, R., Liu, Y., Cao, Y., Zhao, J., Yuan, L., Fan, H.: ArchME: a systems modeling language extension for mechatronic system architecture modeling. AI EDAM **32**(1), 75–91 (2018). https://doi.org/10.1017/S0890060417000245

19. Garlan, D., Monroe, R.T., Wile, D.: ACME: architectural description of component-based systems. In: Foundations of Component-Based Systems, pp. 47–68. Cambridge University Press (2000)

20. Garrido, J.M.: Object orientation. In: Garrido, J.M. (ed.) Object Oriented Simulation, pp. 51–58. Springer, Boston (2009). https://doi.org/10.1007/978-1-4419-0516-1_5

21. Gawthrop, P.J., Bevan, G.P.: Bond-graph modeling. IEEE Control Syst. **27**(2), 24–45 (2007). https://doi.org/10.1109/MCS.2007.338279

22. Geitner, G.: Power flow diagrams using a bond graph library under simulink. In: IECON 2006–32nd Annual Conference on IEEE Industrial Electronics, pp. 5282–5288, November 2006. https://doi.org/10.1109/IECON.2006.347232

23. Golo, G., van der Schaft, A., Breedveld, P.C., Maschke, B.M.: Hamiltonian formulation of bond graphs. In: Johansson, R., Rantzer, A. (eds.) Nonlinear and Hybrid Systems in Automotive Control, pp. 351–372. Springer, London (2003)

24. Karnopp, D., Rosenberg, R.C.: Analysis and Simulation of Multiport Systems: The Bond Graph Approach to Physical System Dynamics. MIT Press, Cambridge (1968)

25. Krötzsch, M., Simancik, F., Horrocks, I.: Description logics. IEEE Intell. Syst. **29**(1), 12–19 (2014). https://doi.org/10.1109/MIS.2013.123

26. Lee, E.A.: Cyber physical systems: design challenges. In: 2008 11th IEEE International Symposium on Object and Component-Oriented Real-Time Distributed Computing (ISORC), pp. 363–369, May 2008. https://doi.org/10.1109/ISORC.2008.25

27. Mersha, A.Y.: On autonomous and teleoperated aerial service robots (2014). https://doi.org/10.3990/1.9789036536585

28. Novák, P., Šindelář, R.: Component-based design of simulation models utilizing bond-graph theory. IFAC Proc. Vol. **47**(3), 9229–9234 (2014). https://doi.org/10.3182/20140824-6-ZA-1003.01167

29. Paynter, H.M., Briggs, P.: Analysis and Design of Engineering Systems: Class Notes for M.I.T. Course 2.751. MIT Press, Cambridge (1961). Massachusetts Institute of Technology

30. Rijgersberg, H., van Assem, M., Top, J.: Ontology of units of measure and related concepts. Semant. Web **4**(1), 3–13 (2013). https://doi.org/10.3233/SW-2012-0069

31. Sampath Kumar, V.R., Shanmugavel, M., Ganapathy, V., Shirinzadeh, B.: Unified meta-modeling framework using bond graph grammars for conceptual modeling. Robot. Auton. Syst. **72**, 114–130 (2015). https://doi.org/10.1016/j.robot.2015.05.003

32. van der Schaft, A., Cervera, J.: Composition of Dirac structures and control of Port-Hamiltonian systems. In: Proceedings of the 15th International Symposium on the Mathematical Theory of Networks and Systems. University of Notre Dame (2002)

33. Scioni, E., et al.: Hierarchical hypergraphs for knowledge-centric robot systems. In: A Composable Structural Meta Model and its Domain Specific Language NPC4 (2016). https://doi.org/10.6092/JOSER_2016_07_01_p55

34. Secchi, C., Bonfe, M., Fantuzzi, C.: On the use of UML for modeling mechatronic systems. IEEE Trans. Autom. Sci. Eng. **4**(1), 105–113 (2007). https://doi.org/10.1109/TASE.2006.879686

35. Sen, S., Vangheluwe, H.: Multi-domain physical system modeling and control based on meta-modeling and graph rewriting. In: 2006 IEEE Conference on Computer Aided Control System Design, 2006 IEEE International Conference on Control Applications, 2006 IEEE International Symposium on Intelligent Control. pp. 69–75, October 2006. https://doi.org/10.1109/CACSD-CCA-ISIC.2006.4776626

36. Song, I.Y., Froehlich, K.: Entity-relationship modeling. IEEE Potentials **13**(5), 29–34 (1995). https://doi.org/10.1109/45.464652

37. Stramigioli, S.: Intrinsically passive control using sampled data system passivity. In: Multi-point Interaction with Real and Virtual Objects, pp. 215–229, July 2005. https://doi.org/10.1007/11429555_14

38. Stramigioli, Stefano: Energy-aware robotics. In: Camlibel, M.Kanat, Julius, A.Agung, Pasumarthy, Ramkrishna, Scherpen, Jacquelien M.A. (eds.) Mathematical Control Theory I. LNCIS, vol. 461, pp. 37–50. Springer, Cham (2015). https://doi.org/10.1007/978-3-319-20988-3_3

39. Turki, S., Soriano, T.: A SysML extension for Bond Graphs support (2005)

40. van der Schaft, A., Jeltsema, D.: Port-Hamiltonian systems theory: an introductory overview. Found. Trends® Syst. Control **1**(2), 173–378 (2014). https://doi.org/10.1561/2600000002

Multilabeled Petri Nets

Kasper Dokter[✉]

Leiden University, Leiden, Netherlands
K.P.C.Dokter@cwi.nl

Abstract. We introduce multilabeled Petri nets as an inherently parallel generalization of constraint automata. Composition of multilabeled nets does not suffer from state-space explosions, which makes them an adequate intermediate representation for code generation. We present also an abstraction operator for multilabeled nets that eliminates internal transitions, which optimizes the execution of multilabeled nets.

1 Introduction

In a concurrent system, multiple components simultaniously interact according to a protocol, which defines all allowed interactions amongst them. Most concurrent programming languages offer syntax to specify the protocol as choreography. In a choreography, components interact directly via locks, semaphores, and queues. As a result, the protocol code blends with the component code, and the protocol is hard to edit, reuse, or analyse. Exogenous coordination languages, such as Reo [2] or BIP [4], offer syntax to specify the protocol as an orchestration. In an orchestration, components interact indirectly via a central protocol component. In this case, the protocol is easy to edit, reuse, and analyse.

Although we desire the protocol specification as an orchestration, we want a protocol implementation as a choreography. Indeed, implementing the protocol with a single central protocol component can easily introduce a performance bottleneck. To prevent the bottleneck, we aim for distributed protocols, i.e., protocols implemented as multiple parallel components.

It is the responsability of the compiler of the coordination language to produce efficient protocol implementations. Jongmans [11] developed a compiler that generates protocol implementations based on constraint automata [3]. A constraint automaton is a pair (P, A) consisting of a set of variables P and a state machine A, whose transition labels are pairs (N, g) consisting of a set of variables $N \subseteq P$ and a constraint g on their values. The elements P, N, and g are respectively called interface, synchronization constraint, and data constraint.

Constraint automata model protocols in a simple and intuitive manner. However, being a state machine, a constraint automaton is inherently sequential. As we desire distributed protocols, the constraint automaton representation of protocols is not completely adequate. State-space explosions for constraint automata serve as evidence for this mismatch. Alternative algorithms to compute the composite constraint automaton have only partial success [13, Fig. 17(a)].

© Springer Nature Switzerland AG 2020
F. Arbab and S.-S. Jongmans (Eds.): FACS 2019, LNCS 12018, pp. 106–126, 2020.
https://doi.org/10.1007/978-3-030-40914-2_6

In contrast to state machines, Petri nets [18] are inherently parallel. Moreover, a state machine can be viewed as a Petri net for which every transition has a single input place and a single output place. It seems natural to generalize constraint automata by replacing the underlying state machine by a Petri net.

In the current paper, we introduce multilabeled Petri nets as an inherently parallel extension to constraint automata. After stating some basic results on monoids and multisets (Sect. 2), we view a constraint automaton as a *multilabeled Petri net* (Sect. 3), which is an ordinary Petri net whose transitions are labeled by multisets of actions. If multiple (not necessarily distinct) transitions in a multilabeled Petri net fire in parallel, the composite transition is labeled by the union of the labels of its constituent transitions.

We generalize constraint automaton composition to composition of multilabeled Petri nets (Sect. 4). While the composition of arbitrary multilabeled Petri nets seems hard to compute, we develop an efficient algorithm that composes *square-free* nets. Intuitively, a multilabeled Petri net is squarefree iff any action occurs at most once in every (parallel) execution step of the net.

The number of places in the composite Petri net grows linearly, which prevents the state-space explosion. Therefore, multilabeled Petri nets are an adequate intermediate representation of protocols. Since a transition-space explosion is still possible, multilabeled Petri nets are not a silver bullet.

Multilabeled Petri nets can contain silent transitions, which have no observable behavior. Such silent transitions can be the result of hiding irrelevant actions. In protocol implementations based on multilabeled Petri nets, silent transitions do not perform any I/O-operation and delay the throughput of the protocol. We define an abstraction operator for multilabeled nets (Sect. 5) that removes silent transitions. We develop an algorithm that computes the abstraction of a multilabeled net.

Finally, we summarize the results (Sect. 6) and point out future work.

Running Example. We illustrate the composition and abstraction of multilabeled Petri nets by an example on a mail server and a client.

Figure 1(a) shows the Petri net of a client that can compose, send, receive, and delete messages. Composed messages are stored as concepts, and received messages end up in the inbox. The client can concurrently send and receive messages, as is the case for a large company with internal mail between different departments. We want every message to be transferred to two recipients (e.g., adding a recipient in CC). We represent this intend labeling the send transition with the expression a^2, which denotes a multiset that contains a twice.

Figure 1(b) shows the Petri net of a server that can transfer messages. A fingerprint of each transferred message is logged. We assume that message transferal is not buffered: a send message is immediately received. We represent this by labeling the transfer transition with the expression ab that denotes the multiset that contains a and b. Note that the order of a and b is irrelevant, as ab and ba denote the same multiset.

Figure 1(c) shows the composition of the client and server, as defined by the composition operator presented in the current paper. Composition of multilabeled nets synchronize transitions that agree on shared actions. Observe that the transfer transition must fire twice in order to agree on a with the send transition. Consequently, the receive transition must fire twice in order to agree on b with the transfer[2] transition. Hence, the send, transfer, and receive transitions synchronize into a single parallel transition send | transfer[2] | receive[2], which we denote as t. The actions c (compose) and d (delete) of the client are not shared with the server. Therefore, the client-server composition allows the client to compose and print a message, without synchronizing with the mail server.

Figure 1(d) shows the abstraction of the composite system, where message transferal (actions a and b) is hidden. The resulting silent transition t is eliminated by replacing t by the sequential composition t; delete, which we denote as s. Note that s deletes only one of the two messages that are sent. The remaining message can be deleted by the usual delete transition.

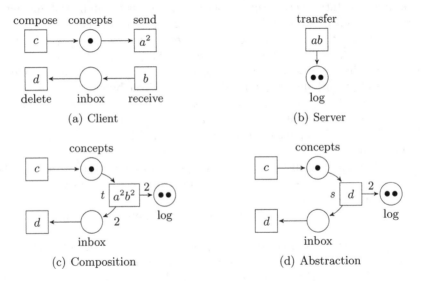

(a) Client

(b) Server

(c) Composition

(d) Abstraction

Fig. 1. Multilabeled nets of a client (a) and a mail server (b). Their composition (c) synchronizes a single send transition with two transfer and two receive transitions (t equals send | transfer[2] | receive[2]). The abstraction (d) hides the a and b actions and eliminates the resulting silent transition (s equals t; delete).

Related Work. The literature offers a wide variety of composition operators on Petri nets, which can be divided in two categories, namely *topological* and *parallel* compositions.

By topological compositions, we mean those operators that 'glue Petri nets along their places and transitions'. For example, Mazurkiewicz [15, Section 8] defines a composition operator that glues transitions with the same name. Bernardinello and De Cindio [5, Part II] define a composition that glues Petri nets on places and transitions. Kotov [14] defines multiple composition operators, including superposition, which glue two Petri nets along shared places and transitions. Hierarchical composition [8] is a topological composition that identifies a complete Petri net with a single transition of another Petri net.

By parallel compositions, we mean those operators that 'synchronize transitions of the constituent Petri nets'. A single transition of one Petri net can synchronize with multiple transitions of the other net. Hence, parallel composition can be seen as duplication plus gluing, which distinguishes it from topological composition.

Parallel composition is called synchronous, if subsystems must 'run at the same speed'. That is, both Petri nets in a composition must simultaneously fire a transition. Examples of these operators are those that are defined as a product in a category of Petri nets [9,10,16,23]. Such composition operators are not always convenient for specification.

In asynchronous parallel composition, subsystems can run at different speeds. That is, one Petri net can fire a transition, while the other net does nothing. Examples of asynchronous compositions include the composition of Petri Boxes [6], Signal Transition Graphs [21], extended safe nets [20, Chapter 4], zero safe nets [7], and general Petri nets [10].

The asynchronous composition operator defined by Anisimov, Golenkov, and Kharitonov [1] comes closest to our composition operator. Their parallel composition $_\alpha\|_\beta$ of Petri nets is relative to some transition (multi)labelings α and β of its operands. Unlike our composition operator, Anisimov et al. suppose a CCS-like synchronization that synchronizes an action a with its conjugate $\bar{a}$. As such, their composition cannot be used as a formal semantics for Reo circuits.

Silent transitions in safe Petri nets can be removed in at least two different ways. Vogler and Wollowski [21] use contraction, which deletes a silent transition and merges all its input and output places. In general, contraction does not yield a safe net.

Wimmel [22] eliminates silent transitions from safe Petri nets in three steps: First, he unfolds a safe Petri net into an occurrence net, whose graph structure is acyclic. Next, he finds the process [17] (i.e., the set of all accepted pomsets) of the occurrence net, while ignoring all silent transitions. Finally, he constructs a suitable finite quotient of the process. By construction, the resulting Petri net has no silent transitions and is pomset-equivalent to the original net.

2 Preliminaries

2.1 Graded Monoids

A **monoid** is a triple $(M, +, 0)$, where $+ : M \times M \to M$ is an associative binary operation, with identity element 0. A **submonoid** of $(M, +, 0)$ is a monoid of the form $(S, +, 0)$, where $S \subseteq M$ is a subset of M that contains $0 \in S$ and is closed under addition.

An element x in a monoid M is **invertible** iff there exists some $y \in M$, such that $x + y = 0$. An element x in a monoid M is **irreducible** iff $x = a + b$ implies that a or b is invertible, for all $a, b \in M$.

A monoid $(M, +, 0)$ is **graded** if there exists a map $g : M \to \mathbb{N}$, such that, for all $x, y \in M$, we have $g(x + y) = g(x) + g(y)$ and if $g(x) = 0$ then x is invertible.

Lemma 1. *Every element in a graded monoid is a sum of irreducibles.*

Proof. Let $x \in M$ be arbitrary. We show by induction on the grading $g(x) \in \mathbb{N}$ that x is a sum of irreducibles.

If $g(x) = 0$, then x is invertible. Hence, $x + y = 0$, for some $y \in M$. If $x = a + b$, for some $a, b \in M$, then $a + (b + y) = 0$ and a is invertible. Thus, x is irreducibile. In particular, x is a sum of irreducibles.

Suppose that every $y \in M$, with $g(y) < g(x)$, is a sum of irreducibles. If x is irreducible, then x is a sum of irreducibles. Suppose that x is reducible, i.e., $x = a + b$, for some non-invertible $a, b \in M$. By definition of the grading, $g(a), g(b) > 0$. Hence, $g(a), g(b) < g(a) + g(b) = g(x)$. By the induction hypothesis, a and b are sums of irreducibles, and so is x.

By induction, every element in a graded monoid is a sum of irreducibles. $\square$

A **right-ideal** of a monoid $(M, +, 0)$ is a subset $I \subseteq M$, such that $x \in I$ and $y \in M$ implies $x + y \in I$, for all $x, y \in M$. A right-ideal I of a monoid M is **proper** iff $0 \notin I$. An element $x \in I$ is **right-irreducible (in I)** iff $x = a + b$ and $a \in I$ implies that b is invertible, for all $a, b \in M$.

Lemma 2. *Let I be a proper right ideal of a graded monoid M. Every element in M is a (possibly empty) sum of right-irreducibles in I plus an element $r \in M \setminus I$.*

Proof. Since the lemma holds trivially for right-irreducibles in I, let $x \in M$ be right-reducible. We show by induction on the grading $g(x) \in \mathbb{N}$ that x is a possibly empty sum of right-irreducibles in I plus an element $r \in M \setminus I$.

If $g(x) = 0$, then x is invertible. Since I is proper, we have $x \in M \setminus I$. Hence, x is an empty sum of right-irreducibles plus $r = x$.

Suppose that every y, with $g(y) < g(x)$, is a possibly empty sum of right-irreducibles in I plus an element $r \in M \setminus I$. Since x is right-reducible, we find some $a \in I$ and some non-invertible b, such that $x = a + b$. Non-invertibility of b implies that $g(b) > 0$. Hence,

$$g(a) < g(a) + g(b) = g(a + b) = g(x).$$

The induction hypothesis shows that $a = (\sum_{i=1}^{n} a_i) + r$ is a sum of $n \geq 0$ right-irreducibles $a_1, \ldots, a_n$ in I plus an element $r \in M \setminus I$. Since $a \in I$ and $r \in M \setminus I$, we have $n \geq 1$. Thus, $\sum_{i=1}^{n} a_i$ is an element of the right-ideal I. As I is proper, $\sum_{i=1}^{n} a_i$ is non-invertible, and $g(\sum_{i=1}^{n} a_i) \geq 1$. This implies that

$$g(r + b) < g(\textstyle\sum_{i=1}^{n} a_i) + g(r + b) = g((\textstyle\sum_{i=1}^{n} a_i) + r + b) = g(x).$$

The induction hypothesis applied to $r + b$ yields right-irreducibles $b_1, \ldots, b_m$ and an element $r' \in M \setminus I$, such that $r + b = (\sum_{j=1}^{m} b_j) + r'$. Hence, $x = a + b = (\sum_{i=1}^{n} a_i) + r + b = (\sum_{i=1}^{n} a_i) + (\sum_{j=1}^{m} b_j) + r'$, which proves the lemma. $\square$

2.2 Multisets

A **multiset** over a set X is an unordered collection of elements with duplicates, which is formally represented as a map $m : X \to \mathbb{N}$ that counts the number of occurrences of each $x \in X$ in the multiset. The set of all multisets over X is denoted as $\mathbb{N}^X$. The **cardinality** $|m|$ of a multiset m is defined as the cardinality of the set $\{(x,k) \mid x \in X, 0 \le k < m(x)\}$. A multiset m is **non-empty** iff $0 < |m|$, and **finite** iff $|m| < \aleph_0$, where $\aleph_0$ is the first infinite cardinal number. The empty multiset is denoted as $\emptyset$. The set of all finite multisets over X is denoted as $\mathbb{N}^{(X)} = \{m : X \to \mathbb{N} \mid |m| < \aleph_0\}$. The **free commutative monoid** is the triple $(\mathbb{N}^{(X)}, \cup, \emptyset)$.

For $k \in \mathbb{N}$, and multisets $m, m' \in \mathbb{N}^X$, the **union** $m \cup m'$, **intersection** $m \cap m'$, **difference** $m \setminus m'$, and **multiplication** km are defined, for $x \in X$, as

$$(m \cup m')(x) = m(x) + m'(x),$$
$$(m \cap m')(x) = \min(m(x), m'(x)),$$
$$(m \setminus m')(x) = m(x) \mathbin{\dot-} m'(x)$$
$$(k \cdot m)(x) = k \cdot m(x),$$

where $\dot-$ is monus, defined as $a \mathbin{\dot-} b = \max(a - b, 0)$, for all $a, b \in \mathbb{N}$. The **subset** relation of multisets is defined as $m \subseteq m'$ iff $m(x) \le m'(x)$, for all $x \in X$.

Multisets $m_0, m_1, m_2 \in \mathbb{N}^X$ satisfy the following identities:

$$m_0 \setminus (m_1 \cup m_2) = (m_0 \setminus m_1) \setminus m_2$$
$$m_0 \cup (m_1 \setminus m_0) = m_1 \cup (m_0 \setminus m_1)$$
$$(m_0 \cup m_1) \setminus m_2 = (m_0 \setminus m_2) \cup (m_1 \setminus (m_2 \setminus m_0))$$

Restriction $m|_Y$ of a multiset m on X to a subset $Y \subseteq X$ is defined, for all $y \in Y$, as $m|_Y(y) = m(y)$.

It is convenient to represent a finite multiset over a set X as (an equivalence class of) a finite sequence of elements from X. Let X^* be the free monoid on X that consists of all finite words of elements from X (including the empty word ϵ). A word $w \in X^*$ induces a multiset $w : X \to \mathbb{N}$, by defining, for all $x \in X$,

$$\epsilon(x) = 0, \quad wx(x) = w(x) + 1, \quad wy(x) = w(x), \text{ for } y \ne x.$$

Note that different words might define the same multiset. For example, xy and yx both denote the multiset wherein both x and y occur once.

3   Multilabeled Petri Nets

Multilabeled Petri nets are Petri nets whose transitions are labeled with a multiset of actions.

Definition 1. *A multilabeled (Petri) net is a tuple (A, P, T, μ_0) with*

1. A a set of actions,

2. P *a set of places,*
3. $T \subseteq \mathbb{N}^P \times \mathbb{N}^A \times \mathbb{N}^P$ *a set of transitions, and*
4. $\mu_0 : P \to \mathbb{N}$ *an initial marking.*

Inspired by Goltz [10], the notation in Definition 1 slightly differs from the usual notation of Petri nets. The advantage of this presentation is that transitions (including its set of input and output places) can be studied in isolation, which allows for parallel and sequential composition of transitions.

For a transition $t = (P, \alpha, Q) \in T$, we write $^\bullet t = P$ for the multiset of input places of t, we write $t^\bullet = Q$ for the multiset of output places of t, and we write $\ell(t) = \alpha$ for the multiset of labels of t.

For the development of the composition operator for multilabeled nets in Sect. 4, we use the standard concurrent semantics of Petri nets, which allows multiple transitions to fire in parallel.

Definition 2. *A* **multitransition** *of a multilabeled net* (A, P, T, μ_0) *is a finite multiset* $\theta : T \to \mathbb{N}$ *of transitions.* $\mathbb{N}^{(T)}$ *denotes the set of all multitransitions.*

A multitransition $\theta \in \mathbb{N}^{(T)}$ of a multilabeled net $N = (A, P, T, \mu_0)$ defines a (concrete) transition $\tau(\theta) = (^\bullet\theta, \ell(\theta), \theta^\bullet)$ in $\mathbb{N}^P \times \mathbb{N}^A \times \mathbb{N}^P$, wherein

$$^\bullet\theta = \bigcup\nolimits_{t \in T} \theta(t) \cdot {}^\bullet t$$
$$\theta^\bullet = \bigcup\nolimits_{t \in T} \theta(t) \cdot t^\bullet$$
$$\ell(\theta) = \bigcup\nolimits_{t \in T} \theta(t) \cdot \ell(t)$$

A multitransition θ of a N is **enabled** at a marking $\mu \in \mathbb{N}^P$ iff $^\bullet\theta \subseteq \mu$. A marking $\mu' \in \mathbb{N}^P$ is **obtained** from a marking $\mu \in \mathbb{N}^P$ via a multitransition θ (denoted $\mu\ [\theta\rangle\ \mu'$) iff θ is enabled at μ, and $\mu' = (\mu \setminus {}^\bullet\theta) \cup \theta^\bullet$.

Definition 3. *The* **concurrent semantics** *of a multilabeled net* (A, P, T, μ_0) *is a pointed, directed, labeled graph* (V, E, μ_0), *consisting of*

1. *vertices* $V = \{\mu : P \to \mathbb{N}\}$, *and*
2. *labeled edges* $E = \{(\mu, \ell(\theta), \mu') \in V \times \mathbb{N}^A \times V \mid \mu\ [\theta\rangle\ \mu', |\theta| > 0\}$.

Note that the empty multitransition $\theta = \emptyset$ does not constitute a valid step in the semantics, as $\emptyset$ allows for internal divergent behavior (by always firing $\emptyset$).

For the development of the abstraction operator in Sect. 5, we rely on the interleaving semantics of nets:

Definition 4. *The* **interleaving semantics** *of a multilabeled net* (A, P, T, μ_0) *is a pointed, directed, labeled graph* (V, E, μ_0), *consisting of*

1. *vertices* $V = \{\mu : P \to \mathbb{N}\}$, *and*
2. *labeled edges* $E = \{(\mu, \ell(t), \mu') \in V \times \mathbb{N}^A \times V \mid \mu\ [t\rangle\ \mu'\}$.

The only difference between the concurrent semantics and interleaving semantics of multilabeled nets is the cardinality of the multitransitions.

We introduce some terminology for a multilabeled net $N = (A, P, T, \mu_0)$. N is called **finite** iff A, P, and T are all finite. For $k \geq 1$, N is called k-**bounded** iff every reachable marking μ satisfies $\mu(p) \leq k$, for all $p \in P$. N is called **safe** iff N is 1-bounded.

A marking $\mu' \in \mathbb{N}^P$ is reachable from a marking $\mu \in \mathbb{N}^P$ via a sequence of transitions $t_1 \cdots t_n \in T^*$, with $n \geq 0$, (denoted $\mu \, [t_1 \cdots t_n\rangle \, \mu'$) iff there exists markings $\mu_1, \ldots, \mu_{n-1} \in \mathbb{N}^P$, such that $\mu \, [t_1\rangle \, \mu_1 \cdots \mu_{n-1} \, [t_n\rangle \, \mu'$. A **firing sequence** of N is a sequence of transitions $t_1 \cdots t_n \in T^*$, with $n \geq 0$ and $\mu_0 \, [t_1 \cdots t_n\rangle \, \mu'$, for some marking μ'. A marking μ is called **reachable** iff μ is reachable from the initial marking μ_0.

A transition t is **dead** in N iff t does not occur on any firing sequence. A transition t is **potentially fireable** in N iff t is not dead in N. For $k \geq 1$, a transition $t \in T$ in a multilabeled net N is k-**live** iff ${}^\bullet t(p) \leq k$, for all $p \in P$. Every potentially fireable transition in a k-bounded net is k-live.

3.1   Constraint Automata

As stated earlier, multilabeled nets generalize constraint automata [3] (without data constraints). If data constraints are ignored, the interpretation of constraint automata as multilabeled nets is rather straightforward. Figure 2 shows the multilabeled nets for some frequently used Reo primitives.

The sync(a, b) protocol in Fig. 2(a) accepts a datum at input a and immediately offers it at output b. Since the sync(a, b) protocol is stateless, its multilabeled net does not contain a place. The fifo$_1(a, b)$ protocol in Fig. 2(b) accepts a datum at input a and stores it. In the next step, it offers the stored datum at output b. The node$_{m,n}(a, \ldots, a_{m+n})$ protocol in Fig. 2(c) accepts a datum at any input a_i, with $1 \leq i \leq m$, and immediately offers a copy of it at every output a_j, with $m < j \leq m+n$. The place in the node$_{m,n}(a, \ldots, a_{m+n})$ protocol does not serve as memory, but encodes the conflict between the transitions. The transformer$_f(a, b)$ protocol accepts a datum d_a from its input a, and simultaneously offers the datum $f(d_a)$ at its output b. The transformation of datum d_a into d_b is modeled by the data constraint $d_b = f(d_a)$.

In general, constraint automata can have non-trivial data constraints, as is the case for transformer$_f(a, b)$. It is certainly possible to extend the definition of multilabeled nets (Definition 1) to include data constraints as well. However, such extension would not add any expressiveness to multilabeled nets, because data constraints can be encoded as fresh actions. For example, the data constraint $d_b = f(d_a)$ can be encoded as a fresh action $\langle d_b = f(d_a)\rangle$, which we call a **data-constraint action**. Freshness ensures that data-constraint actions are not used for synchronization (as is defined in Sect. 4). After composition, data-constraint actions can be decoded back into data constraints. If multiple data-constraint actions end up in the same transition label, their decoded data constraints are combined via conjunction. Figure 2(d) shows the multilabeled net for the transformer channel.

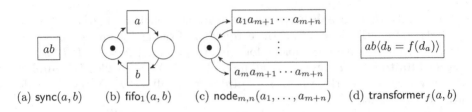

Fig. 2. Multilabeled nets for Reo primitives. The action $\langle d_b = f(d_a) \rangle$ in the transformer is the encoding of a data constraint.

The above trick to encode data constraints as actions can be similarly applied to other semantic models [12].

The multilabeled nets that come from constraint automata are square-free:

Definition 5. *A multitransition θ in a multilabeled net N is* **square-free** *iff $\ell(\theta)(a) \leq 1$, for all $a \in A(N)$. A multilabeled net N is* **square-free** *iff every potentially-fireable multitransition θ in N is square-free.*

It is easy to verify that all multilabeled nets in Fig. 2 are square-free: every transition is square-free and every multitransition θ of size $|\theta| > 1$ is dead.

The main reason for considering square-free nets is that their composition can be easily computed.

4   Composition

For this section, fix two multilabeled nets $N_i = (A_i, P_i, T_i, \mu_{0i})$, for $i \in \{0, 1\}$, which are interpreted according to the concurrent semantics in Definition 3.

We define the composition $N_0 \times N_1$ of N_0 and N_1 that synchronizes N_0 and N_1 on shared actions $A_0 \cap A_1$. We follow a standard approach to define the (parallel) composition of multilabeled nets. First, we generate all combinations of transitions in N_0 and N_1 that can fire in parallel. Next, we restrict to synchronizing combinations that 'agree on shared actions'. Finally, we restrict to a subset of combinations that generate all synchronizing combinations.

Recall that the disjoint union $X + Y$ of two sets X and Y is defined as $(X \times \{0\}) \cup (Y \times \{1\})$.

Definition 6. *A* **global transition** *is a finite multiset $\eta : T_0 + T_1 \to \mathbb{N}$.*

A global transition $\eta \in \mathbb{N}^{(T_0 + T_1)}$ has, for $i \in \{0, 1\}$, a **local component** $\eta|_i \in \mathbb{N}^{(T_i)}$ defined as $\eta|_i(t) = \eta(t, i)$, for all $t \in T_i$.

The set $\mathbb{N}^{(T_0+T_1)}$ of all global transitions, endowed with the union $\cup$ of multiset and the empty multiset $\emptyset$, constitutes a monoid. Moreover, multiset cardinality, $|\cdot|$, defines a grading on $\mathbb{N}^{(T_0+T_1)}$. In particular, any submonoid of $\mathbb{N}^{(T_0+T_1)}$ is a graded monoid.

We now formalize the notion of agreement on shared actions:

Definition 7. *A global transition $\eta \in \mathbb{N}^{(T_0+T_1)}$ is S-compatible, $S \subseteq A_0 \cap A_1$, iff $\ell(\eta|_0)(a) = \ell(\eta|_1)(a)$, for all $a \in S$. The set of S-compatible global transitions is denoted as $C(S) \subseteq \mathbb{N}^{(T_0+T_1)}$. We call η **compatible**, if η is $A_0 \cap A_1$-compatible.*

Intuitively, the local components $\eta|_0$ and $\eta|_1$ of an S-compatible multitransition $\eta \in \mathbb{N}^{(T_0+T_1)}$ agree only on the shared actions in S, while they may disagree on shared actions $a \in (A_0 \cap A_1) \setminus S$ outside of S.

A compatible global transition $\eta \in C(A_0 \cap A_1)$ defines a (concrete) transition $\lambda(\eta) = ({}^\bullet\eta, \ell(\eta), \eta^\bullet) \in \mathbb{N}^{P_0+P_1} \times \mathbb{N}^{A_0 \cup A_1} \times \mathbb{N}^{P_0+P_1}$, with

$$
\begin{aligned}
{}^\bullet\eta(p, i) &= {}^\bullet(\eta|_i)(p) \\
\eta^\bullet(p, i) &= (\eta|_i)^\bullet(p) \\
\ell(\eta)(a) &= \ell(\eta|_i)(a) \qquad \text{if } a \in A_i
\end{aligned}
$$

for all $(t, i) \in T_0 + T_1$ and $a \in A_0 \cup A_1$. Note that $\ell(\eta)(a)$ is well-defined, since $\ell(\eta|_0)(a) = \ell(\eta|_1)(a)$, for $a \in A_0 \cap A_1$.

The empty global transition $\emptyset$ is trivially S-compatible, for all $S \subseteq A_0 \cap A_1$. Furthermore, the union $\alpha \cup \beta$ of two compatible global transitions is again S-compatible, for all $S \subseteq A_0 \cap A_1$. Hence, the set of S-compatibles $C(S)$ is a (graded) submonoid of the graded monoid $\mathbb{N}^{(T_0+T_1)}$. Lemma 1 shows that every compatible global transition can be decomposed into irreducibles.

Any reducible S-compatible global transition $\eta = \alpha \cup \beta$, for some global transitions α and β, is redundant, as $\lambda(\eta)$ can be simulated by firing $\lambda(\alpha)$ and $\lambda(\beta)$ in parallel. Hence, we consider the set $C_0(S) \subseteq C(S)$ of all irreducible S-compatible global transitions.

The image of $C_0(A_0 \cap A_1)$ under the map $\lambda : C(A_0 \cap A_1) \rightarrow \mathbb{N}^{P_0+P_1} \times \mathbb{N}^{A_0 \cup A_1} \times \mathbb{N}^{P_0+P_1}$ defines the set of transitions of the composition:

Definition 8. *The **composition** $N_0 \times N_1$ of the multilabeled nets N_0 and N_1 is a multilabeled net with actions $A_0 \cup A_1$, places $P_0 + P_1$, transitions $\lambda(C_0(A_0 \cap A_1))$, and initial marking μ_0 defined as $\mu_0(p, i) = \mu_{0i}(p)$, for all $(p, i) \in P_0 + P_1$.*

It is laborious but straightforward to verify that the composition of multilabeled Petri nets is associative. The composition is commutative only up to renaming of places, due to the index from the disjoint union. The composition is idempotent only up to semantic equivalence, as is duplicates the places.

Example 1. Consider the mail example in Fig. 1, and suppose that the set of actions of the client equals $\{a, b, c, d\}$ and the set of actions of the server equals $\{a, b\}$. According to Definition 8, the transitions in the composition of the nets in

Fig. 1(a) and (b) consist of irreducible compatible global transitions. The compose transition in Fig. 1(a) and the (implicit) empty multitransition $\emptyset$ in Fig. 1(b) trivially agree on shared actions. Therefore, (compose | $\emptyset$) is a compatible transition. Being of length 1, the compose transition is necessarily irreducible, which implies that the compose transition is a transition of the composition in Fig. 1(c).

Similarly, the global transition η, with components $\eta|_0 = $ send | receive^2 and $\eta|_1 = $ transfer^2, is also a compatible transition. Indeed, the label of $\eta|_0$ and $\eta|_1$ both equal a^2b^2. Clearly, η is irreducible, which shows that $\lambda(\eta)$ is a transition of the composition in Fig. 1(c).   ♣

4.1   Composition Algorithm

Definition 8 only defines the transitions of the composition: it does not suggest a procedure on how these transitions can be found. It seems difficult to compute the composition of arbitrary multilabeled nets. Since the current work is motivated by constraint automata, we develop an algorithm that computes the composition of square-free multilabeled nets (Definition 5).

Lemma 3 shows that square-freeness must be checked only for atomic nets.

Lemma 3. *If N_0 and N_1 are square-free, then so is $N_0 \times N_1$.*

Proof. If a global transition $\eta \in \mathbb{N}^{(T_0+T_1)}$ is potentially fireable, then so are its local components $\eta|_0$ and $\eta|_1$. For $i \in \{0,1\}$, square-freeness of N_i implies that $\ell(\eta|_i)(a) \leq 1$ and $a \in A_i$. By construction, $\ell(\eta)(a) \leq 1$, for all $a \in A_0 \cup A_1$. Hence, $N_0 \times N_1$ is square-free.   □

By Definition 8, the composition $N_0 \times N_1$ can contain dead transitions. Since these dead transitions do not contribute to the behaviour of the multilabeled net, it is no problem if our composition algorithm does not generate them. As the composition, $N_0 \times N_1$, is square-free, we consider only square-free transitions:

Definition 9. *A global transition $\eta : T_0 + T_1 \to \mathbb{N}$ is **square-free** if $\lambda(\eta)$ is square-free. For $S \subseteq A_0 \cap A_1$, we denote the set of all square-free, irreducible S-compatible global transitions as $\overline{C_0}(S) \subseteq C_0(S)$.*

We compute the composition of square-free nets N_0 and N_1 by recursion on the number of shared actions. This procedure is conveniently expressed with the following terminology:

Definition 10. *The **difference** $d_a(\eta)$ of a global transition $\eta \in \mathbb{N}^{(T_0+T_1)}$ at a shared action $a \in A_0 \cap A_1$ is defined as the integer*

$$d_a(\eta) = \ell(\eta|_0)(a) - \ell(\eta|_1)(a).$$

The set of all square-free, irreducible, S-compatible global transitions with difference $d \in \mathbb{Z}$ is denoted as $\overline{C_0^d}(S)$.

It is straightforward to verify that $d_a(\alpha \cup \beta) = d_a(\alpha) + d_a(\beta)$, for global transitions $\alpha, \beta \in \mathbb{N}^{(T_0+T_1)}$.

Since every global transition is $\emptyset$-composite, $C(\emptyset) = \mathbb{N}^{(T_0+T_1)}$ and $C_0(\emptyset) = T_0 + T_1$, where each $(t, i) \in T_0 + T_1$ is viewed as a singleton multiset on $T_0 + T_1$.

Lemma 4 expresses square-free, irreducible S-compatibles in terms of square-free, irreducible S'-compatibles, with $S' \subseteq S$.

Lemma 4. *If $S \subseteq A_0 \cap A_1$, and $a \in (A_0 \cap A_1) \setminus S$ then*

$$\overline{C_0}(S \cup \{a\}) \subseteq \overline{C_0^0}(S) \cup \{\alpha_{-1} \cup \alpha_1 \mid \alpha_d \in \overline{C_0^d}(S)\} \subseteq C(S \cup \{a\}).$$

Proof. For the first inclusion, let $\eta \in C_0(S \cup \{a\})$. Since every $S \cup \{a\}$-compatible is also S-compatible, we have that $\eta \in C(S)$. As $C(S)$ is a graded monoid, Lemma 1 shows that, for some $n \geq 1$ and $\beta_1, \ldots, \beta_n \in C_0(S)$, we have

$$\eta = \beta_1 \cup \cdots \cup \beta_n$$

Since η is square-free, we have, for every $1 \leq k \leq n$, that

$$\ell(\beta_k)(a) \leq \sum_{i=1}^n \ell(\beta_i)(a) = \ell(\bigcup_{i=1}^n \beta_i)(a) = \ell(\eta)(a) \leq 1,$$

which shows that every β_i, $1 \leq i \leq n$, is square-free. In particular, the difference of β_i at a satisfies $d_a(\beta_i) \in \{-1, 0, 1\}$. We distinguish two cases:

Case 1: Suppose that $d_a(\beta_1) = 0$. Then, β_1 and $\beta_2 \cup \ldots \cup \beta_n$ are both S-compatible. Irreducibility of η shows that $n = 1$. Hence, $\eta = \beta_1 \in \overline{C_0^0}(S)$.

Case 2: Suppose that $d_a(\beta_1) = \pm 1$. Since η is S-compatible, we have that

$$\sum_{i=1}^n d_a(\beta_i) = d_a(\bigcup_{i=1}^n \beta_i) = d_a(\eta) = 0$$

Since $d_a(\beta_i) \in \{-1, 0, 1\}$, for $1 \leq i \leq n$, we find some $1 < k \leq n$, such that $d_a(\beta_k) = -d_a(\beta_1) = \mp 1$. Without loss of generality, we assume that $k = 2$. From $d_a(\beta_1 \cup \beta_2) = d_a(\beta_1) + d_a(\beta_2) = 0$, it follows that $\beta_1 \cup \beta_2$ and $\beta_3 \cup \ldots \cup \beta_n$ are both S-compatible. Irreducibility of η and non-emptyness of β_i, for $1 \leq i \leq n$, shows that $n = 2$, which implies that $\eta = \beta_1 \cup \beta_2$, with $\beta_i \in \overline{C_0^{\pm 2i \mp 3}}(S)$, for $i \in \{1, 2\}$. In both cases, $\eta \in \overline{C_0^0}(S) \cup \{\alpha_{-1} \cup \alpha_1 \mid \alpha_d \in \overline{C_0^d}(S)\}$.

For the second inclusion, let $\eta \in \overline{C_0^0}(S) \cup \{\alpha_{-1} \cup \alpha_1 \mid \alpha_d \in \overline{C_0^d}(S)\}$. Suppose that $\eta \in \overline{C_0^0}(S)$. By construction, $d_a(\eta) = 0$, which shows that $\eta \in C(S \cup \{a\})$. Every decomposition of η in $C(S \cup \{a\})$ is also a decomposition in $C(S)$. Hence, irreducibility of η in $C(S)$ implies that $\eta \in \overline{C_0}(S \cup \{a\})$.

Suppose $\eta \in \{\alpha_{-1} \cup \alpha_1 \mid \alpha_d \in \overline{C_0^d}(S)\}$. Then, we have $d_a(\eta) = d_a(\alpha_{-1} \cup \alpha_1) = d_a(\alpha_{-1}) + d_a(\alpha_1) = -1 + 1 = 0$. Hence, $\eta \in C(S \cup \{a\})$. $\square$

Algorithm 1 computes a set $C \subseteq C(A_0 \cap A_1)$ of compatible global transitions of two square-free multilabeled nets N_0 and N_1, including all square-free, irreducible global transitions (i.e., $C \supseteq \overline{C_0}(A_0 \cap A_1)$). We conjecture that Algorithm 1 actually generates $C = \overline{C_0}(A_0 \cap A_1)$, which means that Algorithm 1 produces only irreducible global transitions.

For clarity, we present distribution (Algorithm 1) in its simplest form. Distribution can be optimized by using an appropriate data structure for set C for efficient constructions of the subsets $C^d = \{\eta \in C \mid d_a(\eta) = d\}$, for $d \in \{-1, 0, 1\}$.

Algorithm 1. Distribution

Require: Two finite, square-free multilabeled nets N_0 and N_1.
Ensure: $\overline{C_0}(A_0 \cap A_1) \subseteq C \subseteq C(A_0 \cap A_1)$.
1: $C \leftarrow T_0 + T_1 \subseteq \mathbb{N}^{(T_0+T_1)}$
2: **for** $a \in A_0 \cap A_1$ **do**
3:     **for** $d \in \{-1, 0, 1\}$ **do**
4:         $C^d \leftarrow \{\eta \in C \mid d_a(\eta) = d\}$
5:     **end for**
6:     $C \leftarrow C^0 \cup \{\alpha \cup \beta \mid (\alpha, \beta) \in C^{-1} \times C^1, \alpha \cup \beta \text{ square-free}\}$
7: **end for**

Theorem 1. *Distribution (Algorithm 1) is totally correct.*

Proof. By Lemma 4, after $S \subseteq A_0 \cap A_1$ iterations, set C consists of all square-free, irreducible S-compatibles. Finiteness of N_0 and N_1 ensures that $A_0 \cap A_1$ is finite, which implies termination. $\qquad\square$

Example 2 (Alternator). Consider the $\mathsf{alternator}_n$ protocol, $n \geq 2$, defined as the following composition of node, sync, $\mathsf{syncdrain}$, and fifo_1 components:

$$\mathsf{alternator}_n = \prod_{i=1}^n \left(\mathsf{node}_{1,3}(a_i; a_i^1, a_i^2, a_i^3) \times \mathsf{node}_{2,1}(b_i^1, b_i^2; b_i) \times \mathsf{sync}(a_i^2, b_i^1)\right)$$
$$\times \prod_{i=1}^{n-1} \left(\mathsf{syncdrain}(a_i^3, a_{i+1}^1) \times \mathsf{fifo}_1(b_{i+1}, b_i^2)\right),$$

where the multilabeled net for $\mathsf{syncdrain}$ and sync are identical. For those familiar with the syntax of Reo, $\mathsf{alternator}_5$ has the following diagram:

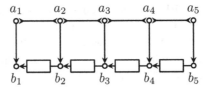

Using Algorithm 1 and Theorem 1, we can compute the multilabeled net of $\mathsf{alternator}_5$. The composition is shown in Fig. 3(a). For readability, we do not draw the places induced by the node components. Their contents remains the same throughout the execution of the $\mathsf{alternator}_5$. We also hide all internal actions by removing from the transition labels all actions other than b_1 and $a_1, \ldots, a_5$. We formalize this procedure in Definition 11.

The multilabeled net in Fig. 3(a) can be used for the implementation of the $\mathsf{alternator}_5$ protocol. In a naive implementation, a place is implemented as a variable, and a transition is implemented as a thread. Each thread reads and writes to these variables according to the flow relation, and performs I/O-operations according to the transition label. In particular, transitions with an empty label do not perform any I/O-operation. Of course, care must be taken for variables that are shared amongst different threads.   ♣

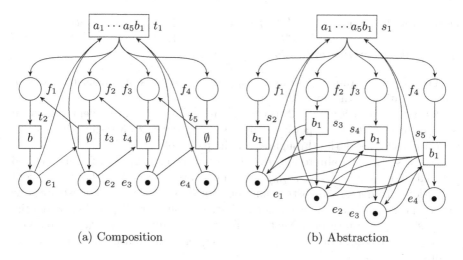

(a) Composition   (b) Abstraction

Fig. 3. Composition and abstraction of the alternator$_5$ protocol.

The number of places in a composition $N_0 \times N_1$ is the sum of the places in N_0 and N_1, which shows that the composition operator in Definition 8 does not suffer from state-space explosions. The total number of transitions in a composition $N_0 \times N_1$ equals the number of irreducible compatible subsets of N_0 and N_1, which can potentially grow large. The following result shows that, for 'nice compositions', the number of transitions in the composition does not blow up.

Corollary 1. *The composition $N_0 \times N_1$ has at most $|T_0| + |T_1|$ transitions, if for every $a \in A_0 \cap A_1$ there exists an $i \in \{0, 1\}$ with $|\{t \in T_i \mid \ell(t)(a) > 0\}| \leq 1$.*

Proof. The condition ensures that, in Algorithm 1, $|C^{-1}| = 1$ or $|C^1| = 1$. Hence, the size of C does not increase, which shows that the upper bound holds. □

Example 3. As shown in [13, Fig. 17(a)], the constraint automaton representation of the alternator$_n$ protocol does not scale well in $n \geq 1$. However, for multilabeled Petri nets the situation is completely different. Applying Corollary 1 to the compositions in alternator$_n$ in Example 2, it can be shown that, for every $n \geq 1$, the number of transitions of the alternator$_n$ is equal to n. As such, the multilabeled net representation of the alternator$_n$ protocol does not suffer from a state-space or a transition-space explosion. ♣

5   Abstraction

The definition of multilabeled Petri nets allows for *silent* transitions, i.e., transitions t with an empty label $\ell(t) = \emptyset$. Such silent transitions can be introduced by hiding internal actions:

Definition 11 (Hiding). *Hiding an action $a \in A$ in a multilabeled net N yields the net $\exists_a N = (B, P, \{({}^\bullet t, \ell(t)|_B, t^\bullet) \mid t \in T\}, \mu)$, where $B = A \setminus \{a\}$ and $\ell(t)|_B$ is the restriction of $\ell(t)$ to B.*

The hiding operator $\exists_a$ simply drops all occurrences of a from the label of every transition in the given Petri net.

As indicated in Example 2, the naive implementation of a silent transition is a thread that does not perform any I/O-operations. As a result, these silent transitions can delay the throughput of the protocol.

In this section, we aim to improve the generated code by transforming a fixed multilabeled net $N = (A, P, T, \mu_0)$ into an equivalent net ∂N without any silent transitions. To define the abstraction operator ∂, we follow the same strategy as for composition of multilabeled nets. First, we consider all possible sequential compositions of transitions. Next, we restrict to sequences of transitions with observable behavior. Finally, we restrict to sequences of transitions that generate all observable traces.

5.1   Sequential Compositions

Consider the set T^* of all firing sequences of N, i.e., all finite sequences of transitions of N. Following Mazurkiewicz [15], strictly different firing sequences can be considered identical up to permutation of independent transitions.

Definition 12. *The **dependency relation** $D \subseteq T \times T$ is defined as*

$$(s, t) \in D \quad \Leftrightarrow \quad s^\bullet \cap {}^\bullet t \neq \emptyset \text{ or } t^\bullet \cap {}^\bullet s \neq \emptyset \text{ or } \ell(s) \neq \emptyset \neq \ell(t)$$

Intuitively, transitions s and t are dependent iff one transition takes the output of the other as input, or if both transitions are observable. Note that conflicting transitions are not necessarily dependent.

The dependency relation D induces a **trace equivalence** $\equiv \subseteq T^*$ defined as the smallest congruence on T^*, such that $st \equiv ts$, for all $(s, t) \notin D$. An equivalence class $[x] = \{y \in T^* \mid y \equiv x\}$ of a firing sequence x is called a **trace**.

The **trace monoid** $T^*/\equiv$ is the set $\{[x] \mid x \in T^*\}$ of all traces, endowed with composition, defined, for all $x, y \in T^*$, as $[x][y] = [xy]$. Since $\equiv$ is a congruence, composition of traces is well-defined.

Observable behavior of traces is a map $o : T^*/\equiv \to (\mathbb{N}^A \setminus \{\emptyset\})^*$ defined, for all $x \in T^*$, as

$$o([\epsilon]) = \epsilon, \qquad o([xt]) = \begin{cases} o([x])\ell(t) & \text{if } \ell(t) \neq \emptyset \\ o([x]) & \text{otherwise} \end{cases}$$

Since observable transitions do not commute (Definition 12), o is well-defined.

Next, we define map $\sigma : T^*/\equiv \to \mathbb{N}^P \times \mathbb{N}^A \times \mathbb{N}^P$ that maps every trace w to a concrete transition $\sigma(w) \in \mathbb{N}^P \times \mathbb{N}^A \times \mathbb{N}^P$. The definition of this map relies on sequential composition of transitions:

Definition 13. *The* **sequential composition** $s;t$ *of transitions* s, t *in a multi-labeled net* N *is defined as* $(^\bullet(s;t), \ell(s;t), (s;t)^\bullet)$, *where*

$$^\bullet(s;t) = {}^\bullet s \cup ({}^\bullet t \setminus s^\bullet), \qquad (s;t)^\bullet = t^\bullet \cup (s^\bullet \setminus {}^\bullet t), \qquad \ell(s;t) = o([st])$$

For a sequential composition $s;t$, transition t can use the tokens produced by s. Hence, t consumes only the tokens from the multiset difference ${}^\bullet t \setminus s^\bullet$. Therefore, the sequential composition $s;t$ consumes only the tokens in the multiset union ${}^\bullet s \cup ({}^\bullet t \setminus s^\bullet)$.

Example 4. Figure 4(b) shows some sequential compositions of transitions s and t in Fig. 4(a). Intuitively, the sequential composition $s;t$ performs t after s. The token generated in place q by s is immediately consumed by t. Therefore, $s;t$ does not have q as input place or output place.

Note, however, that $t;s$ has q both as input place and output place, because the token produced at place q by s comes too late.   ♣

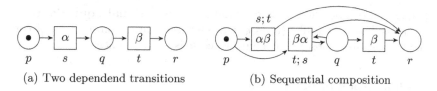

(a) Two dependend transitions   (b) Sequential composition

Fig. 4. Sequential composition of transitions with multilabels α and β.

Sequential composition of traces is a map $\sigma : T^*/{\equiv} \rightarrow \mathbb{N}^P \times \mathbb{N}^A \times \mathbb{N}^P$ defined, for all traces $w \in T^*/{\equiv}$ and transitions $t \in T$, as

$$\sigma([\epsilon]) = (\emptyset, \emptyset, \emptyset), \qquad \sigma(wt) = \sigma(w);t. \tag{1}$$

Lemmas 5 and 6 show that σ is a well-defined homomorphism, that is, $\sigma(u)$ does not depend on the representative of $u \in T^*/{\equiv}$, and $\sigma(uv) = \sigma(u);\sigma(v)$, for all traces $u, v \in T^*/{\equiv}$.

Lemma 5. *If* $(s,t) \notin D$, *then* $s;t = t;s$.

Proof. If $(s,t) \notin D$, then $s^\bullet \cap {}^\bullet t = t^\bullet \cap {}^\bullet s = \emptyset$ and either $\ell(s) \neq \emptyset$ or $\ell(t) \neq \emptyset$. The latter condition implies that $\ell(s;t) = o([st]) = o([ts]) = \ell(t;s)$. The former condition implies that ${}^\bullet t \setminus s^\bullet = {}^\bullet t$ and ${}^\bullet s \setminus t^\bullet = {}^\bullet s$, which implies

$$^\bullet(s;t) = {}^\bullet s \cup ({}^\bullet t \setminus s^\bullet) = ({}^\bullet s \setminus t^\bullet) \cup {}^\bullet t = {}^\bullet(t;s)$$

Similarly, it follows that $(s;t)^\bullet = (t;s)^\bullet$, which proves $s;t = t;s$.   □

By construction, sequential composition of traces σ parses each trace as a left-associative composition. Thus, for traces u and v in $T^*/{\equiv}$, $\sigma(uv)$ and $\sigma(u);\sigma(v)$ could evaluate to different transitions. Lemma 6 shows that these expressions are equal, and that sequential composition of traces is a homomorphism.

Lemma 6. *Sequential composition of transitions is associative.*

Proof. The identities in Sect. 2 show, for all transitions $r, s, t \in T$, that

$$
\begin{aligned}
{}^{\bullet}(r;(s;t)) &= {}^{\bullet}r \cup ({}^{\bullet}(s;t) \setminus r^{\bullet}) \\
&= {}^{\bullet}r \cup (({}^{\bullet}s \cup ({}^{\bullet}t \setminus s^{\bullet})) \setminus r^{\bullet}) \\
&= {}^{\bullet}r \cup ({}^{\bullet}s \setminus r^{\bullet}) \cup (({}^{\bullet}t \setminus s^{\bullet}) \setminus (r^{\bullet} \setminus {}^{\bullet}s)) \\
&= {}^{\bullet}(r;s) \cup ({}^{\bullet}t \setminus (s^{\bullet} \cup (r^{\bullet} \setminus {}^{\bullet}s))) \\
&= {}^{\bullet}(r;s) \cup ({}^{\bullet}t \setminus (r;s)^{\bullet}) \\
&= {}^{\bullet}((r;s);t)
\end{aligned}
$$

Similarly $(r;(s;t))^{\bullet} = ((r;s);t)^{\bullet}$. Since concatenation is associative,

$$
\ell(x;(y;z)) = o(x)o(y)o(z) = \ell((x;y);z),
$$

which shows that $x;(y;z)$ and $(x;y);z$. $\qquad\square$

Sequential composition of traces induces **observational equivalence** $\approx \subseteq T^{*}/\equiv$ defined as $v \approx w$ iff $\sigma(v) = \sigma(w)$. Since sequential composition of traces is a homomorphism, observable equivalence $\approx$ is a congruence.

5.2   Elimination of Silent Traces

Consider the set $O = \{w \in T^{*}/\equiv \mid o(w) \neq \epsilon\}$ of observable traces, which constitutes a proper right-ideal in the trace monoid $T^{*}/\equiv$. Lemma 2 shows that every observable trace w, with $o(w) \neq \epsilon$, can be decomposed as a sequence of right-irreducibles in O followed by a silent trace v, with $o(v) = \epsilon$.

Every trace w whose observable behavior $o(w)$ has length greater than one admits a decomposition uv, with $o(u) \neq \epsilon$ and $v \not\equiv \epsilon$. Hence, the observable behavior $o(w)$ of a right-irreducible trace $w \in O$ must have length one.

Nevertheless, the length $|w|$ of the trace w can be arbitrarily large. To shrink the set of traces that generate all observable behavior, we consider two additional conditions.

Definition 14. *A trace w is* **acyclic** *iff $|w| = \min_{w' \approx w} |w'|$.*

Every contiguous subtrace of an acyclic trace is acyclic.

Lemma 7. *If uvw is acyclic, then v is acyclic.*

Proof. If v is not acyclic, then there exists a trace $v' \approx v$, such that $|v'| < |v|$. Then, $|uv'w| < |uvw|$, and $\sigma(uv'w) = \sigma(u);\sigma(v');\sigma(w) = \sigma(u);\sigma(v);\sigma(w) = \sigma(uvw)$. Hence, $uv'w \approx uvw$, and uvw is not acyclic. $\qquad\square$

Definition 15. *A trace w is k-live iff $\sigma(w')$ is k-live, for every suffix w' of w.*

Every contiguous subtrace of a k-live trace is k-live.

Algorithm 2. Abstraction

Require: A finite, k-bounded multilabeled net $N = (A, P, T, \mu_0)$.
Ensure: $H = \{x \in T^* \mid [x]$ right-irreducible, acyclic, and k-live$\}$.
1: $H \leftarrow \{t \mid t \in T, \ \ell(t) \neq \emptyset\}$
2: **repeat**
3:     $h \leftarrow |H|$
4:     $H \leftarrow H \cup \{sx \mid x \in H, \ s \in T, \ \ell(s) = \emptyset, \ sx \not\equiv xs, \ \sigma([sx]) \notin \sigma(H), \ [sx] \ k\text{-live}\}$
5: **until** $h = |H|$

Lemma 8. *If uvw is k-live, then v is k-live.*

Proof. If v is not k-live, then v can be decomposed as $v = v_0 v_1$, such that $\sigma(v_1)$ is not k-live. Then, uvw can be decomposed as $(uv_0)(v_1 w)$. From ${}^\bullet \sigma(v_1 w) = {}^\bullet(\sigma(v_1); \sigma(w)) = {}^\bullet\sigma(v_1) \cup ({}^\bullet\sigma(w) \setminus \sigma(v_1)^\bullet) \supseteq {}^\bullet\sigma(v_1)$, it follows that $\sigma(v_1 w)$ is also not k-live, and that uvw is not k-live. $\qquad\square$

The results in Lemmas 2, 7 and 8 show that the trace $w = [t_1 \ldots t_n]$ of every firing sequence $t_1 \ldots t_n$, $n \geq 1$, can be generated from right-irreducible, acyclic, k-live traces modulo observational equivalence $\approx$. To see this, note that w comes from a firing sequence, which means that w is k-live by construction. Now, select some acyclic $w' \approx w$ (i.e., w' is of minimal length). Lemma 2 applied for the right-ideal $O = \{u \in T^*/\equiv \mid o(u) \neq \epsilon\}$ yields a decomposition $w' = a_1 \ldots a_n b$, where $a_1, \ldots, a_n$ are right-irreducible in O, and $o(b) = \epsilon$ is silent. Since w is acylcic and k-live, Lemmas 7 and 8 show that the traces $a_1, \ldots, a_n$ are right-irreducible, acyclic, and k-live.

Definition 16. *The* **abstraction** *∂N of the multilabeled net N is defined as* $(A, P, \{\sigma(w) \mid w \text{ right-irreducible, acyclic, and } k\text{-live}\}, \mu)$.

Example 5. Consider the composite net N in Fig. 1(c). Hiding actions a and b results in a net $N' = \exists_a \exists_b N$, where send | transfer^2 | receive^2 is a silent transition. From the current marking, the net N' can eventually produce observable behavior. However, to do so, N' must first firing a silent transition.

Application of the abstraction operator yields the $\partial N'$, as shown in Fig. 1(d). In contrast with N', the abstraction $\partial N'$ can directly fire an observable transition in the current marking. As a result, the abstraction $\partial N'$ produces observable behavior faster than N', and the executable code generated from $\partial N'$ potentially optimizes the code generated from N'.   ♣

5.3 Abstraction Algorithm

The transitions of the abstraction ∂N of a multilabeled net N can be computed by recursion on the length of the underlying traces. Algorithm 2 shows a straightforward tree search that starts from the shortest possible acyclic, k-live, right-irreducible traces, namely the observable transitions.

Theorem 2. *Algorithm 2 is totally correct.*

Proof. It is routine to check that, after $n \geq 0$ iterations, H contains all acyclic, k-live, right-irreducible traces of length $n + 1$. For termination, observe that a finite, k-bounded net has only finitely many k-live transitions. □

Table 1. Abstraction of the multilabeled net in Fig. 3(a) using Algorithm 2.

w	s	$^\bullet w$	$w^\bullet$	$\ell(w)$	acyclic	1-live	irreducible
t_1	s_1	$e_1 e_2 e_3 e_4$	$f_1 f_2 f_3 f_4$	$a_1 \cdots a_5 b_1$	y	y	y
t_2	s_2	f_1	e_1	b	y	y	y
t_3		$e_1 f_2$	$e_2 f_1$	$\emptyset$	y	y	y
t_4		$e_2 f_3$	$e_3 f_2$	$\emptyset$	y	y	y
t_5		$e_3 f_4$	$e_4 f_3$	$\emptyset$	y	y	y
$t_3 t_1$		$e_1^2 e_3 e_4 f_2$	$f_1^2 f_2 f_3 f_4$	$a_1 \cdots a_5 b_1$	y		y
$t_4 t_1$		$e_1 e_2^2 e_4 f_3$	$f_1 f_2^2 f_3 f_4$	$a_1 \cdots a_5 b_1$	y		y
$t_5 t_1$		$e_1 e_2 e_3^2 f_4$	$f_1 f_2 f_3^2 f_4$	$a_1 \cdots a_5 b_1$	y		y
$t_3 t_2$	s_3	$e_1 f_2$	$e_1 e_2$	b	y	y	y
$t_4 t_2$		$e_2 f_1 f_3$	$e_1 e_3 f_2$	b	y	y	
$t_5 t_2$		$e_3 f_1 f_4$	$e_1 e_4 f_3$	b	y	y	
$t_3 t_3 t_2$		$e_1^2 f_2^2$	$e_1 e_2^2 f_1$	b	y		y
$t_4 t_3 t_2$	s_4	$e_1 e_2 f_3$	$e_1 e_2 e_3$	b	y	y	y
$t_5 t_3 t_2$		$e_1 e_3 f_2 f_4$	$e_1 e_2 e_4 f_3$	b	y	y	
$t_3 t_4 t_3 t_2$		$e_1^2 f_2 f_3$	$e_1 e_2 e_3 f_1$	b	y		y
$t_4 t_4 t_3 t_2$		$e_1 e_2^2 f_3^2$	$e_1 e_2 e_3^2 f_2$	b	y		y
$t_5 t_4 t_3 t_2$	s_5	$e_1 e_2 e_3 f_4$	$e_1 e_2 e_3 e_4$	b	y	y	y
$t_3 t_5 t_4 t_3 t_2$		$e_1^2 e_3 f_2 f_4$	$e_1 e_2 e_3 e_4 f_1$	b	y		y
$t_4 t_5 t_4 t_3 t_2$		$e_1 e_2^2 f_3 f_4$	$e_1 e_2 e_3 e_4 f_2$	b	y		y
$t_5 t_5 t_4 t_3 t_2$		$e_1 e_2 e_3^2 f_4^2$	$e_1 e_2 e_3 e_4^2 f_3$	b	y		y

Example 6. We use Algorithm 2 to eliminate the silent transitions alternator$_5$ protocol as shown in Fig. 3(a). Table 1 shows the intermediate steps

Figure 3(b) shows the resulting multilabeled net ∂alternator$_5$. In ∂alternator$_5$, a reader component at the output b_1 of alternator$_5$ does not need to wait for silent transitions to move the data into the fifo$_1$ buffer between b_2 and b_1. Instead, the reader can immediately take the data from the first non-empty fifo$_1$ buffer. As such, the naive implementation of the abstraction ∂alternator$_5$ should have higher throughput than alternator$_5$. We did not yet verify this claim experimentally. ♣

6   Conclusion

We introduce multilabeled Petri nets, i.e., Petri nets whose transitions are labeled by a multiset of actions. We define a binary composition operator for multilabeled nets, and construct an algorithm to compute the composition. We define a unary abstraction operator that removes silent transitions from multilabeled nets, and construct an algorithm to compute the abstraction.

The composition algorithm Algorithm 1 assumes that the multilabeled nets are square-free. Although the assumption of square-freeness is justified in the context of constraint automata, it would be very useful to be able to automatically compose more general multilabeled nets (such as the running example).

While the definition of the composition operator relies on the concurrent semantics of nets, the definition of the abstraction operator relies on interleaving semantics. As a result, the composition and abstraction operator are not interoperable, in the sense that $\partial(N_0 \times N_1) \neq \partial N_0 \times \partial N_1$, for multilabeled nets N_0 and N_1. Such an identity would allow for simplification of intermediate compositions, which potentially speeds up the construction of the composition.

References

1. Anisimov, N.A., Golenkov, E.A., Kharitonov, D.I.: Compositional petri net approach to the development of concurrent and distributed systems. Program. Comput. Softw. **27**(6), 309–319 (2001). https://doi.org/10.1023/A:1012758417962
2. Arbab, F.: Puff, the magic protocol. In: Agha, G., Danvy, O., Meseguer, J. (eds.) Formal Modeling: Actors, Open Systems, Biological Systems. LNCS, vol. 7000, pp. 169–206. Springer, Heidelberg (2011). https://doi.org/10.1007/978-3-642-24933-4_9
3. Baier, C., Sirjani, M., Arbab, F., Rutten, J.J.M.M.: Modeling component connectors in Reo by constraint automata. Sci. Comput. Program. **61**(2), 75–113 (2006). https://doi.org/10.1016/j.scico.2005.10.008
4. Basu, A., Bozga, M., Sifakis, J.: Modeling heterogeneous real-time components in BIP. In: Fourth IEEE International Conference on Software Engineering and Formal Methods (SEFM 2006), 11–15 September 2006, Pune, India, pp. 3–12. IEEE Computer Society (2006). https://doi.org/10.1109/SEFM.2006.27
5. Bernardinello, L., De Cindio, F.: A survey of basic net models and modular net classes. In: Rozenberg, G. (ed.) Advances in Petri Nets 1992. LNCS, vol. 609, pp. 304–351. Springer, Heidelberg (1992). https://doi.org/10.1007/3-540-55610-9_177
6. Best, E., Devillers, R., Hall, J.G.: The box calculus: a new causal algebra with multi-label communication. In: Rozenberg, G. (ed.) Advances in Petri Nets 1992. LNCS, vol. 609, pp. 21–69. Springer, Heidelberg (1992). https://doi.org/10.1007/3-540-55610-9_167
7. Bruni, R., Montanari, U.: Zero-safe nets, or transition synchronization made simple. Electr. Notes Theor. Comput. Sci. **7**, 55–74 (1997). https://doi.org/10.1016/S1571-0661(05)80466-9
8. Fehling, R.: A concept of hierarchical Petri nets with building blocks. In: Rozenberg, G. (ed.) ICATPN 1991. LNCS, vol. 674, pp. 148–168. Springer, Heidelberg (1993). https://doi.org/10.1007/3-540-56689-9_43
9. van Glabbeek, R., Vaandrager, F.: Petri net models for algebraic theories of concurrency. In: de Bakker, J.W., Nijman, A.J., Treleaven, P.C. (eds.) PARLE 1987. LNCS, vol. 259, pp. 224–242. Springer, Heidelberg (1987). https://doi.org/10.1007/3-540-17945-3_13

10. Goltz, U.: On representing CCS programs by finite petri nets. In: Chytil, M.P., Koubek, V., Janiga, L. (eds.) MFCS 1988. LNCS, vol. 324, pp. 339–350. Springer, Heidelberg (1988). https://doi.org/10.1007/BFb0017157

11. Jongmans, S.S.T.Q.: Automata-theoretic protocol programming. Ph.D. thesis, Centrum Wiskunde & Informatica (CWI), Faculty of Science, Leiden University (2016)

12. Jongmans, S.T.Q., Arbab, F.: Overview of thirty semantic formalisms for Reo. Sci. Ann. Comp. Sci. **22**(1), 201–251 (2012). https://doi.org/10.7561/SACS.2012.1.201

13. Jongmans, S.T.Q., Kappé, T., Arbab, F.: Constraint automata with memory cells and their composition. Sci. Comput. Program. **146**, 50–86 (2017). https://doi.org/10.1016/j.scico.2017.03.006

14. Kotov, V.E.: An algebra for parallelism based on petri nets. In: Winkowski, J. (ed.) MFCS 1978. LNCS, vol. 64, pp. 39–55. Springer, Heidelberg (1978). https://doi.org/10.1007/3-540-08921-7_55

15. Mazurkiewicz, A.: Introduction to trace theory. In: Diekert, V., Rozenberg, G. (eds.) The Book of Traces, Chap. 1, pp. 3–41. World Scientific, Singapore (1995). https://doi.org/10.1142/9789814261456_0001

16. Meseguer, J., Montanari, U.: Petri nets are monoids. Inf. Comput. **88**(2), 105–155 (1990). https://doi.org/10.1016/0890-5401(90)90013-8

17. Pratt, V.: Modeling concurrency with partial orders. Int. J. Parallel Program. **15**(1), 33–71 (1986). https://doi.org/10.1007/BF01379149

18. Reisig, W.: Understanding Petri Nets - Modeling Techniques, Analysis Methods. Case Studies. Springer, Heidelberg (2013). https://doi.org/10.1007/978-3-642-33278-4

19. Rozenberg, G. (ed.): Advances in Petri Nets 1992, The DEMON Project, LNCS, vol. 609. Springer, Heidelberg (1992). https://doi.org/10.1007/3-540-55610-9

20. Taubner, D.A.: Finite Reresentations of CCS and TCSP Programs by Automata and Petri Nets. LNCS, vol. 369. Springer, Heidelberg (1989). https://doi.org/10.1007/3-540-51525-9

21. Vogler, W., Wollowski, R.: Decomposition in asynchronous circuit design. In: Cortadella, J., Yakovlev, A., Rozenberg, G. (eds.) Concurrency and Hardware Design, Advances in Petri Nets. LNCS, vol. 2549, pp. 152–190. Springer, Heidelberg (2002). https://doi.org/10.1007/3-540-36190-1_5

22. Wimmel, H.: Eliminating internal behaviour in petri nets. In: Cortadella, J., Reisig, W. (eds.) ICATPN 2004. LNCS, vol. 3099, pp. 411–425. Springer, Heidelberg (2004). https://doi.org/10.1007/978-3-540-27793-4_23

23. Winskel, G.: Petri nets, algebras, morphisms, and compositionality. Inf. Comput. **72**(3), 197–238 (1987). https://doi.org/10.1016/0890-5401(87)90032-0

A Service-Oriented Approach for Decomposing and Verifying Hybrid System Models

Timm Liebrenz[1]([✉]), Paula Herber[2], and Sabine Glesner[1]

[1] Software and Embedded Systems Engineering Group, TU Berlin, Berlin, Germany
{timm.liebrenz,sabine.glesner}@tu-berlin.de
[2] Embedded Systems Group, University of Münster, Münster, Germany
paula.herber@uni-muenster.de

Abstract. The design of fault free hybrid control systems, which combine discrete and continuous behavior, is a challenging task. Their hybrid behavior and further factors make their design and verification challenging: These systems can consist of multiple interacting components, and commonly used design languages, like MATLAB Simulink do not directly allow for the verification of hybrid behavior. By providing *hybrid contracts*, which formally define the interface behavior of hybrid system components in differential dynamic logic ($d\mathcal{L}$), and providing a decomposition technique, we enable compositional verification of Simulink models with interacting components. This enables us to use the interactive theorem prover KeYmaera X to prove the correctness of hybrid control systems modeled in Simulink, which we demonstrate with an automotive industrial case study.

Keywords: Hybrid systems · Compositional verification · Theorem proving · Model-driven development

1 Introduction

Hybrid control systems are applied in many different domains and their functionality is steadily increasing. They combine both discrete and continuous behavior. Model-driven development languages like Matlab Simulink are often used to cope with the complexity of hybrid control systems. Simulink enables modeling and simulation of hybrid systems and provides mature tool support for graphical editing, simulation, and automated code generation. Models in Simulink can contain interacting components and the development environment can be used to simulate their interactions. However, simulation can typically not be applied to all input scenarios. Thus, especially in safety-critical areas (e.g. automotive or medical systems), where faulty behavior can cause injuries or threatens human life, formal verification is highly desirable. However, due to the state-space explosion problem and the infinite space of continuous values, formal verification does not scale for hybrid control systems.

© Springer Nature Switzerland AG 2020
F. Arbab and S.-S. Jongmans (Eds.): FACS 2019, LNCS 12018, pp. 127–146, 2020.
https://doi.org/10.1007/978-3-030-40914-2_7

In this paper, we present a novel approach for the decomposition and formal verification of hybrid control systems modeled in Simulink, by exploiting the use of components in the system design. The key ideas are to decompose a given system into services, and to use the concept of *hybrid contracts* to abstract from the inner structure of each service for system verification. Our definition of a service in Simulink [17] enriches a Simulink component with variability and extends it by a *hybrid contract* to provide a defined interface. Note that [17] only provides the ideas for contracts and does not provide means for their creation and verification.

We define hybrid contracts in differential dynamic logic ($d\mathcal{L}$). This enables us to abstractly capture the dynamic behavior of hybrid services that combine discrete as well as time-continuous behavior in a formally well-founded manner. To formally verify that a service adheres to its hybrid contract, we use our transformation from Simulink to $d\mathcal{L}$ [16] to transform Simulink services as well as complete systems into a formal $d\mathcal{L}$ representation and use the interactive theorem prover KeYmaera X. Previously, the transformed models of larger systems could not be verified, due to the size of the resulting model. To enable compositional verification of hybrid control systems, we extend our previous transformation [16] with an abstraction mechanism, where we replace services with their hybrid contracts. We cope with concurrently executed contracts as well as with their correct embedding into hybrid programs.

The remainder of this paper is structured as follows: In Sect. 2, we introduce Simulink, $d\mathcal{L}$, and briefly summarize our transformation from Simulink to $d\mathcal{L}$ as presented in [16]. We discuss related work in Sect. 3. We introduce our service-oriented decomposition and verification approach for hybrid system models in Sect. 4. We provide a proof sketch for the soundness of our embedding to ensure that safety properties are preserved by our abstraction technique. We demonstrate the applicability of our approach with an industrial case study provided by our partners from the automotive industry in Sect. 5. We conclude in Sect. 6.

2   Preliminaries

In this section, we introduce Simulink and $d\mathcal{L}$ as modeling languages for hybrid systems. Furthermore, we present our existing transformation of Simulink models into $d\mathcal{L}$.

2.1   Simulink

MATLAB/Simulink [18] is a data flow oriented modeling language and integrated modeling tool. It enables the design and simulation of hybrid system. The basic building elements are blocks with input and output ports and signals that connect the ports of blocks. Blocks perform calculations on their input values and write the results to their outputs. Signals transfer the values of outputs to inputs. Blocks can be assigned to different groups, e.g., direct feed-through, time-discrete, time-continuous, and control flow.

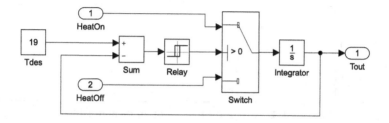

Fig. 1. Temperature control system in Simulink

Figure 1 shows a temperature control system in Simulink. It takes a heating (*HeatOn*) and a cooling (*HeatOff*) value as inputs, to keep the temperature in range of a desired value and the resulting temperature is provided at its output (*Tout*). The desired temperature is given by a constant (*Tdes*). The current temperature is determined by an *Integrator* block, which integrates the incoming signal over time. Whether the heating or cooling value is used, is determined by a *Switch* block. The *Switch* block is controlled by a *Relay* block, which outputs 1 if the current temperature is 0.5 °C under *Tdes* and 0 if the temperature is 0.5 °C above *Tdes*. Note that the turn on and turn off point are not shown in the figure. These values are given as parameters of the *Relay* block. For temperatures in between, the *Relay* keeps its current output.

2.2   KeYmaera X and the Differential Dynamic Logic (d$\mathcal{L}$)

KeYmaera X [12] is an interactive theorem prover for deductive verification of hybrid systems. A system is modeled in differential dynamic logic (d$\mathcal{L}$) as a hybrid program. A *sequential execution* $\alpha; \beta$ describes that the hybrid program α is executed before β. A *nondeterministic choice* $\alpha ++ \beta$ means that either α or β can be executed. With a *nondeterministic repetition* α^*, an arbitrary number of executions of α is modeled. A *discrete assignment* $x := \theta$ assigns a value θ to a variable x. A *nondeterministic assignment* $x := *$ assigns an arbitrary value to a variable x. In a *continuous evolution* $\{x'_1 = \theta_1, ..., x'_n = \theta_n \& H\}$ the variables $x_1, ..., x_n$ evolve continuously over time along their respective derivatives $\theta_1, ..., \theta_n$. This evolution is executed for an arbitrary amount of time as long as the conditions given by the evolution domain H hold. A test formula $?\phi$ continues the execution if ϕ holds and aborts the execution otherwise.

Listing 1.1 shows a hybrid program that models a person on an escalator [21]. The variable x denotes the position of the person and v the speed of the escalator. The person can step down ($x := x - 1$) if he is not already at the bottom ($?(x > 1)$), or move up with speed v for an arbitrary time ($x' = v$). This can be repeated an arbitrary number of times ($\{\cdot\}^*$). The specification $[\cdot]$ requires that under the precondition $x \geq 2 \land v \geq 0$, after all possible executions, $x \geq 0$.

```
1   x >= 2 & v >= 0 ->
2   { [ {
3        ?(x > 1); x := x − 1;
4        ++
5        {x' = v}
6   }* ] x >= 0 }
```

Listing 1.1. A hybrid program of an escalator [21]

2.3 Simulink to d$\mathcal{L}$ Transformation

In [16], we have presented a transformation of hybrid Simulink models into the formal semantics of d$\mathcal{L}$. The transformation takes a Simulink model as input and creates a representation in d$\mathcal{L}$ according to transformation rules for each individual block. The transformation only considers atomic blocks. Subsystems are flattened, i.e. internal blocks are extracted from subsystems and put in the outer system. Note that with this approach, our previously presented transformation is not able to make use of subsystems to encapsulate inner block structures.

A transformation rule is defined as a tuple $\{m, \beta\}$, where m is a replacement *macro* and β is a hybrid program that also can be empty. A macro $m = [id \leftarrow \alpha]$ is a replacement mechanism that can be applied to a hybrid program to replace a placeholder id with a hybrid program α. A *conditional macro* $m_c = [id \Leftarrow CM]$, where id is the identifier that should be replaced and CM is a set of conditional replacements (α_i, c_i) The expansion of a conditional macro is defined as follows:

$$\beta[id \Leftarrow CM] = \beta[id \Leftarrow \{(\alpha_1, c_1), ..., (\alpha_n, c_n)\}]$$
$$= \{?(c_1); \beta[id \leftarrow \alpha_1, c_1]; + + ... + +?(c_n); \beta[id \leftarrow \alpha_n, c_n]; \}$$

A conditional macro can be applied to hybrid program β. If β contains the identifier id, the conditional macro replaces β by a nondeterministic choice. The nondeterministic choice splits β into multiple cases (one case for each condition c_i). In each case with condition c_i, the identifier id inside the hybrid program is replaced by the corresponding α_i. This causes an exponential growth of the target d$\mathcal{L}$ model in the number of control flow changes. Therefore, this transformation does not scale well and larger models that contain multiple components cannot be handled with this simple transformation approach.

```
1   Variable Declarations
2   Preconditions & Initializations −>
3   [ {β;γ}* ] Postcondition
```

Listing 1.2. Structure of a transformed model

```
1    {smallStep:=0.0; TOut:=Integrator; HeatOn:=∗; HeatOff:=∗;
2      { ?(Tdes−Integrator>=0.5); Relay:=1.0;
3      ++
4        ?(Tdes−Integrator<=−0.5); Relay:=0.0;
5      ++
6        ?((Tdes−Integrator<0.5) ∧ (Tdes−Integrator>−0.5));
7      } { ?((Tdes−Integrator>=0.5) ∧ (Relay>0.0));
8        {simTime' = 1.0, Integrator' = HeatOn, smallStep' = 1.0
9           & ((Tdes−Integrator>=0.5) ∧ (Relay>0.0)) ∨ (smallStep<=smallStepSize)}
10     ++
11       ?((Tdes−Integrator>=0.5) ∧ (Relay<=0.0));
12       {..., Integrator' = HeatOff, ...
13          & ((Tdes−Integrator>=0.5) ∧ (Relay<=0.0)) ∨...}
14     ++
15     ... }
16   }∗
```

Listing 1.3. Simulation loop of a temperature control system in d$\mathcal{L}$

The transformation handles all blocks in a given model, collects macros and builds a hybrid program that respects all data, control and timing dependencies. The general structure of a resulting d$\mathcal{L}$ model is depicted in Listing 1.2. The behavior of the transformed system is captured within a *global simulation loop* $\{\cdot\}^*$, while the behavior of single blocks is captured in a sequential composition of discrete assignments β and continuous evolutions γ. *Preconditions* and *postcondition* may be defined for verification. The transformation supports models that contain both discrete and continuous blocks and is not restricted to a fixed step size solver. A parameter *smallStepSize* can be used to set an upper bound on the time after a Zero-Crossing before the next signal evaluation step.

The d$\mathcal{L}$ representation of the temperature control system from Fig. 1 is shown in Listing 1.3. Lines 2–6 represent the Relay block, i.e., the output is set to 1, to 0, or does not change. An excerpt of the continuous behavior of the integrator block is given by Lines 7–14. In a nondeterministic choice, the current control flow is determined by a test formula. In the temperature control system, there are six possible control flow combinations, three states of the relay block (on, off and hold-state) times two switch states (upper or lower input signal is written to its output). Two of them are shown in Lines 7 and 11, the others are skipped here for brevity of presentation. Note that the control flow condition determines the evolution domain to stop the continuous evolutions if the control flow changes. The system in d$\mathcal{L}$ allows the evolution to go a *smallStep* over the evolution domain, to model small deviations due to numerical errors.

3   Related Work

In this section, we discuss approaches for the verification of Simulink models, and work that addresses the verification of hybrid systems in general.

Simulink Verification. To model hybrid control systems, Simulink is widely used. However, the semantics of Simulink is only informally defined. The Simulink Design Verifier [19] provides means to use model checking and abstract interpretation techniques to explore the state space of a Simulink model. However, only discrete system behavior is considered and the scalability is limited [14]. In [2], a transformation of Simulink into the input language for the deductive verification system Why3 [10] is presented. In [14], Simulink models are transformed in an input representation of UCLID [15], which enables Satisfiability Modulo Theory (SMT) solving for the verification of safety properties. In [26], Simulink models are transformed into Boogie [4], which uses the SMT solver Z3 [9] for verification. While all of these approaches enable formal verification, they only consider a discrete subset of Simulink and are therefore not applicable for hybrid systems. In [6] a transformation of Simulink models into sequential programs is presented, which then are used for contract-based verification. However, while this approach supports multi-rate systems, it is only applicable for time-discrete systems. Some approaches extend the Simulink block library to provide wider means to handle hybrid systems. In [27], a toolbox for hybrid equations in Simulink is presented. However, these blocks only extend the simulation capabilities and do not enable verification. The tool CheckMate [7] enables the modeling and verification of Simulink models using hybrid automata. Special blocks to model polyhedral invariant hybrid automata (PIHA) are provided, which are then used for verification. However, this approach is only applicable for a special class of hybrid systems. A transformation of hybrid Simulink models in a specific hybrid automata dialect (SpaceEx) is presented in [20]. This enables the use of reachability algorithms to check safety properties of the system. Concurrency is modeled with parallel composition of hybrid automata. This causes the state space to increase exponentially in the number of concurrent blocks, and thus again does not scale well. The tool CLawZ [23] is used to automatically prove code that is provided by code generation of a Simulink model. By automatically transforming a Simulink model into a Z representation and associating the variables with the generated code, the model can be verified. Code generation only considers a discrete subset of Simulink. This approach cannot be used to check the interplay of the controller and its environment in the Simulink model.

Hybrid System Verification. Hybrid systems model the interactions between discrete state changes and continuous evolutions. A widely used modeling technique for such systems is hybrid automata [1]. Nodes represent system states that are defined by differential equations, which represent the evolution of continuous variables. The edges represent discrete jumps between system states. To cope with concurrent systems, approaches that are based on hybrid automata usually use some kind of parallel composition [3,11,13]. However, since the state space increases exponentially in the number of concurrent processes of a parallel composition, these approaches suffer from the state space explosion problem and therefore do not scale well. In [8], the authors present a compositional analysis for hybrid systems: The systems are represented by continuous dynamics and discrete systems that are defined over finite alphabets. The approach considers stability, passivity and norms on inputs and outputs. However, no timing properties are considered. In [5], the authors use the tool *Ariadne* to perform assume-guarantee reasoning to verify nonlinear hybrid automata. The approach uses approximations to calculate the reachable states of components and check whether components fulfill their contracts. This approach does not mention timing in its component contracts. In [24], Platzer has introduced the *differential dynamic logic* together with a compositional proof calculus. The idea of the proof calculus is to reduce properties of hybrid programs to properties of their parts. In [22], Platzer et al. embedded their differential dynamic logic into a contract-based approach for the compositional verification of hybrid systems. We adopt the idea that system parts can be described by contracts defined in d$\mathcal{L}$ to enable compositional verification, and make heavy use of the underlying compositionality of the proof calculus implemented in the KeYmaera X theorem prover. Note that in [22], the authors assume that the system under verification has a classical controller-plant structure, where each component has their own controller. In Simulink and other data flow or signal flow oriented languages, however, we have to cope with hierarchical subsystems and the resulting cascading control flow through multiple subsystems, that is, the controller might be distributed over multiple subsystems or services. This can be elegantly captured by adding discrete control conditions to the evolution domains of a service, respectively its hybrid contract. Compared to the approaches presented in [22], our approach mainly introduces two novel contributions: First, we introduce the concept of services instead of components, and second, we support the designer in the definition of contracts by providing approaches to (1) systematically add observer variables to a given service and (2) to systematically define its dynamic properties. The service-oriented approach provides us with an elegant way to cope with the data flow oriented nature of Simulink models, as we do not reason about input and output states, but about the properties of the input and output trajectories. The concept of a service is, in our opinion, better suited to capture this than a classical component-based approach, as the service can be considered as something influencing the flow between its input and output port.

4   A Service-Oriented Verification Approach for Hybrid System Models

Our key idea to enable scalable verification of hybrid control system is to decompose a given Simulink model into services, and provide *hybrid contracts* to describe their dynamic interface behavior. Our overall approach is depicted in Fig. 2. First, the designer identifies components that can be considered to be independent and configurable services [17] and creates hybrid contracts for them. The designer determines the components in the model that are used as services. The best candidates for such services are subsystem, which already encapsulate functionality and have an interface via their input and output ports. It is also possible to choose a group of connected blocks as service. Second, we transform

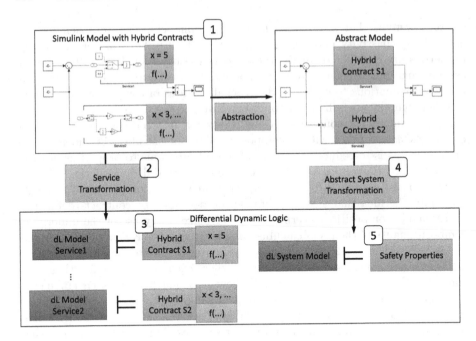

Fig. 2. Our service-oriented verification approach

each service individually into a behaviorally equivalent formal d$\mathcal{L}$ representation using our transformation defined in [16]. Third, we use the interactive theorem prover KeyMaera X to verify that the service fulfills its contract. Fourth, we use the contracts in an abstract system transformation. A hybrid contract abstracts from the inner structure of a given service, but describes the interface behavior in sufficient detail to represent their behavior in the resulting d$\mathcal{L}$ model. Fifth, we use the abstracted d$\mathcal{L}$ representation to semi-automatically verify safety properties using KeyMaera X. Note that the verified properties can be represented as a hybrid contract of the overall system. Thus, our approach can be used hierarchically, i.e., a large system can be divided into services that can be individually verified, and each service can again be divided into smaller services. As a consequence, our approach scales well even for large and heterogeneous hybrid systems if adequate contracts can be found for all subsystems. We use the term service in Simulink [17] to define a submodel that provides its interface behavior via hybrid contracts. These services can be reused in other designs and their contracts can be reused for the verification. We demonstrate the applicability and scalability of our approach with a case study from the automotive industry in Sect. 5. Our approach is sound in the sense that safety properties are preserved by our abstraction using hybrid contracts. Soundness is guaranteed by the verification of individual services and by correctly embedding services into the overall d$\mathcal{L}$ representation of a system.

In the following, we first present assumptions we impose on a given Simulink model. Then, we present our concept of hybrid contracts, briefly discuss the verification of individual services, and then the verification of the overall system using the abstraction described above, i.e., the replacement of services by their hybrid contracts. Finally, we briefly discuss soundness and provide a proof sketch to reason about correctness of our embedding of hybrid contracts into the $d\mathcal{L}$ representation of a given Simulink model.

4.1 Assumptions

The transformation from Simulink to $d\mathcal{L}$ requires that a given Simulink model fulfills the assumptions defined in [16]:

1. No algebraic loops are used.
2. No S-function blocks are used.
3. No external scripts or libraries are used.

Note that feedback loops are generally allowed, only *algebraic* loops, that is, loops where only feed-through blocks are used, are prohibited. The use of algebraic loops is generally discouraged, therefore this is not a strong limitation. S-functions enable the use of system-functions, which can be written in various programming languages. Currently, S-functions could only be over-approximated in the transformation by assuming that the output could be every possible value at every time. Note that there are some additional assumptions on the concrete block set defined in [16] due to the state of the implementation of the transformation. The implementation supports representatives of the most relevant block classes, namely arithmetic, logic, discrete, continuous, and control flow blocks. If extensions to the supported block set do not change the general structure of the resulting $d\mathcal{L}$ model, the concepts of the approach that we present in this paper are not affected.

4.2 Hybrid Contracts

Hybrid contracts capture the hybrid behavior of a Simulink system, which means that they are restricted neither to purely discrete nor to purely continuous system behavior. To enable the integration of contracts in the system verification, we represent hybrid contracts in $d\mathcal{L}$. A *hybrid contract* is a tuple $c = (\Phi_{in}, \Phi_{out})$, where Φ_{in} is a set of assumptions, and Φ_{out} is a set of guarantees. The assumptions describe behavior of input variables of a service. The guarantees provide information about the trajectories and values of output variables. Assumptions and guarantees are modeled in $d\mathcal{L}$. The use of $d\mathcal{L}$ to define contracts enables us to describe a variety of possible behavior and we are not limited to classical discrete system properties, for example, range and timing properties. Instead, we inherit the full expressiveness of $d\mathcal{L}$, which enables us to also express dynamic properties, i.e., differential equations together with discrete control signals, as continuous evolutions and assignments. However, to systematically make use of the high expressiveness of d$\mathcal{L}$ requires a high expertise by the designer. To ease the contract definition process, we provide (1) *extension functions*, which enable the designer to systematically introduce observer variables into the design under verification and (2) *property templates*, i.e. templates for range properties, timing properties, and dynamic properties, which can be used as design patterns for typical requirements or contract definitions.

Extension Functions. To enable designers to systematically define requirements and contracts, we introduce functions that extend hybrid programs by observer variables. The introduced variables may not change the behavior of the underlying program. Observer variables can be used to store the values of signals for use in contracts, e.g., to compare the value of a signal at the start of a run of the simulation loop with its value at the end.

Discrete Observer Variables. We introduce a function $u_D(\alpha, v, \theta, F)$ that prepends a conditional discrete assignment to a hybrid program. Under the condition F, the term θ is assigned to variable v. It is required that v is not bound in α. The expression θ may read variables from α.

$$u_D(\alpha, v, \theta, F) == \{?(F); v := \theta; ++?(\neg F); \}\alpha$$

Example. We use u_D to keep track how a variable changes during the execution of a simulation loop. Therefore, we introduce a variable x_{last} that stores the value of x at the beginning of the simulation loop.

$$\alpha_{new} = u_D(\alpha, x_{last}, x, true)$$

We set the condition to *true* to perform this assignment at every execution of the simulation loop.

Continuous Observer Variables. We introduce a function $u_C(\alpha, v' = \theta)$ that adds a continuously evolving variable $v' = \theta$ to the continuous evolutions of a given hybrid program. We define u_C for hybrid programs as follows:

$$u_C(\alpha; \beta, v' = \theta) == u_C(\alpha, v' = \theta); u_C(\beta, v' = \theta)$$
$$u_C(\alpha ++ \beta, v' = \theta) == u_C(\alpha, v' = \theta) ++ u_C(\beta, v' = \theta)$$
$$u_C(?(F), v' = \theta) ==?(F)$$
$$u_C((x_1' = \mu_1, ..., x_n' = \mu_n \& F), v' = \theta) == (x_1' = \mu_1, ..., x_n' = \mu_n, v' = \theta \& F)$$

Other parts of a hybrid program are not changed by u_C. To ensure that u_C does not change the behavior of the underlying system α, we require that v is a fresh variable, i.e., it is not used anywhere in α.

Example. To add an observer variable that stores the elapsed simulation time, we use u_C as follows:

$$\alpha_{new} = u_C(\alpha, simTime' = 1)$$

To introduce a clock variable *timer* that is reset under the condition $flag = true$, we use u_C and u_D in combination:

$$\alpha_{new} = u_C(u_D(\alpha, timer, 0, flag = true), timer' = 1)$$

Property Templates. To ease the contract definition process, we provide three property templates, namely range properties, timing properties, and dynamic properties, which can be used as design patterns for the contract definition. These templates can be used by a designer to create guarantees of a contract.

Range Properties. Range properties define bounds on the values of a given signal.

$$[\{\beta; \gamma\}^*] \, x \sim r_1 \wedge y \sim r_2 \wedge \ldots$$

Where x, y are signals, $' \sim' \in \{=, \neq, <, >, >=, <=\}$, and $r_1, r_2 \in \mathbb{R}$. Note that we typically assume that range properties should always hold for all possible runs of a given hybrid program, i.e. we use the [] operator.

Example. In the temperature control system, we can describe the requirement to keep the output temperature $Tout$ in desired bounds as a range property:

$$[\{\beta; \gamma\}^*](TOut \leq Tdes + \delta \wedge TOut \geq Tdes - \delta)$$

where $Tdes$ represents the desired temperature, $\delta > 0$ denotes the allowed deviation, $TOut$ is the controlled temperature.

Timing Properties. To ensure that the system reacts to a special system state or signal before or after a given time, a timing property can be used. To refer to arbitrary starting points of the simulation time, we propose to use *clock* variables. A clock variable has a derivative of 1 and can be reset in the d$\mathcal{L}$ representation of the system, for example, if a special system state is reached or if a given signal exceeds a given value. The general form of a timing property is:

$$[\{u_D(\beta, clock, 0, true); u_C(\gamma, clock' = 1)\}^*](stateCondition \rightarrow clock \sim r)$$

where *clock* is a clock variable, and *stateCondition* describes a special system state that triggers a timing condition $clock \sim r$.

Example. In the temperature control system, we use a timing property to ensure that no rapid switching occurs. With additional clock variables *relayOnTime* and *relayOffTime* that keep track of the time elapsed between switching, we can formulate a property that states that no rapid switching occurs, i.e. at least a time *MIN* elapses before the state of the switch changes.

$$[u_D(u_D(\beta, relayOnTime, 0, 19.0 - Integrator >= 0.5),$$
$$relayOffTime, 0, 19.0 - Integrator <= -0.5);$$
$$\{u_C(u_C(\gamma, \; relayOnTime' = 1), relayOffTime' = 1)\}^*]$$
$$(Relay = 0.0 \rightarrow relayOnTime \geq MIN$$
$$\wedge \; Relay = 1.0 \rightarrow relayOffTime \geq MIN)$$

This property describes that depending on the state of variable *Relay* the values of the clocks *relayOnTime* and *relayOffTime* reached at least a value of *MIN*.

Dynamic Properties. Dynamic properties define arbitrary relations of values before and after a run of the simulation loop. We use observer variables to keep track of the change of values and the elapsed time.

Example. To ensure that a dynamic change of the temperature is always kept in bounds that are given by δ, we define the following dynamic property:

$$[\{u_D(u_D(\beta, clock, 0, true), Integrator_{last}, Integrator, true); u_C(\gamma, clock' = 1)\}^*]$$
$$(Integrator_{last} - Integrator \leq \delta \cdot clock \; \&$$
$$Integrator_{last} - Integrator \geq -\delta \cdot clock)$$

where *clock* is a clock variable and $Integrator_{last}$ an observer variable, which stores the last value of the temperature (*Integrator*) at the start of each simulation loop. The property ensures that $Integrator_{last}$ can not deviate from *Integrator* by more than $\delta \cdot clock$ in all runs of the hybrid program, for $\delta > 0$.

4.3   Service Verification

To ensure that a given contract holds for a service, we transform the service into $d\mathcal{L}$. If the service only contains atomic Simulink blocks, we can use our Simulink to $d\mathcal{L}$ transformation [16]. If a service contains other subsystems, we can either flatten the system by replacing all subsystems by their inner blocks or consider these subsystems as services and use contracts to represent them (see Sect. 4.4).

The transformation provides us with a hybrid program α. To ensure that a contract $(\Phi_{pre}, \Phi_{post})$ holds for this system, we create the following $d\mathcal{L}$ formula.

$$(\Phi_{pre}) \rightarrow [\alpha](\Phi_{post})$$

where Φ_{pre} is a conjunction of preconditions, α is the transformed system as hybrid program and Φ_{post} are the postconditions. We need to ensure that under the assumption that Φ_{pre} holds, all possible runs of the hybrid program α need to fulfill the guarantee Φ_{post}. We can verify such contracts using KeyMaera X.

The creation of contracts is done by the designer. Guarantees can conceptually be derived from the system requirements and the desired behavior of the system, e.g. no division by zero. Often, assumptions can be inferred during the proof of the service. Open goals can point out missing properties of input signals that should be added as assumption.

```
1    {smallStep := 0.0;
2     S_HeatOn := 29 − S_Tout + 15 − S_Tout;
3     S_HeatOff := 15 − S_Tout;
4     S_Tout := *;
5     ?(((S_HeatOn ≥ 0) ∧ (S_HeatOn ≤ 10) ∧ (S_HeatOff ≤ 0) ∧ (S_HeatOff ≥ −10))
6     −>
7     ((19.0 − 2.0 ≤ S_Tout) ∧ (S_Tout ≤ 19.0 + 2.0)));
8     ScopeInput0 := S_Tout;
9    }*
```

Listing 1.4. Compositional verification example: abstract system

4.4 Service-Oriented System Verification

To overcome scalability issues, we present a service-oriented verification approach that uses hybrid contracts to abstract the behavior of services in system verification. The key idea is to decompose the system into services, define a hybrid contract for each service, and separately verify that (1) each service adheres to its hybrid contract, and that (2) the overall system with services replaced by their hybrid contracts satisfies the requirements. A service is represented as:

$$s = (B, S, I, O, C)$$

Its inner block structure is given by its blocks B and the signal lines S that connect the blocks. The interface of the service is given by its input ports $I \subset B$ and its output ports $O \subset B$. The set of hybrid contracts is given by C, where a contract $c \in C = (\Phi_{pre}, \Phi_{post})$ contains assumptions Φ_{pre} and guarantees Φ_{post}. We use the following rules to transform a service by its contract instead of its inner structure:

1. We introduce fresh variables for all input ports in I and output ports in O.
2. We connect its input ports, i.e., we assign the input variables of the service with the respective expressions from the surrounding Simulink model.
3. We nondeterministically choose the output signals of the service such that it satisfies the contract, i.e., we use a nondeterministic assignment and restrict the resulting value with Φ_{out} whenever Φ_{in} holds.
4. If a contract refers to inner variables, these are added to the $d\mathcal{L}$ model as fresh variables.

Temperature Control System. An illustrative embedding of our temperature control service into a simple environment is shown in Listing 1.4. Lines 2–3 show assignments to the input ports of the service given by the environment. They depend on the service output as it is embedded into a feedback loop in the overall system. In Lines 4–8, the output signal is assigned according to the contract.

4.5   Soundness

In this section, we discuss the soundness of our abstraction, which replaces services by their contracts and abstract their inner block structure. Our approach is sound in the sense that properties that hold for the abstract system are guaranteed to be preserved in the concrete system. Our key idea to ensure soundness is threefold: First, we assume that variables written by a service are not written somewhere else in the system. This is ensured by the syntactical structure of a Simulink model, as all inner blocks of a service can only be accessed through input signals. Second, we assume that a contract provides an over-approximation of the service behavior. The over-approximation of contracts is ensured by the service verification, where we semi-automatically prove in KeYmaera X that a service adheres to its contract. Third, and finally, we have to ensure that our proposed abstraction of service by its contract during system transformation does not restrict the behavior that would be created by the system transformation when using the inner block structure of the service. This enables us to use the contract as replacement for the service and still be sure that properties of the abstracted model also hold in the original model.

To show the latter, we use the dynamic and static semantics of $d\mathcal{L}$, described in more detail in [25]. $BV(\alpha)$ denotes the set of bound variables, whose values are changed in α. Furthermore, we look at the transition semantics of hybrid programs. For each interpretation I, a hybrid program is interpreted as binary transition relation $I[\![\alpha]\!] \subseteq S \times S$ on states. Where a transition is defined by its source state ν and its target state ω, and each state $\nu, \omega \in S$ is a mapping of all variables to a value in $\mathbb{R}$. E.g., $I[\![x := \theta]\!] = \{(\nu, \omega) :$ where $\nu = \omega$ except $\omega(x)$ has the value of term θ in state $\nu\}$.

We introduce two extension functions for hybrid programs: $store(BV(\alpha))$ represents a sequential execution of all discrete assignments, where the current values of the bound variables of α are assigned to observer variables. Furthermore, $BV(\alpha) := *$ assigns arbitrary values to all bound variables of α.

Let sys be a system that contains a service ser. Furthermore, the inner service ser fulfills the contract (X_A, X_G). The behavior of sys is given by the hybrid program $\alpha_{sys}{}^*$, where α_{sys} represent the behavior of sys for one execution of the simulation loop. Note that the hybrid program is created via the transformation of a Simulink model, i.e. the original Simulink model of sys contains a service ser. Therefore, the behavior that is represented in α_{sys} contains the behavior of the service ser. The behavior the service ser is represented by the hybrid program $\alpha_{ser}{}^*$, where α_{ser} represents the transformed behavior of the Simulink service ser. Note that the α_{ser} only contains the behavior of ser. To verify that the system sys fulfills the property Q under the assumptions P, we show the following:

$$\vdash P \rightarrow [(\alpha_{sys})^*]Q$$

The assumptions P are properties that hold for the whole system, like initial values and properties of the environment.

In the following, we consider hybrid programs that are created by our transformation, see Sect. 2.3. These programs have the form $\alpha = \beta; \gamma$, where $beta$

contains the discrete behavior and γ contains the continuous evolutions. We can split the concrete system behavior α_{sys} to obtain $\beta_0; \beta_{ser}; \beta_1; \gamma_{sys}$, where β_0 and β_1 are discrete parts that do not contain functionality of the service, β_{ser} denotes the discrete part of the service, and γ_{sys} represents the continuous part of the system. For the abstracted system, we write

$$\beta_0; store(BV(\beta_{ser})); BV(\beta_{ser}) := *; ?(X'_A \rightarrow X_G); \beta_1;$$
$$store(BV(\gamma_{ser})); BV(\gamma_{ser}) := *; \gamma'_{sys}; ?(X'_A \rightarrow X_G)$$

Where γ'_{sys} denotes the continuous evolutions of the system, where all evolutions and evolution domains of γ_{ser} are removed, and X'_A denotes the contract assumptions where all occurrences of bound variables of α_{ser} are replaced with their respective observer variables. We use this to show that the runs of the concrete model are a subset of the abstracted model.

Since the system parts β_0 and β_1 provide identical transitions in the concrete and abstracted model, we only need to show that the two abstracted parts for the discrete and continuous parts provide at least the runs of their concrete representation. Namely, we need to show:

(1) $I[\![\beta_{ser}]\!] \subseteq I[\![store(BV(\beta_{ser})); BV(\beta_{ser}) := *; ?(X'_A \rightarrow X_G);]\!]$

(2) $I[\![\gamma_{sys}]\!] \subseteq I[\![store(BV(\gamma_{ser})); BV(\gamma_{ser}) := *; \gamma'_{sys}; ?(X'_A \rightarrow X_G);]\!]$

Note that (2) needs to consider γ_{sys}, since all continuous evolutions evolve concurrently. For (1), it suffices syntactically to consider β_{ser} because we have introduced fresh variables for all input and output ports. We briefly discuss a proof sketch for (2), (1) can be shown analogously, without continuous evolutions.

For the abstracted system, we obtain the following runs:

$$I[\![store(BV(\gamma_{ser})); BV(\gamma_{ser}) := *; \gamma'_{sys}; ?(X'_A \rightarrow X_G);]\!]$$
$$= I[\![store(BV(\gamma_{ser}))]\!] \circ I[\![BV(\gamma_{ser}) := *]\!] \circ I[\![\gamma'_{sys}]\!] \circ I[\![?(X'_A \rightarrow X_G)]\!]$$
$$= \{(\nu, \omega) : \omega = \nu \text{ with the following exceptions:}$$

  (i) all observer variables are assigned with the observed values

  (ii) all variables $v \in BV(\gamma_{ser})$ have an arbitrary value that fulfills $(X'_A \rightarrow X_G)$

  (iii) all variables in γ'_{sys} evolve according to their evolution domains$\}$

For the concrete system the following states are reachable:

$$I[\![\gamma_{sys}]\!] = \{(\nu, \omega) : \omega = \nu \text{ except for all bound variables in } \gamma_{sys},$$
$$\text{which evolve according to their evolution domains}\}$$

As (i) observer variables are guaranteed by definition to only read variables from the original system, and (ii) our service verification has shown that the contract $(X'_A \rightarrow X_G)$ provides an over-approximation of the possible service behavior, i.e. allows more values for $v \in BV(\gamma_{ser})$ than the concrete service, we can conclude that (2) is satisfied, and thus our embedding of contracts to abstract from services is sound.

Table 1. Verification times in hh:mm

		Concrete system	Abstracted system	Service
Temperature control two services	Correct temperature range	1:00	0:02	0:07
	No rapid switching	-	0:05	0:16
Distance warner	No overflows	Not loadable	0:30	0:19
	Correct gain behavior	Not loadable	0:24	7:00

5   Evaluation

We evaluate our approach with two case studies, namely a temperature control system with two interacting services, and a multi-object distance warner model provided by our partners from the automotive industry. All experiments were run with KeYmaera X version 4.6.1 on a 2.6 Ghz 4-core machine running Windows 7. The fully-automatic transformation of the given Simulink models into $d\mathcal{L}$ is performed in a few seconds, which is neglectable compared to the verification times. The verification is done interactively. In particular, the contracts that are verified for each service and later on used in the abstracted system are created by hand and some invariants were manually added during the verification process.

Temperature Control System with Two Interacting Services. As first case study, we have used a model with two interacting temperature control services.[1] The service, which is used in this system, is the model presented in Sect. 2.1. One service controls the output temperature of the heating unit, while the other controls the room temperature. We have automatically transformed this system into $d\mathcal{L}$ and verified in KeyMaera X that the temperature is kept in a given range and that no rapid switching occurs. To verify these properties, we have manually added two invariants to the simulation loop. For the temperature bounds, we use range properties as invariants to describe the upper and lower temperature bound. To show that no rapid switching occurs, the invariant contains two clock variables that keep track of the time since the relay switched its state. The minimum time between changes is described by a timing property. Furthermore, we use a second set of timing properties that capture that the heating or cooling amount since the last state switch is bounded. The verification results are shown in the upper two rows of Table 1. We can see that the verification time for the system with contracts is significantly reduced compared to the flattened concrete system. Even the verification time of system and service together is less than the time for the flattened system. Since we only need to verify the contract for both services once, we can reuse the verification result.

---

[1] The $d\mathcal{L}$ models and proofs of the temperature control system are available online: https://www.sese.tu-berlin.de/ServiceVerification/parameter/en/.

For the rapid switching property of the concrete system, we could not even finish a proof, since KeYmaera X aborted due to an internal error, likely caused by the size of the model. With our service-oriented verification approach, we were able to verify both properties in less than half an hour.

Distance Warner. Our second case study is an industrial example of an automotive hybrid control system, namely a multi-object distance control system. It measures the relative speed of up to two leading vehicles and computes the relative distances. Furthermore, continuous integrators are used to compute the relative distance as a dynamic function of the relative speed. The system contains 263 blocks and 364 signal lines. To enable contract-based verification, we have encapsulated a core component of the distance warner, namely the distance calculator, which is used twice in the overall system, as a service. As showcase, we checked the following safety properties: (1) the distance warner does not produce overflows, and (2) the distance warner performs a correct gain behavior, where the gain of the outgoing discrete signal is depending on the incoming continuous signal. To show the overflow property, we added a range property as invariant. For the property that considers the gain behavior, we added observer variables that keep track of the current sign of the input signal. The main property of the gain behavior is formulated as dynamic property that compares the last output state to the current one. The verification times are shown in the two bottom rows of Table 1. Note that we first detected a bug in the original system, where an overflow was possible at an integrator in the system. We detected this bug within 20 min and subsequently used a patched version of the system where the integrator is saturated. For the corrected system, we were not able to verify any properties on the concrete system, since KeYmaera X crashed when starting a new proof. For the abstracted system with the distance calculator as a service, we were able to show both desired properties for all possible input scenarios in less than 8 h.

6   Conclusion

In this paper, we have presented a service-oriented verification approach for hybrid control systems modeled in Simulink. As our first main contribution, we have introduced *hybrid contracts*, which abstractly capture the dynamic behavior of Simulink subsystems that are encapsulated as services. To ease the definition of hybrid contracts, we have presented *extension functions*, which enable the designer to systematically insert observer variables into a given hybrid system, and *property templates*, which can be used as design patterns to support the designer in defining contracts based on commonly used properties, i.e., range, timing, and dynamic properties. Our existing transformation from Simulink into the formally well-defined differential dynamic logic ($d\mathcal{L}$) [16] enables us to formally verify that a service adheres to its hybrid contract using the interactive theorem prover KeYmaera X. As our second main contribution, we have presented a service-oriented verification approach for hybrid control systems modeled in Simulink. The key idea is that we use our hybrid contracts to replace services in a given hybrid system by their contracts, and thus enable the hierarchical abstraction from implementation details of services. To ensure the correct embedding of services into concrete systems, we have provided a proof sketch for the soundness of this abstraction technique. We have demonstrated the applicability of our approach with experimental results. In particular, we have used an automotive industrial case study of a hybrid distance warner, which could neither be verified using the Simulink Design Verifier nor KeYMaera X on the concrete system. With our service-oriented verification approach, we have been able to detect a bug in the original system (namely an overflow), and we have been able to verify functional and non-functional properties of a patched version of the hybrid distance warner within less than 8 h.

In future work, we plan to increase the automation of our approach, for example, by automatically generating contracts for common error classes, e.g. no division-by-zero or overflow detection, and by providing Simulink-specific tactics to ease the verification process. Furthermore, we intend to extend the applicability of our services and hybrid contracts to other modeling languages. By abstracting the inner structure of services and providing their interface behavior via hybrid contracts, they can be used in other representations that only consider their hybrid contracts. By extending our technique to other design languages than Simulink and developing transformation to d$\mathcal{L}$, Services can be designed in any supported modeling language. The services can be used in new designs and the design language of the new system can be different than the services that can be used in the system. It is only necessary to provide the interface ports and their hybrid contracts. In the new design, the service is then used as black box. Appropriate transformations can be used for these black box services to verify properties of the system containing such a service.

References

1. Alur, R., Courcoubetis, C., Henzinger, T.A., Ho, P.-H.: Hybrid automata: an algorithmic approach to the specification and verification of hybrid systems. In: Grossman, R.L., Nerode, A., Ravn, A.P., Rischel, H. (eds.) HS 1991-1992. LNCS, vol. 736, pp. 209–229. Springer, Heidelberg (1993). https://doi.org/10.1007/3-540-57318-6_30
2. Araiza-Illan, D., Eder, K., Richards, A.: Formal verification of control systems' properties with theorem proving. In: 2014 UKACC International Conference on Control (CONTROL), pp. 244–249. IEEE (2014)
3. Aştefănoaei, L., Bensalem, S., Bozga, M.: A compositional approach to the verification of hybrid systems. In: Ábrahám, E., Bonsangue, M., Johnsen, E.B. (eds.) Theory and Practice of Formal Methods. LNCS, vol. 9660, pp. 88–103. Springer, Cham (2016). https://doi.org/10.1007/978-3-319-30734-3_8
4. Barnett, M., Chang, B.-Y.E., DeLine, R., Jacobs, B., Leino, K.R.M.: Boogie: a modular reusable verifier for object-oriented programs. In: de Boer, F.S., Bonsangue, M.M., Graf, S., de Roever, W.-P. (eds.) FMCO 2005. LNCS, vol. 4111, pp. 364–387. Springer, Heidelberg (2006). https://doi.org/10.1007/11804192_17
5. Benvenuti, L., Bresolin, D., Collins, P., Ferrari, A., Geretti, L., Villa, T.: Assume-guarantee verification of nonlinear hybrid systems with ariadne. Int. J. Robust Nonlinear Control 24(4), 699–724 (2014)

6. Boström, P.: Contract-based verification of Simulink models. In: Qin, S., Qiu, Z. (eds.) ICFEM 2011. LNCS, vol. 6991, pp. 291–306. Springer, Heidelberg (2011). https://doi.org/10.1007/978-3-642-24559-6_21

7. Chutinan, A., Krogh, B.H.: Computational techniques for hybrid system verification. In: IEEE Transactions on Automatic Control, vol. 48, pp. 64–75. IEEE (2003)

8. Cubuktepe, M., Ahmadi, M., Topcu, U., Hencey, B.: Compositional analysis of hybrid systems defined over finite alphabets. IFAC-PapersOnLine **51**(16), 115–120 (2018)

9. De Moura, L., Bjørner, N.: Z3: an efficient SMT solver. In: Ramakrishnan, C.R., Rehof, J. (eds.) TACAS 2008. LNCS, vol. 4963, pp. 337–340. Springer, Heidelberg (2008). https://doi.org/10.1007/978-3-540-78800-3_24

10. Filliâtre, J.-C., Paskevich, A.: Why3 — where programs meet provers. In: Felleisen, M., Gardner, P. (eds.) ESOP 2013. LNCS, vol. 7792, pp. 125–128. Springer, Heidelberg (2013). https://doi.org/10.1007/978-3-642-37036-6_8

11. Frehse, G.: PHAVer: algorithmic verification of hybrid systems past hytech. In: Morari, M., Thiele, L. (eds.) HSCC 2005. LNCS, vol. 3414, pp. 258–273. Springer, Heidelberg (2005). https://doi.org/10.1007/978-3-540-31954-2_17

12. Fulton, N., Mitsch, S., Quesel, J.-D., Völp, M., Platzer, A.: KeYmaera X: an axiomatic tactical theorem prover for hybrid systems. In: Felty, A.P., Middeldorp, A. (eds.) CADE 2015. LNCS (LNAI), vol. 9195, pp. 527–538. Springer, Cham (2015). https://doi.org/10.1007/978-3-319-21401-6_36

13. Henzinger, T.A., Ho, P.H., Wong-Toi, H.: HyTech: a model checker for hybrid systems. Int. J. Softw. Tools Technol. Transf. **1**, 110–122 (1997)

14. Herber, P., Reicherdt, R., Bittner, P.: Bit-precise formal verification of discrete-time MATLAB/Simulink models using SMT solving. In: 2013 Proceedings of the International Conference on Embedded Software (EMSOFT), pp. 1–10. IEEE (2013)

15. Lahiri, S.K., Seshia, S.A.: The UCLID decision procedure. In: Alur, R., Peled, D.A. (eds.) CAV 2004. LNCS, vol. 3114, pp. 475–478. Springer, Heidelberg (2004). https://doi.org/10.1007/978-3-540-27813-9_40

16. Liebrenz, T., Herber, P., Glesner, S.: Deductive verification of hybrid control systems modeled in Simulink with KeYmaera X. In: Sun, J., Sun, M. (eds.) ICFEM 2018. LNCS, vol. 11232, pp. 89–105. Springer, Cham (2018). https://doi.org/10.1007/978-3-030-02450-5_6

17. Liebrenz, T., Herber, P., Göthel, T., Glesner, S.: Towards service-oriented design of hybrid systems modeled in Simulink. In: 2017 IEEE 41st Annual Computer Software and Applications Conference (COMPSAC), vol. 2, pp. 469–474. IEEE (2017)

18. MathWorks: MATLAB Simulink. www.mathworks.com/products/simulink.html

19. MathWorks: White Paper: Code Verification and Run-Time Error Detection Through Abstract Interpretation. Technical report (2008)

20. Minopoli, S., Frehse, G.: SL2SX translator: from Simulink to SpaceEx models. In: 19th International Conference on Hybrid Systems: Computation and Control, pp. 93–98. ACM (2016)

21. Mitsch, S., Platzer, A.: The KeYmaera X proof IDE: concepts on usability in hybrid systems theorem proving. In: 3rd Workshop on Formal Integrated Development Environment. Electronic Proceedings in Theoretical Computer Science, vol. 240, pp. 67–81. Open Publishing Association (2017)

22. Müller, A., Mitsch, S., Retschitzegger, W., Schwinger, W., Platzer, A.: Change and delay contracts for hybrid system component verification. In: Huisman, M., Rubin, J. (eds.) FASE 2017. LNCS, vol. 10202, pp. 134–151. Springer, Heidelberg (2017). https://doi.org/10.1007/978-3-662-54494-5_8

23. O'Halloran, C.: Automated verification of code automatically generated from Simulink®. Autom. Softw. Eng. **20**(2), 237–264 (2013)

24. Platzer, A.: Differential dynamic logic for hybrid systems. J. Autom. Reason. **41**(2), 143–189 (2008)

25. Platzer, A.: A complete uniform substitution calculus for differential dynamic logic. J. Autom. Reason. **59**(2), 219–265 (2017)

26. Reicherdt, R., Glesner, S.: Formal verification of discrete-time MATLAB/Simulink models using boogie. In: Giannakopoulou, D., Salaün, G. (eds.) SEFM 2014. LNCS, vol. 8702, pp. 190–204. Springer, Cham (2014). https://doi.org/10.1007/978-3-319-10431-7_14

27. Sanfelice, R., Copp, D., Nanez, P.: A toolbox for simulation of hybrid systems in Matlab/Simulink: Hybrid Equations (HyEQ) toolbox. In: 16th International Conference on Hybrid Systems: Computation and Control, pp. 101–106. ACM (2013)

Compositional Liveness-Preserving Conformance Testing of Timed I/O Automata

Lars Luthmann$^{(\boxtimes)}$ (ID), Hendrik Göttmann (ID), and Malte Lochau (ID)

Real-Time Systems Lab, TU Darmstadt, Darmstadt, Germany
{lars.luthmann,malte.lochau}@es.tu-darmstadt.de,
h.goettmann@stud.tu-darmstadt.de

Abstract. I/O conformance testing theories (e.g., *ioco*) are concerned with formally defining when observable output behaviors of an implementation conform to those permitted by a specification. Thereupon, several real-time extensions of *ioco*, usually called *tioco*, have been proposed, further taking into account permitted delays between actions. In this paper, we propose an improved version of *tioco*, called *live timed ioco* (*ltioco*), tackling various weaknesses of existing definitions. Here, a reasonable adaptation of quiescence (i.e., observable absence of any outputs) to real-time behaviors has to be done with care: *ltioco* therefore distinguishes safe outputs being allowed to happen, from live outputs being enforced to happen within a certain time period thus inducing two different facets of quiescence. Furthermore, *tioco* is frequently defined on Timed I/O Labeled Transition Systems (TIOLTS), a semantic model of Timed I/O Automata (TIOA) which is infinitely branching and thus infeasible for practical testing tools. Instead, we extend the theory of zone graphs to enable *ltioco* testing on a finite semantic model of TIOA. Finally, we investigate compositionality of *ltioco* with respect to parallel composition including a proper treatment of silent transitions.

Keywords: Real-time testing · Timed Automata · Input/output conformance testing · Compositionality

1 Introduction

Model-based testing constitutes a practically emerging, yet theoretically founded technique for automated quality assurance of software systems [16]. In particular, input/output conformance testing theories formalize notions of observable conformance between an implementation under test and a specification, where the **ioco** theory [42] constitutes one of the most prominent examples. The **ioco** relation requires both the input/output-behaviors of the specification and the implementation to be represented as input/output labeled transition systems (IOLTS),

L. Luthmann and M. Lochau—This work was funded by the Hessian LOEWE initiative within the Software-Factory 4.0 project.

© Springer Nature Switzerland AG 2020
F. Arbab and S.-S. Jongmans (Eds.): FACS 2019, LNCS 12018, pp. 147–169, 2020.
https://doi.org/10.1007/978-3-030-40914-2_8

where the IOLTS of the implementation is unknown (black-box assumption) [11]. For an implementation to satisfy **ioco**, all its possible output behaviors must be permitted by the specification. To rule out trivial implementations never showing any output, **ioco** employs the notion of *quiescence* to explicitly permit starvation. In order to ensure proper test-execution semantics, **ioco** requires *input-enabled* implementations, never blocking any (test-)inputs. Hence, **ioco** is concerned with the correct *ordering* of (or causality among) input/output (re-)actions, whereas quantified *time delays* between action occurrences are not considered. However, reasoning about real-time behaviors becomes more and more crucial and various real-time extensions of **ioco**, so-called **tioco**, have been recently proposed [14,24,26,27,39]. Based on timed extensions of IOLTS (so-called TIOLTS), a system run progresses by either actively performing discrete, instantaneous actions or by inactively letting a quantified amount of time pass. Nevertheless, existing definitions of **tioco** suffer from several weaknesses which we tackle in this paper by proposing an improved version called *live timed ioco* (**ltioco**). Our contributions can be summarized as follows.

- Recent adoptions of quiescence in a timed setting also show several weaknesses: most recent versions of **tioco** either do not incorporate any notion of quiescence at all [24,26,28,39], or define quiescence in terms of (either infinite or bounded) time intervals without observable output actions [14,39]. Both fail to distinguish the *enabling* of output actions (i.e., an output is allowed to occur in a time interval to constitute *safe* behavior) from *enforced* output actions (i.e., an output must occur in a certain time interval to meet *liveness* requirements). To this end, **ltioco** distinguishes safe outputs from live outputs thus explicitly incorporating the two different facets of timed quiescence. We prove correctness of **ltioco** with respect to TIOLTS semantics and we show that **ltioco** is strictly more discriminating than most recent versions of **tioco**.
- We investigate compositionality properties of **ltioco** with respect to (synchronous) parallel composition including silent transitions.
- Finally, all recent versions of *tioco* are defined on TIOLTS, constituting a semantic model of Timed I/O Automata (TIOA) which is infinitely branching and thus infeasible for practical testing tools. Instead, we extend the notion of zone graphs to effectively check *ltioco* on a finite semantic model of TIOA using so-called *span traces*. Thereupon, we developed a tool for online testing using **tioco** (see https://www.es.tu-darmstadt.de/ltioco).

The remainder of this paper is structured as follows. We first give a formal introduction into TIOA and parallel composition of TIOA in Sect. 2. Then, we discuss existing notions of **tioco** and point out their weaknesses in Sect. 3 which we address in the subsequent Sect. 4. Furthermore, we give an intuition on how to apply zone graphs for an efficient implementation of our approach in Sect. 5 and we summarize related work in Sect. 6. Please note that all proofs are provided in the accompanying technical report [31].

2   Timed Input/Output Automata

We first recall foundations of *Timed Automata (TA)* [2,3], extension of TA by input/output labels [17,35,36] and their composition involving silent transitions [10].

TA are labeled finite state-transition graphs with states being called *locations* and transitions being called *switches*. A TA is further defined with respect to a finite set $\mathcal{C}$ of *clocks* over a numerical *clock domain* $\mathbb{T}$ (e.g., $\mathbb{T} = \mathbb{N}_0$ for *discrete time* and $\mathbb{T} = \mathbb{R}_+$ with $\mathbb{R}_+ := \{r \mid r \in \mathbb{R} \wedge r \geq 0\}$ for *dense time*). Clocks constitute constantly and synchronously increasing, yet independently resettable variables over $\mathbb{T}$ for measuring and restricting time intervals (durations/delays) between action occurrences. Note that we consider $\mathbb{T} = \mathbb{N}_0$ in all examples for the sake of readability. In particular, we consider *Timed Safety Automata* [23] in which time-critical behaviors are expressed by *clock constraints* as *guards* for switches and *invariants* for locations. Guards restrict time intervals in which a switch is enabled while residing in its source location, whereas invariants restrict time intervals in which a TA run is permitted to reside in a location. Alternative TA definitions may incorporate distinguished *acceptance locations* thus employing Büchi acceptance semantics on *infinite runs* [2,23] which is out of the scope of this paper as model-based testing is inherently limited to *finite* test runs.

Timed Input/Output-labeled Automata (TIOA) extend TA for timed interface specifications (e.g., for model-based conformance testing of time-critical components or systems [35,36]). The *label alphabet* $\Sigma = \Sigma_I \cup \Sigma_O$ of a TIOA consists of two disjoint subsets of (externally controllable, internally observable) *input actions* Σ_I and (externally observable, internally controllable) *output actions* Σ_O. The special symbol $\tau \notin \Sigma$ summarizes *internal actions* of silent switches being neither externally controllable nor visible, and we write $\Sigma_\tau = \Sigma \cup \{\tau\}$ for short.

Definition 1 (TIOA). *A TIOA $\mathcal{A}$ is a tuple $(L, \ell_0, \Sigma_I, \Sigma_O, \rightarrow, I)$, where*

- *L is a finite set of locations with initial location $\ell_0 \in L$,*
- *Σ_I and Σ_O are sets of input actions and output actions with $\Sigma_I \cap \Sigma_O = \emptyset$,*
- *$\rightarrow \subseteq L \times \mathcal{B}(\mathcal{C}) \times \Sigma_\tau \times 2^\mathcal{C} \times L$ is a relation defining switches, with a set $\mathcal{B}(\mathcal{C})$ of clock constraints φ inductively defined as*

$$\varphi := x \sim r \mid x - y \sim r \mid \neg\varphi \mid \varphi \wedge \varphi \mid \text{true},$$

  where $x, y \in \mathcal{C}$, $r \in \mathbb{Q}_+$, and $\sim \in \{<, \leq, =, \geq, >\}$, and
- *$I : L \rightarrow \mathcal{B}(\mathcal{C})$ is a function assigning location invariants.*

We write $\ell \xrightarrow{g, \sigma, R} \ell'$ to denote switches from location ℓ to ℓ' with guard g, action σ and set $R \subseteq \mathcal{C}$ of clocks being reset. Without loss of generality, we assume each location invariant being unequal to true to be *downward-closed* (i.e., with clauses $x \leq r$ or $x < r$) [8]. The operational semantics of TIOA may be defined as *Timed Input/Output Labeled Transition System (TIOLTS)* [22]. A TIOLTS state $\langle \ell, u \rangle$ is a pair consisting of a location $\ell \in L$ and a *clock valuation* $u \in \mathcal{C} \rightarrow \mathbb{T}$. A

TIOLTS defines two kinds of transitions: (1) passage of time while inactively residing in a location, and (2) instantaneous switches between locations due to action occurrences (including τ). Given a clock valuation u, $u + d$ denotes the clock valuation mapping each clock $c \in \mathcal{C}$ to the updated clock value $u(c) + d$ with $d \in \mathbb{T}$. For a subset $R \subseteq \mathcal{C}$ of clocks, $[R \mapsto 0]u$ denotes the clock valuation mapping every clock in R to 0 while preserving the values of all other clocks in $\mathcal{C} \setminus R$. Finally, $u \in g$ denotes that clock valuation u satisfies clock constraint $g \in \mathcal{B}(\mathcal{C})$. We further distinguish between *strong* and *weak* transitions, depending on whether silent transitions are visible or not.

Definition 2 (TIOLTS). *The TIOLTS of TIOA $(L, \ell_0, \Sigma_I, \Sigma_O, \rightarrow, I)$ is a tuple $(S, s_0, \Sigma_I, \Sigma_O, \twoheadrightarrow)$, where*

- $S = L \times (\mathcal{C} \rightarrow \mathbb{T})$ *is a set of* states *with initial state* $s_0 = \langle \ell_0, [\mathcal{C} \mapsto 0]u_0 \rangle \in S$,
- $\hat{\Sigma}_\tau = \Sigma_I \cup \Sigma_O \cup \{\tau\} \cup \Delta$ *is a set of* labels *with* $\Delta = \mathbb{T}$, $\Sigma_\tau \cap \Delta = \emptyset$, *and*
- $\twoheadrightarrow \subseteq S \times \hat{\Sigma}_\tau \times S$ *is a set of* (strong) transitions *being the least relation satisfying the rules:*
  - $\langle \ell, u \rangle \xrightarrow{d} \langle \ell, u + d \rangle$ *if* $u \in I(\ell)$ *and* $(u + d) \in I(\ell)$ *for* $d \in \mathbb{T}$, *and*
  - $\langle \ell, u \rangle \xrightarrow{\sigma} \langle \ell', u' \rangle$ *if* $\ell \xrightarrow{g, \sigma, R} \ell'$, $u \in g$, $u' = [R \mapsto 0]u$, $u' \in I(\ell')$, $\sigma \in \Sigma_\tau$.

By $\Rightarrow \subseteq S \times \hat{\Sigma} \times S$ we further denote a set of (weak) transitions *being the least relation satisfying the rules:*

- $s_0 \xrightarrow{\tau^n} s_n$ *if* $\exists s_1, \ldots s_{n-1} \in S : s_0 \xrightarrow{\tau} s_1 \xrightarrow{\tau} \ldots \xrightarrow{\tau} s_n$ *with* $n \in \mathbb{N}_0$,
- $s \xrightarrow{\sigma} s'$ *if* $\exists s_1, s_2 \in S : s \xrightarrow{\tau^n} s_1 \xrightarrow{\sigma} s_2 \xrightarrow{\tau^m} s'$ *with* $n, m \in \mathbb{N}_0$,
- $s \xrightarrow{d} s'$ *if* $s \xrightarrow{d} s'$,
- $s \xrightarrow{0} s'$ *if* $s \xrightarrow{\tau^n} s'$ *with* $n \in \mathbb{N}_0$,
- $s_0 \xrightarrow{\sigma_1 \cdots \sigma_n}$ *if* $\exists s_1, \ldots s_n \in S : s_0 \xrightarrow{\sigma_1} s_1 \xrightarrow{\sigma_2} \ldots \xrightarrow{\sigma_n} s_n$ *with* $n \in \mathbb{N}_0$, *and*
- $s \xrightarrow{d + d'} s'$ *if* $\exists s'' \in S : s \xrightarrow{d} s''$ *and* $s'' \xrightarrow{d'} s'$.

We only consider *strongly convergent* TIOA (i.e., having TIOLTS without infinite τ-sequences). By $[\![\mathcal{A}]\!]^x_S$, $x \in \{w, s\}$, we refer to the (either weak or strong) TIOLTS semantics of TIOA $\mathcal{A}$, where we omit parameter x if not relevant. The weak semantics is obtained by replacing all occurrences of $\rightarrow$ by $\Rightarrow$ in all definitions. We recall three essential properties for strong TIOLTS semantics of any given TIOA [1,17].

Proposition 1. *Let $(S, s_0, \Sigma_I, \Sigma_O, \twoheadrightarrow)$ be a TIOLTS of a TIOA.*

- *(Time Add)* $\forall s_1, s_3 \in S, \forall d_1, d_2 \in \Delta : s_1 \xrightarrow{d_1 + d_2} s_3 \Leftrightarrow \exists s_2 : s_1 \xrightarrow{d_1} s_2 \xrightarrow{d_2} s_3$
- *(Time Reflex)* $\forall s_1, s_2 \in S : s_1 \xrightarrow{0} s_2 \Rightarrow s_1 = s_2$
- *(Time Determ)* $\forall s_1, s_2, s_3 \in S : s_1 \xrightarrow{d} s_2$ *and* $s_1 \xrightarrow{d} s_3$ *then* $s_2 = s_3$

In contrast, the weak semantics obviously obstructs all three properties.

Furthermore, by $traces(s_0) = \{\omega \mid s_0 \xrightarrow{\omega}\}$ we denote the set of all *traces* $\omega = \alpha_1 \alpha_2 \cdots \alpha_k \in (\Sigma \cup \Delta)^*$ corresponding to some path $s_0 \xrightarrow{\alpha_1} s_1 \xrightarrow{\alpha_2} \cdots \xrightarrow{\alpha_k} s_k$

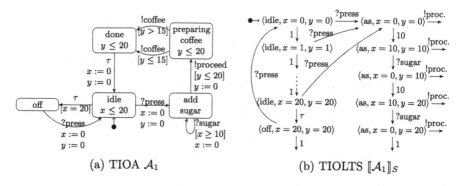

(a) TIOA $\mathcal{A}_1$ (b) TIOLTS $[\![\mathcal{A}_1]\!]_S$

Fig. 1. TIOA for a simple vending machine [4,8] and extract from TIOLTS

of TIOLTS s. Given a TIOA $\mathcal{A}$, the TIOLTS $[\![\mathcal{A}]\!]_S$ defines all possible *(timed)* *runs* $s_0 = \langle \ell_0, u_0 \rangle \xrightarrow{d_1, \sigma_1} \langle \ell_1, u_1 \rangle \xrightarrow{d_2, \sigma_2} \cdots$ of $\mathcal{A}$ in terms of sequences of *(timed)* *steps* $s \xrightarrow{d,\sigma} s''$ denoting $\exists s' \in S : s \xrightarrow{d} s' \xrightarrow{\sigma} s''$ [39]. We refer to the set of weak/strong traces of state s by $traces(s)^x$, $x \in \{w, s\}$, respectively.

Example 1. Figure 1a shows a (simplified) TIOA $\mathcal{A}_1$ of a vending machine with two clocks, x and y, and Fig. 1b depicts an extract from its TIOLTS. Switches are labeled with actions (prefixes "?" for inputs and "!" for outputs), guards (e.g., $x \leq 20$), and (possibly empty) clock resets. We label locations by their names (e.g., initial location *idle*) and their location invariants. Clock constraints being equal to true are omitted. Each (timed) *run* of the machine starts in initial location *idle*, where a user may *press* a button to switch to location *add sugar*. If no button is pressed for 20 time units (e.g., seconds), the machine is turned *off* via a silent switch and may be switched to *idle*, again, by pressing a button. In location *add sugar*, sugar may be repeatedly selected, where at least 10 s must pass between two consecutive requests and the machine proceeds to location *preparing coffee* at most 20 s after input *press*. Here, coffee is dispensed for at most 20 s and the machine finally returns to *idle*. The machine either produces small coffees (finishing after less than 15 s) or large coffees (requiring more than 15 s). This example illustrates the semantic differences between guards and invariants: guards restrict time intervals in which a switch is *allowed* to be taken, whereas invariants define time intervals after which a location is *enforced* to be left (e.g., it is allowed to perform *!proceed* to leave location *add sugar* while $y \leq 20$ holds, whereas it is enforced to leave location *preparing coffee* in case of $y = 20$). Hence, guards express *safety* conditions, whereas invariants express *liveness* conditions of timed runs.

A TIOA is supposed to specify one particular part of an arbitrary complex system composed of several concurrently interacting *components*. We define CCS-like *parallel composition* of TIOA with synchronous communication via shared input/output actions, becoming internal τ-actions [17]. As a prerequisite

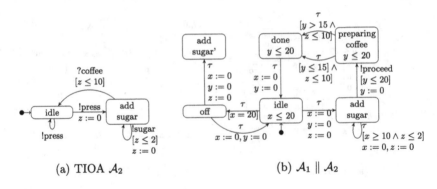

Fig. 2. Sample TIOA composition

for composing two TIOA $\mathcal{A}_1$ and $\mathcal{A}_2$, denoted as $\mathcal{A}_{1\|2} = \mathcal{A}_1 \parallel \mathcal{A}_2$, we require both to be *composable* (i.e., all shared actions have opposed directions).

Definition 3 (TIOA Composition). *Let* $(L_j, \ell_{0_j}, \Sigma_{I_j}, \Sigma_{O_j}, \to_j, I_j)$ *with* $j \in \{1,2\}$ *be TIOA with* $\Sigma_{I_1} \cap \Sigma_{I_2} = \emptyset$, $\Sigma_{O_1} \cap \Sigma_{O_2} = \emptyset$ *and* $C_1 \cap C_2 = \emptyset$. *Their* parallel composition *is a TIOA* $(L_1 \times L_2, (\ell_{0_1}, \ell_{0_2}), \Sigma_{I_{1\|2}}, \Sigma_{O_{1\|2}}, \to_{1\|2}, I_{1\|2})$ *over* $C_{1\|2} = C_1 \cup C_2$ *with* $\Sigma_{I_{1\|2}} = (\Sigma_{I_1} \cup \Sigma_{I_2}) \backslash (\Sigma_{O_1} \cup \Sigma_{O_2})$, $\Sigma_{O_{1\|2}} = (\Sigma_{O_1} \cup \Sigma_{O_2}) \backslash (\Sigma_{I_1} \cup \Sigma_{I_2})$, $I_{1\|2}(\ell_1, \ell_2) = I_1(\ell_1) \wedge I_2(\ell_2)$, *and* $\to_{1\|2}$ *is the least relation satisfying the rules:*

(1) $(\ell_1, \ell_2) \xrightarrow{g_1, \sigma, R_1}_{1\|2} (\ell'_1, \ell_2)$ *if* $\ell_1 \xrightarrow{g_1, \sigma, R_1}_1 \ell'_1$ *and* $\sigma \in (\Sigma_1 \setminus \Sigma_2) \cup \{\tau\}$

(2) $(\ell_1, \ell_2) \xrightarrow{g_2, \sigma, R_2}_{1\|2} (\ell_1, \ell'_2)$ *if* $\ell_2 \xrightarrow{g_2, \sigma, R_2}_2 \ell'_2$ *and* $\sigma \in (\Sigma_2 \setminus \Sigma_1) \cup \{\tau\}$

(3) $(\ell_1, \ell_2) \xrightarrow{g_1 \wedge g_2, \tau, R_1 \cup R_2}_{1\|2} (\ell'_1, \ell'_2)$ *if* $\ell_1 \xrightarrow{g_1, \sigma, R_1}_1 \ell'_1, \ell_2 \xrightarrow{g_2, \sigma, R_2}_2 \ell'_2$ *and*
$$\sigma \in (\Sigma_1 \cap \Sigma_2).$$

Example 2. Consider TIOA $\mathcal{A}_1$, $\mathcal{A}_2$ and their parallel composition $\mathcal{A}_1 \parallel \mathcal{A}_2$ (cf. Figs. 1a, 2a, and b). A customer $\mathcal{A}_2$ may *press* a button, add sugar and wait for coffee. In $\mathcal{A}_1 \parallel \mathcal{A}_2$, shared actions are performed synchronously only if being enabled in both $\mathcal{A}_1$ and $\mathcal{A}_2$, thus resulting in a τ-step. For instance, the synchronized switch from *idle* to *add sugar* is labeled with τ and clocks x, y (from $\mathcal{A}_1$) and z (from $\mathcal{A}_2$) being reset. Similarly, the *sugar* loop also becomes a τ-step, while clock resets are unified and guards are conjugated. In contrast, switch *proceed* does not become internal as this output is not observed by $\mathcal{A}_2$ (but instead transmitted to some administration component). Location *preparing coffee* has two switches labeled with τ as both *coffee* switches of $\mathcal{A}_1$ are synchronized with the *coffee* switch of $\mathcal{A}_2$. Location *off* has a τ-step to *add sugar'* as the switch of $\mathcal{A}_1$ from *off* to *idle* may also be synchronized with the switch of $\mathcal{A}_2$ from *idle* to *add sugar*. Here, *add sugar'* does not have any outgoing transitions as *add sugar* ($\mathcal{A}_2$) has no actions shared with *idle* ($\mathcal{A}_1$). The *sugar* loop and the switch from *preparing coffee* to *done* guarded by $y > 15 \wedge z \leq 10$ are semantically incompatible as their guards are unsatisfiable in all runs.

3   Timed Input/Output Conformance

TIOLTS have been considered as a formal basis for conformance testing theories of time-critical input/output behaviors [39]. Timed conformance relations are usually defined in the flavor of **ioco** testing, as initially proposed on input/output labeled transition systems (IOLTS) for untimed behaviors [42].

Intuitively, IOLTS *im* representing an implementation under test *input/output-conforms* to IOLTS *sp* representing a specification, denoted *im* **ioco** *sp*, if for all input behaviors specified in *sp*, the observable output behaviors of *im* for those input behaviors are permitted by *sp*. Input behaviors may be only partially specified (i.e., only for relevant/intended environmental input sequences, the expected output behaviors are explicitly captured in *sp*), whereas implementation *im* is supposed to be *input-enabled* (i.e., to never block any input action). Timed adaptations of **ioco**, so-called **tioco**, consider both *im* and *sp* to be represented as TIOLTS as checking timed input/output conformance directly on TIOA is unfeasible due to non-observability of clock resets in timed runs. For instance, in the example in Fig. 1a, it is unknown if it is allowed to wait for 20 time units in *idle* if we reach this location from *done* as resets of x and y are not observable. Similar to the untimed case, TIOLTS *im* is supposed to be input-enabled (i.e., *im* must always—*at any time*—be able to instantaneously accept all possible inputs). In addition, for *im* to specify realistic behaviors, we further impose the *independent-progress* property: In each state, *im* is able to either wait for an infinite amount of time or to eventually perform an output action thus preventing *forced inputs* [17,39].

Definition 4. *Let* $(S, s_0, \Sigma_I, \Sigma_O, \twoheadrightarrow)$ *be a TIOLTS.*

- *(Input-Enabledness) State* $s \in S$ *is* weak input-enabled *iff* $\forall i \in \Sigma_I : s \overset{i}{\twoheadrightarrow}$.
- *(Independent Progress) State* $s \in S$ *of a TIOLTS enables* weak independent progress *iff* $\forall d \in \Delta : s \overset{d}{\twoheadrightarrow}$ *or* $\exists d \in \Delta, \exists o \in \Sigma_O : s \overset{d}{\twoheadrightarrow} \overset{o}{\twoheadrightarrow}$.

A TIOLTS is *(weak) input-enabled* iff all states are (weak) input-enabled and it enables *(weak) independent progress* if all states do (for the strong versions of both properties, we replace $\twoheadrightarrow$ by $\rightarrow$). Similarly to **ioco**, we assume weak input-enabledness and independent progress for all implementations under test, whereas specifications may be underspecified. This is required for practical testing where an implementation should always at least accept (and then potentially ignore) every input. Conversely, the environment (i.e., a tester) should not be enforced by the implementation to provide a particular input in order to guarantee any progress. For instance, consider Fig. 1a: location *off* is not input-enabled as there is no switch for input *sugar*. However, if there would be such a switch, then also location *idle* would be weak input-enabled as output *off* may be reached by a τ-step. In contrast, all locations in Fig. 1a enable (weak) independent progress.

We now revisit two major definitions of **tioco** from recent literature. We first consider the (notationally slightly adapted) definition of Krichen and Tripakis [26] which we will refer to as **tioco**$_\Delta$. It is based on the assumption that, in addition to timed traces consisting of sequences of timed steps (d, o) including output actions $o \in \Sigma_O$, also all possible delays $d \in \Delta$ permitted to elapse in states $s \in S$ are observable in isolation.

Definition 5 (tioco$_\Delta$). *Let* im, sp *be a TIOLTS over* $\Sigma = \Sigma_I \cup \Sigma_O$, $s \in S$, $S' \subseteq S$, *and* $\xi \in (\Delta \times \Sigma)^*$.

- $s \, \mathbf{after} \, \xi := \{s' \mid s \overset{\xi}{\twoheadrightarrow} s'\}$,
- $\mathbf{out}_\Delta(s) := \{o \mid o \in \Sigma_O, s \overset{o}{\twoheadrightarrow}\} \cup \{d \mid d \in \Delta, s \overset{d}{\twoheadrightarrow}\}$,
- $\mathbf{out}_\Delta(S') := \bigcup_{s \in S'} \mathbf{out}_\Delta(s)$,

- $ttraces(s) := \{\xi \mid s \overset{\xi}{\twoheadrightarrow}\}$, *and*
- $im \, \mathbf{tioco}_\Delta \, sp :\Leftrightarrow \forall \xi \in ttraces(sp) : \mathbf{out}_\Delta(im \, \mathbf{after} \, \xi) \subseteq \mathbf{out}_\Delta(sp \, \mathbf{after} \, \xi)$

We may use the name of the whole TIOLTS and the name of its initial state interchangeably as frequently done in **ioco**-based theories (e.g., by $im \, \mathbf{after} \, \xi$ we refer to the set of states being reachable by ξ from the initial state of im). The second version of **tioco**, which we will denote as **tioco$_\delta$**, does not rely on observability of arbitrary delays, but instead incorporates a notion of *timed quiescence* [39]. Quiescence constitutes another fundamental concept of (untimed) **ioco**: IOLTS state s is *quiescent*, denoted $\delta(s)$, if no output or internal action is enabled in s thus requiring an input to proceed a (suspended) run reaching s. By making quiescence observable by a special output δ, **ioco** rejects trivial implementations im never showing any outputs as this must be explicitly permitted by the specification. In the timed case, state s of a TIOLTS may be considered quiescent if no output action is *ever* (or, at least not until some fixed maximum delay M [14]) enabled in s. To this end, the notion of *timed suspension traces* (*tstraces*) extends *traces* of TIOLTS by timed observable quiescence. The most common definition of **tioco$_\delta$** may be given as follows.

Definition 6 (tioco$_\delta$). *Let* im, sp *be a TIOLTS over* $\Sigma = \Sigma_I \cup \Sigma_O$, $s, s' \in S$, $S' \subseteq S$ *and* $\xi \in (\Delta \times (\Sigma \cup \{\delta\}))^*$.

- s *is* quiescent, *denoted by* $\delta(s)$, *iff* $\forall \mu \in \Sigma_O, \forall d \in \Delta : s \overset{(d, \mu)}{\not\longrightarrow}$,

- $s \, \mathbf{after} \, \xi := \{s' \mid s \overset{\xi}{\twoheadrightarrow} s'\}$,

- $\mathbf{out}(s) := \{(d, o) \mid o \in \Sigma_O, d \in \Delta, s \overset{(d,o)}{\Longrightarrow}\} \cup \{\delta \mid \delta(s)\}$,
- $\mathbf{out}(S') := \bigcup_{s \in S'} \mathbf{out}(s)$,

- $tstraces(s) := \{\xi \mid s \overset{\xi}{\twoheadrightarrow}\}$, *where* $s' \overset{\delta}{\twoheadrightarrow} s'$ *iff* $\delta(s')$, *and*
- $im \, \mathbf{tioco}_\delta \, sp :\Leftrightarrow \forall \xi \in tstraces(sp) : \mathbf{out}(im \, \mathbf{after} \, \xi) \subseteq \mathbf{out}(sp \, \mathbf{after} \, \xi)$

Example 3. Figure 3 provides a collection of small examples illustrating **tioco$_\delta$**. In Fig. 3a, it holds that $[\![A_0]\!]_S \, \mathbf{tioco}_\delta \, [\![A_1]\!]_S$ as the required inclusion relation holds for all possible **out** sets, for instance, $\mathbf{out}([\![A_0]\!]_S \, \mathbf{after} \, \epsilon) = \{(1, o), (2, o)\} \subseteq \mathbf{out}([\![A_1]\!]_S \, \mathbf{after} \, \epsilon) = \{(1, o), (2, o), (3, o)\}$. Note, that this is also true for output behaviors enabled after 3 time units as $[\![A_0]\!]_S$ does not permit to wait for

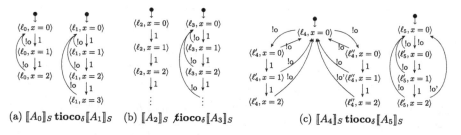

Fig. 3. Examples for **tioco**$_\delta$ on TIOLS

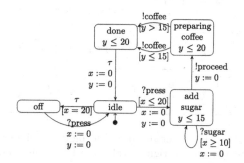

Fig. 4. Example for a candidate implementation $\mathcal{A}'_1$ of $\mathcal{A}_1$ (cf. Fig. 1a)

3 time units, such that the respective **out** set is empty. Hence, **tioco** permits implementations to show less output behavior than the specification allows. Figure 3b depicts a further example where $[\![A_2]\!]_S$ **tioco**$_\delta$ $[\![A_3]\!]_S$ does not hold as **out**($[\![A_2]\!]_S$ **after** ϵ) = $\{\delta\} \not\subseteq$ **out**($[\![A_3]\!]_S$ **after** ϵ) = $\{(1,o),(2,o),\ldots\}$ (i.e., implementation $[\![A_2]\!]_S$ is quiescent but specification $[\![A_3]\!]_S$ is not). The TIOLTS in Fig. 3c illustrates how non-determinism is handled by **tioco**$_\delta$. For specification ($[\![A_5]\!]_S$), it holds that **out**($[\![A_5]\!]_S$ **after** $(0,o)$) = $\{(0,o),(1,o),(2,o),(2,o')\}$, and the same holds for **out**($[\![A_4]\!]_S$ **after** $(0,o)$) and, particularly, $(2,o)$ *and* $(2,o')$. This is due to the fact that in **tioco**$_\delta$ only outputs of those states are considered being reachable via some trace of the specification, but not necessarily of *any* state of the respective TIOLTS. Therefore, it holds that $[\![A_4]\!]_S$ **tioco**$_\delta$ $[\![A_5]\!]_S$.

Next, we apply **tioco**$_\delta$ to our running example to illustrate the differences to **tioco**$_\Delta$.

Example 4. Consider TIOA $\mathcal{A}'_1$ depicted in Fig. 4 to be a candidate implementation of TIOA $\mathcal{A}_1$ in Fig. 1a. First, the guard $y \leq 20$ of the switch labeled with *proceed* is not contained in $\mathcal{A}'_1$, and instead, location *add sugar* has an invariant $y \leq 15$. Considering only this difference, we have $[\![\mathcal{A}'_1]\!]_S$ **tioco**$_\delta$ $[\![\mathcal{A}_1]\!]_S$ as well as $[\![\mathcal{A}'_1]\!]_S$ **tioco**$_\Delta$ $[\![\mathcal{A}_1]\!]_S$ as we forbid output *proceed* for $15 < y \leq 20$ and waiting in *add sugar* for an arbitrary amount of time. In contrast, omitting the switch labeled *!proceed* in $\mathcal{A}'_1$ would lead to a violation of **tioco**$_\delta$ as location *add sugar* would become quiescent (whereas **tioco**$_\Delta$ still holds as it does not

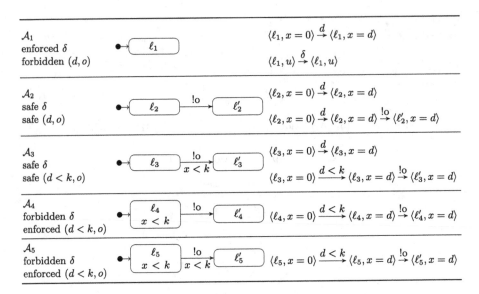

Fig. 5. Examples of allowed/safe, enforced and forbidden actions (TIOA depicted on the left, TIOLTS depicted on the right). Table 1 lists for all combinations of TIOA whether **tioco$_\delta$** holds or not.

Table 1. Combinations of TIOA of Fig. 5 whether **tioco$_\delta$** holds.

im	sp				
	$[\![\mathcal{A}_1]\!]_S$	$[\![\mathcal{A}_2]\!]_S$	$[\![\mathcal{A}_3]\!]_S$	$[\![\mathcal{A}_4]\!]_S$	$[\![\mathcal{A}_5]\!]_S$
$[\![\mathcal{A}_1]\!]_S$	✓	✗	✗	✗	✗
$[\![\mathcal{A}_2]\!]_S$	✗	✓	✗	✗	✗
$[\![\mathcal{A}_3]\!]_S$	✗	✓	✓	✓	✓
$[\![\mathcal{A}_4]\!]_S$	✗	✓	✓	✓	✓
$[\![\mathcal{A}_5]\!]_S$	✗	✓	✓	✓	✓

check for quiescence). Second, the invariant $x \leq 20$ of location *idle* in $\mathcal{A}_1$ is not contained in $\mathcal{A}_1'$ but, instead, becomes a guard to the switch labeled with *?press*. As a result, $[\![\mathcal{A}_1']\!]_S$ **tioco$_\delta$** $[\![\mathcal{A}_1]\!]_S$ still holds as delays in timed runs are only observable by **tioco$_\delta$** if paired with a subsequent output action. In contrast, $[\![\mathcal{A}_1']\!]_S$ **tioco$_\Delta$** $[\![\mathcal{A}_1]\!]_S$ does not hold as in **tioco$_\Delta$** delays of any possible duration are observable, even if no subsequent outputs will ever occur.

Weaknesses of Existing Definitions of Timed Input/Output Conformance. As a result, **tioco$_\Delta$** and **tioco$_\delta$** are incomparable. In addition, observability capabilities required for effectively checking **tioco$_\Delta$** are unrealistic and therefore only of theoretical interest, but infeasible in practice. In contrast, **tioco$_\delta$** is more realistic but fails to guarantee liveness requirements as the notion of quiescence does not properly reflect the differences between allowed and enforced outputs in TIOA specifications. To further illustrate this problem, consider the five TIOA, $\mathcal{A}_1$ to $\mathcal{A}_5$, and their TIOLTS in Fig. 5 (where Table 1 lists for all combinations whether $\textbf{tioco}_\delta$ holds or not). According to Definition 6, location ℓ_1 of $\mathcal{A}_1$ is quiescent, whereas none of the locations ℓ_2 to ℓ_5 of $\mathcal{A}_2$ to $\mathcal{A}_5$ are quiescent as output o is eventually enabled. The table in Fig. 5 shows all possible comparisons of all five TIOA under $\textbf{tioco}_\delta$. Here, the fact that $[\![\mathcal{A}_3]\!]_S$ $\textbf{tioco}_\delta$ $[\![\mathcal{A}_4]\!]_S$ and $[\![\mathcal{A}_3]\!]_S$ $\textbf{tioco}_\delta$ $[\![\mathcal{A}_5]\!]_S$ hold is particularly undesirable (as highlighted in the table): $\mathcal{A}_3$ *may* either produce output o within interval $0 \leq x < k$, or it *may* behave quiescent, whereas $\mathcal{A}_4$ and $\mathcal{A}_5$ *must* produce output o within interval $0 \leq x < k$ and therefore *must not* be quiescent. In contrast, $\mathcal{A}_2$ and $\mathcal{A}_3$ are allowed to be quiescent, by residing for unlimited durations in ℓ_2 and ℓ_3.

We summarize the most important weaknesses of existing versions of **tioco**.

- *(Live Timed Behaviors)* **tioco** either relies on a (unrealistically) strong notion of observability including arbitrary delays, or on a (unnecessarily) weak notion of quiescence not distinguishing allowed from enforced outputs.
- *(Compositionality)* To the best of our knowledge, there only exists one work investigating compositionality properties of **tioco** so far which does not take any notion of quiescence into account [5].
- *(Infinite TIOLTS)* **tioco** is defined on TIOLTS, an infinitely-branching state-transition graph being intractable for realistic testing practices and tools. However, a sound characterization of **tioco** directly on TIOA is also not feasible as timed (suspension) traces are not directly derivable from TIOA.

We next propose an improved version of **tioco** to tackle these weaknesses.

4 Improved Timed Input/Output Conformance

In this section, we tackle the weaknesses of existing versions of **tioco** as described in the previous section.

4.1 Safe vs. Enforced Quiescence

Existing definitions of **tioco** either do not have any notion of quiescence at all [26], or quiescence includes both (1) states that, if no input is provided, will delay forever with no output and (2) states that may eventually produce an output (cf. Fig. 5) [39]. We instead consider two different facets of quiescence: state s is *enforced quiescent* if each run *must* wait in this state for an input for an arbitrary duration to proceed. This coincides with quiescence of $\textbf{tioco}_\delta$. In contrast, state s is *safe quiescent* if a run *may* wait in this state for an input for an arbitrary duration, but *may* also proceed by eventually producing an output. Consequently, state s is *not quiescent*, if a run *must* eventually proceed from this state by producing an output. Hence, s is *live* if it is neither safe quiescent nor enforced quiescent.

Definition 7 (Safe/Enforced Quiescence). *Let* $(S, s_0, \Sigma_I, \Sigma_O, \rightarrow)$ *be a TIOLTS.*

- $s \in S$ *is* safe-quiescent, *denoted* $\delta_S(s)$, *iff* $\forall d \in \Delta : s \overset{d}{\nrightarrow}$.
- $s \in S$ *is* enforced-quiescent, *denoted* $\delta_E(s)$, *iff* $\forall \mu \in \Sigma_O, \forall d \in \Delta : s \overset{(d,\mu)}{\nrightarrow}$.

Intuitively, we may assume enforced-quiescent states to be also safe-quiescent. However, as a counter-example, assume a TIOLTS with one state $\langle \ell, x = 0 \rangle$ (corresponding to a TIOA with one location ℓ and $I(\ell) = x \leq 0$): here, no outputs are possible and no delays are allowed thus obstructing the intuition.

Lemma 1. *Let* $(S, s_0, \Sigma_I, \Sigma_O, \rightarrow)$ *be a TIOLTS. If* $s \in S$ *enables independent progress, then* $\delta_E(s) \Rightarrow \delta_S(s)$.

We add δ_S and δ_E to **out** to distinguish both types of quiescence and adjust *tstraces*, accordingly. This allows us to define *live timed ioco* (**ltioco**$_S$) by extending **tioco**$_\delta$ with outputs δ_S and δ_E. Hence, **ltioco**$_S$ not only guarantees output behaviors of implementation *im* to be *safe* (i.e., allowed to occur within the observed time interval as specified in *sp*), but also requires *im* to be *live* (i.e., to progress with an output within a time interval if enforced by *sp*).

Definition 8. *Let* im, sp *be TIOLTS over* $\Sigma = \Sigma_I \cup \Sigma_O$, $s, s' \in S$, $S' \subseteq S$, $\xi \in (\Delta \times (\Sigma \cup \{\delta_\gamma\}))$.

- s **after** $\xi := \{s' \mid s \overset{\xi}{\twoheadrightarrow} s'\}$,
- **out**$_S(s) := \{(d, o) \mid d \in \Delta, o \in \Sigma_O, s \overset{(d,o)}{\Longrightarrow}\} \cup \{\delta_\gamma \mid \delta_\gamma(s)\}$,
- **out**$_S(S') := \bigcup_{s \in S'} $**out**$_S(s)$,
- *tstraces*$_L(s) := \{\xi \mid s \overset{\xi}{\twoheadrightarrow}\}$, *where* $s' \overset{\delta_\gamma}{\twoheadrightarrow} s'$ *iff* $\delta_\gamma(s')$, *and*
- im **ltioco**$_S$ $sp :\Leftrightarrow \forall \xi \in tstraces_L(sp) : $**out**$_S(im$ **after** $\xi) \subseteq $**out**$_S(sp$ **after** $\xi)$

Obviously, using two different quiescence symbols does not increase complexity of conformance checking as compared to **tioco**$_\delta$ in Definition 6.

Example 5. State $\langle \ell_1, x = 0 \rangle$ in Fig. 5 is quiescent, whereas $\langle \ell_2, x = 0 \rangle$ and $\langle \ell_3, x = 0 \rangle$ are not. With our improved definition, $\langle \ell_1, x = 0 \rangle$ is enforced-quiescent, whereas $\langle \ell_2, x = 0 \rangle$ and $\langle \ell_3, x = 0 \rangle$ are safe-quiescent. States $\langle \ell_4, x = 0 \rangle$ and $\langle \ell_5, x = 0 \rangle$ are neither safe-quiescent nor enforced-quiescent due to the invariants of ℓ_4 and ℓ_5. Hence, **ltioco**$_S$ is now able to reject ℓ_3 as incorrect implementation of ℓ_4 and ℓ_5 as both ℓ_4 and ℓ_5 are not quiescent, whereas ℓ_3 is safe-quiescent. For all other cases, **ltioco**$_S$ yields the same results as listed in Fig. 5.

Lemma 2. **ltioco**$_S$ *is a preorder on the set of input-enabled TIOLTS.*

Furthermore, we can prove that **ltioco**$_S$ is *sound* (i.e., strictly more discriminating) with respect to **tioco**$_\delta$ in the sense that im **ltioco**$_S$ $sp \Rightarrow im$ **tioco**$_\delta$ sp (but not vice versa).

Theorem 1 (Correctness of ltioco$_S$). *Let im and sp be TIOLTS with im being input-enabled and enabling independent progress.*

- *im* ltioco$_S$ *sp* $\Rightarrow$ *im* tioco$_\delta$ *sp*
- *im* ltioco$_S$ *sp* $\Rightarrow$ *im* tioco$_\Delta$ *sp*
- *im* tioco$_\delta$ *sp* $\Rightarrow$ *im* ltioco$_S$ *sp does, in general, not hold.*

Additionally, let sp also be input-enabled.

- *im* ltioco$_S$ *sp* $\Rightarrow$ *traces*^{w}(*im*) $\subseteq$ *traces*^{w}(*sp*)

Note, that *im* tioco$_\Delta$ *sp* $\Rightarrow$ *im* ltioco$_S$ *sp* does not hold as tioco$_\Delta$ has no notion of quiescence, and *im* ltioco$_S$ *sp* $\Rightarrow$ *traces*^{s}(*im*) $\subseteq$ *traces*^{s}(*sp*) does not hold as ltioco$_S$ is limited to observable (weak) steps of timed (suspension) traces.

4.2 Compositionality

For investigating compositionality of ltioco$_S$, we first define parallel composition of TIOA also at the level of TIOLTS.

Definition 9 (TIOLTS Composition). *Let* $(S_j, s_{0_j}, \Sigma_{I_j}, \Sigma_{O_j}, \twoheadrightarrow_j)$ *with* $j \in \{1, 2\}$ *be TIOLTS of composable TIOA. The parallel product is a TIOLTS* $(S_1 \times S_2, (s_{0_1}, s_{0_2}), \Sigma_{I_{1\|2}}, \Sigma_{O_{1\|2}}, \twoheadrightarrow_{1\|2})$, *where* $\Sigma_{I_{1\|2}}$ *and* $\Sigma_{O_{1\|2}}$ *are defined according to Definition 3 and* $\twoheadrightarrow_{1\|2}$ *is the least relation satisfying the rules:*

(1) $(s_1, s_2) \xrightarrow{\sigma}_{1\|2} (s_1', s_2)$ *if* $s_1 \xrightarrow{\sigma}_1 s_1'$, $\sigma \in (\Sigma_1 \setminus \Sigma_2) \cup \{\tau\}$,

(2) $(s_1, s_2) \xrightarrow{\sigma}_{1\|2} (s_1, s_2')$ *if* $s_2' \xrightarrow{\sigma}_2 s_2'$, $\sigma \in (\Sigma_2 \setminus \Sigma_1) \cup \{\tau\}$,

(3) $(s_1, s_2) \xrightarrow{\tau}_{1\|2} (s_1', s_2')$ *if* $s_1 \xrightarrow{\sigma}_1 s_1'$, $s_2' \xrightarrow{\sigma}_2 s_2'$ *and* $\sigma \in (\Sigma_1 \cap \Sigma_2)$, *and*

(4) $(s_1, s_2) \xrightarrow{d}_{1\|2} (s_1', s_2')$ *if* $s_1 \xrightarrow{d}_1 s_1'$, $s_2' \xrightarrow{d}_2 s_2'$ *and* $d \in \Delta$.

Rules (1) and (2) preserve transitions of non-shared (i.e., unsynchronized) actions from both TIOLTS, whereas rule (3) introduces silent transitions for input/output action pairs synchronized between both TIOLTS. Rule (4) preserves (synchronous) delay steps of length d enabled by both TIOLTS. Rule (5) handles inputs leading to the failure state in one of the components, where our notion of composable TIOA ensures that those actions leading to the failure state are not shared. We conclude the following properties.

Lemma 3. *Let* $\mathcal{A}_1$ *and* $\mathcal{A}_2$ *be composable TIOA.*

1. traces($[\![\mathcal{A}_{1\|2}]\!]_S$) = *traces*($[\![\mathcal{A}_1]\!]_S \parallel [\![\mathcal{A}_2]\!]_S$), *and*
2. if $\mathcal{A}_1$ *and* $\mathcal{A}_2$ *are input-enabled and enable independent progress, then this also holds for* $\mathcal{A}_{1\|2}$.

Property (1) ensures parallel composition on TIOA and TIOLTS to commute with respect to timed-traces semantics such that a composed specification can be effectively built from the (finite) TIOA representations of its components. Property (2) ensures that input-enabled and independent-progress enabling TIOA are closed under parallel composition. We now prove compositionality of ltioco$_S$.

Theorem 2. *Let* im_1, im_2, sp_1, *and* sp_2 *be input-enabled and independent progress enabling TIOLTS of composable TIOA. Then it holds that*

$$(im_1 \text{ ltioco}_S \, sp_1) \wedge (im_2 \text{ ltioco}_S \, sp_2) \Rightarrow (im_1 \parallel im_2) \text{ ltioco}_S (sp_1 \parallel sp_2).$$

4.3 Symbolic Live Timed Input/Output Conformance Testing

Concerning the practical intractability of infinitely branching TIOLTS, *zone graphs* have been proposed as *finite* representation of TA semantics [18]. A zone graph $(\mathcal{Z}, \rightsquigarrow)$ of TIOA $\mathcal{A}$ consists of a *transition relation* $\rightsquigarrow$ on a set $\mathcal{Z}$ of *symbolic states* by means of pairs $\langle \ell, \varphi \rangle$ of locations $\ell \in L$ and *zones* $\varphi \in \mathcal{B}(\mathcal{C})$. A zone represents a (potentially infinite) maximum set D of clock valuations satisfying clock constraint φ, where we assume zones in *canonical form* by requiring D to be *closed under entailment* (i.e., φ cannot be strengthened without changing D). We may write D as a synonym for φ and use the notations $D^\uparrow = \{u + d \mid u \in D, d \in \mathbb{T}\}$ and $R(D) = \{[R \mapsto 0]u \mid u \in D\}$. Although zone graphs $(\mathcal{Z}, \rightsquigarrow)$ are, again, not necessarily finite, an equivalent, finite zone graph $(\mathcal{Z}, \rightsquigarrow_k)$ can be obtained with $\rightsquigarrow_k$, (1) by constructing an equivalent *diagonal-free* TA only containing atomic clock constraints of the form $x \sim r$ [10], and (2) by constructing for this TA a *k-bounded* zone graph with all zones being bound by a maximum global *clock ceiling* k using k-normalization [37,38]. Here, the basic idea of k-normalization is to set the value of k to the greatest constant appearing in any clock constraint in the TA. Then, we replace each difference constraint by a difference greater than k (i.e., a difference constraint stating that the difference is greater than k).

As zone-graph constructions from TA ignore switch labels, they are likewise applicable to TIOA. However, in order to lift **ltioco**$_S$ to zone graphs of specifications $\mathcal{A}_{sp}$ and implementations $\mathcal{A}_{im}$ given as TIOA, actions related to TIOA switches (including τ) must be also included as labels for the respective transitions between the corresponding symbolic states. In contrast, symbolic transitions not corresponding to switches of the TIOA are labeled with the special void symbol $\epsilon \notin \Sigma$. We define input/output-labeled zone graph (IOLZG) representations of TIOA as follows.

Definition 10 (IOLZG). *An IOLZG of TIOA $\mathcal{A} = (L, \ell_0, \Sigma_I, \Sigma_O, \rightarrow, I)$ is a tuple $(\mathcal{Z}, z_0, \Sigma_I, \Sigma_O, \rightsquigarrow)$, where*

- $\mathcal{Z} = L \times \mathcal{B}(\mathcal{C})$ *is a set of* symbolic states *with initial state $z_0 = \langle \ell_0, D_0 \rangle \in \mathcal{Z}$,*
- $\Sigma_\tau = \Sigma_I \cup \Sigma_O \cup \{\tau\}$ *is a set of labels, and*
- $\rightsquigarrow \subseteq \mathcal{Z} \times (\Sigma_\tau \cup \{\epsilon\}) \times \mathcal{Z}$ *is a symbolic transition relation being the least relation satisfying the following rules:*
  - $\langle \ell, D \rangle \xrightarrow{\epsilon} \langle \ell, D^\uparrow \wedge I(\ell) \rangle$ *and*
  - $\langle \ell, D \rangle \xrightarrow{\sigma} \langle \ell', R(D \wedge g) \wedge I(\ell') \rangle$ *if $\ell \xrightarrow{g, \sigma, R} \ell'$.*

Let $\langle \ell, D \rangle \in \mathcal{Z}$ be a symbolic state. We further use the following notations.

- $\langle \ell, D \rangle \xrightarrow{d} \langle \ell', R(D \wedge g) \wedge I(\ell') \rangle$ *if $\exists u \in D : u \in g \wedge ([R \mapsto 0](u + d)) \in R(D \wedge g) \wedge I(\ell')$,*
- $\langle \ell, D \rangle \xrightsquigarrow{(d, \sigma)}$ *if $\exists \langle \ell'', D'' \rangle \in \mathcal{Z} : \langle \ell, D \rangle \xrightarrow{d} \langle \ell'', D'' \rangle \xrightarrow{\sigma} \langle \ell', D' \rangle$,*
- $\langle \ell, D \rangle \xrightsquigarrow{(d_1, \sigma_1) \cdots (d_n, \sigma_n)}$ *if $\exists \langle \ell_1, D_1 \rangle, \ldots, \langle \ell_n, D_n \rangle \in \mathcal{Z} : \langle \ell, D \rangle \xrightsquigarrow{(d_1, \sigma_1)} \langle \ell_1, D_1 \rangle \xrightsquigarrow{(d_2, \sigma_2)} \ldots \xrightsquigarrow{(d_n, \sigma_n)} \langle \ell_n, D_n \rangle$ with $n \in \mathbb{N}_0$,*

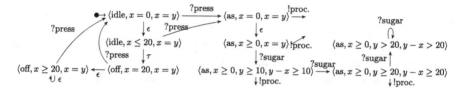

Fig. 6. Example for a k-normalized IOLZG

- $\langle \ell, D \rangle$ is input-enabled *iff* $\forall i \in \Sigma_I, \forall d \in D : \exists \langle \ell', D' \rangle \in \mathcal{Z} : \langle \ell, D \rangle \xrightarrow{(d,i)} \langle \ell', D' \rangle \wedge d \in D'$, and
- $\langle \ell, D \rangle$ *enables* independent progress *iff* $(\forall d \in \Delta : \langle \ell, D \rangle \xrightarrow{d})$ or $\exists d \in \Delta, \exists o \in \Sigma_O : \langle \ell, D \rangle \xrightarrow{d,o}$.

An IOLZG is *input-enabled* and enables *independent progress* if all its state do. Again, we obtain weak steps by replacing $\rightsquigarrow$ by $\Rrightarrow$, where in both relations, ϵ-steps are treated as unobservable. By $[\![\mathcal{A}]\!]^x_{\mathcal{Z}}$, $x \in \{w, s\}$, we refer to the weak/strong IOLZG of TIOA $\mathcal{A}$, again, by possibly omitting x. In fact, k-normalization also applies to IOLZG, where switch labels may cause duplications of transitions but, however, do not affect the set of symbolic states. Hence, the correctness claim for zone graphs of TA (cf. [8]) also holds for IOLZG of TIOA.

Theorem 3. *Let* $s_0 = \langle \ell_0, u_0 \rangle$ *be the initial state of TIOLTS* $[\![\mathcal{A}]\!]_S$ *of TIOA* $\mathcal{A}$ *and* $\langle \ell, \{u_0\} \rangle$ *be the initial state of IOLZG* $[\![\mathcal{A}]\!]_{\mathcal{Z}}$.

- *(Soundness)* $\langle \ell_0, \{u_0\} \rangle \xrightarrow{\xi}_k \langle \ell, D \rangle$ *implies* $\langle \ell_0, u_0 \rangle \xrightarrow{\xi} \langle \ell, u \rangle$ *for all* $u \in D$.
- *(Completeness)* $\langle \ell_0, u_0 \rangle \xrightarrow{\xi} \langle l, u \rangle$ *implies* $\langle \ell_0, \{u_0\} \rangle \xrightarrow{\xi}_k \langle \ell, D \rangle$ *such that* $u \in D$.

Example 6. Figure 6 shows an extract from the (k-normalized) IOLZG of the TIOA in Fig. 1a. Here, $k = 20$ is the largest constant appearing in all clock constraints such that every value of clocks x larger than 20 falls into zone $x > 20$. The initial zone restricts all clock values to 0. Symbolic state $\langle \text{idle}, x \leq 20, x = y \rangle$ comprises all TIOLTS states being in location *idle* as long as $x \leq 20$ holds, and, similarly, for the symbolic states with location *off*. On reaching location *as* (*add sugar*), all clocks are reset. Symbolic state $\langle \text{as}, x \leq 10, y \leq 20, y - x \geq 10 \rangle$ thus aggregates all clock constraints of related TIOLTS runs.

As all TIOLTS states comprised in a symbolic state share the same visible behaviors (up to different clock valuations), IOLZG can be used as a basis for checking **ltioco**$_S$ between respective TIOA. In particular, if a zone of a symbolic state is downward-closed, outputs of that state are enforced as runs may not starve in that state. Correspondingly, we can lift all auxiliary definitions of **ltioco**$_S$ from TIOLTS to IOLZG (marked by index $\mathcal{Z}$). For **out**$_{\mathcal{Z}}$, we have to check for a given symbolic state reached by some *tstrace* whether it is possible to extend the *tstrace* by an output of that symbolic state such that the resulting extended *tstrace* is still a valid *tstrace*. For instance, assume a simple IOLZG with

$\langle \ell_0, x \geq 5 \rangle \overset{lo}{\rightsquigarrow} \langle \ell_1, x < 3 \rangle$: state $\langle \ell_0, x \geq 5 \rangle$ has output o which is only enabled as long as $x < 3$ holds as the state reached by that output is $\langle \ell_1, x < 3 \rangle$. As the set of all valid extensions of *tstraces* by means of pairs of delays and subsequently enabled output actions of one symbolic state is, in general, infinite, they do not provide a reasonable basis for effectively checking **ltioco**$_S$ on zone-graph representations of TIOA. However, a symbolic solution (i.e., comparing the timing constraints for output-action occurrences of symbolic states) is also not feasible for checking **ltioco**$_S$ due to the (generally) unrelated names of locations and clocks of the two different TIOA under consideration. To solve this problem, we instead employ the notion of *spans* [20]: the span of clock c in zone D denotes the minimum time interval containing the minimum and maximum valuations of c enabled in D. We use ∞ to denote upward-open intervals (i.e., $d < \infty$ for all $d \in \mathbb{T}$).

Definition 11 (Span). *Let D be a zone and $c \in C$.*

- span$(c, D) = (lo, up) \in \mathbb{T}_C \times (\mathbb{T}_C \cup \{\infty\})$ *is the minimal interval s.t.* $\forall u \in D : u(c) \geq lo \wedge u(c) \leq up$.
- $(lo, up) \preceq (lo', up') \Leftrightarrow lo \geq lo' \wedge up \leq up'$.
- span$(D) = (lo, up) \Leftrightarrow \forall c \in C: (lo, up) \preceq$ span$(c, D) \wedge \exists c', c'' \in C:$ span$(c', D) = (lo, up') \wedge$ span$(c'', D) = (lo'', up)$.

Given a span $sp = (lo, up)$, we write $d \in sp$ for short if $d \geq lo$ and $d \leq up$ hold. Based on the notion of spans, we are able define *span traces* $(sp_1, \sigma_1), \ldots, (sp_n, \sigma_n)$ as sequences of pairs of spans and action occurrences denoting (maximum) sets of all valid timed traces $(d_1, \sigma_1), \ldots, (d_n, \sigma_n)$ of a given TIOA with equal untimed traces $\sigma_1, \ldots, \sigma_n$ and $d_i \in sp_i$ for $1 \leq i \leq n$.

Example 7. A span trace of the TIOA in Fig. 1a is, for instance, given as $spt = ((20, \infty), \textit{?press}), ((0, 20), \textit{?press}), ((10, \infty), \textit{?sugar})$. This span trace comprises all timed traces that first perform the invisible τ-switch leading to location *off* after exactly 20 time units. The first visible step, performing output action *?press*, then corresponds to the switch leading from location *off* back to location *idle* after at least 20 time units (due to the constraint of the τ-switch). The second occurrence of output action *?press* corresponding to the switch leading from location *idle* to location *add sugar* has to be performed at least 0 and at most 20 time units after the previous step. Afterwards, for the self-switch of location *add sugar* labeled *?sugar* to be enabled, at least 10 time units must elapse.

Please note that the set of valid timed traces of a given untimed trace may not be representable by a single span trace (e.g., in case of non-deterministic TIOA). The minimal, yet complete set of span traces comprising all valid timed traces of a given TIOA $\mathcal{A}$ can be defined with respect to the corresponding IOLZG representation of $\mathcal{A}$ as follows.

Definition 12 (Span Trace). *Let $\mathcal{A} = (L, \ell_0, \Sigma_I, \Sigma_O, \rightarrow, I)$ be a TIOA with IOLZG $(\mathcal{Z}, z_0, \Sigma_I, \Sigma_O, \rightsquigarrow)$. By $\Psi_{\mathcal{Z}}$ we denote the set of span traces of $\mathcal{A}$ being the least set such that $(sp_1, \sigma_1), \ldots, (sp_n, \sigma_n) \in \Psi_{\mathcal{Z}} \Leftrightarrow z_0 \overset{(d_1, \sigma_1) \cdots (d_n, \sigma_n)}{\rightsquigarrow}$, where $d_i \in sp_i, 1 \leq i \leq n$.*

We can show that the set of span traces derived from the IOLZG representation of a TIOA exactly comprises the set of timed traces of the respective TIOLTS representation of the TIOA.

Lemma 4. *Let* $\mathcal{A} = (L, \ell_0, \Sigma_I, \Sigma_O, \rightarrow, I)$ *be a TIOA with TIOLTS* $(S, s_0, \Sigma_I, \Sigma_O, \rightarrow)$. *Then it holds that* $(sp_1, \sigma_1), \ldots, (sp_n, \sigma_n) \in \Psi_{\mathcal{Z}} \Leftrightarrow s_0 \xrightarrow{d_1, \sigma_1} \ldots \xrightarrow{d_n, \sigma_n}$, *where* $d_i \in sp_i, 1 \leq i \leq n$.

Based on this result, we are able to lift *ltioco* from TIOLTS (see Definition 8) to the level of IOLZG and span traces. First, defining the two different notions of quiescence on symbolic states of IOLZG is straightforward. In contrast, the **after**$_{\mathcal{Z}}$ set has now to be redefined in a recursive manner to consecutively traverse span traces ξ instead of timed traces. In particular, the set of symbolic states $\langle \ell, D \rangle$ reachable after ξ is given as the set of symbolic states reachable by all possible sequences of timed steps comprised in ξ. In a similar way, the set of *suspension span traces* (*sptraces*) can be defined for a symbolic state $\langle \ell, D \rangle$ of an IOLZG as the least set of span traces comprising all possible timed traces. Those traces are additionally equipped by special quiescence output symbols δ_E and δ_S to mark occurrences of (enforced or safe) suspension. Thereupon, the **out**$_{\mathcal{Z}}$ set can be defined as the set of all output behaviors (i.e., pairs (sp, o) of spans sp and output actions o including quiescence) being enabled in all symbolic states reachable from state $\langle \ell, D \rangle$ via span trace ξ such that $\xi \cdot (sp, o)$, again, forms a valid span trace. We further define the set $\widehat{\mathbf{out}}_{\mathcal{Z}}(\mathcal{Z}', \xi)$ to contain the **out**$_{\mathcal{Z}}$ sets reachable from sets $\mathcal{Z}'$ of symbolic states via span trace ξ. In case of multiple output behaviors (e.g., (sp, o) and (sp', o)) with equal output actions o, but different spans sp, sp', we implicitly unify overlapping spans by requiring the set $\widehat{\mathbf{out}}_{\mathcal{Z}}(\mathcal{Z}', \xi)$ to be minimal. Finally, we are able to define **ltioco**$_{\mathcal{Z}}$ almost in the usual way, where $\subsetneqq$ is used instead of $\subseteq$ to state that all output behaviors (i.e., sets *spa* of pairs (sp, o) of spans and output actions) of the implementation are subsumed by those of the specification.

Definition 13. *Let* sp, im *be IOLZG over* $\Sigma = \Sigma_I \cup \Sigma_O$, $\gamma \in \{S, E\}$, $\langle \ell, D \rangle \in \mathcal{Z}$, $\mathcal{Z}' \subseteq \mathcal{Z}$, *and* $\xi \in ((\mathbb{T}_C \times (\mathbb{T}_C \cup \{\infty\})) \times (\Sigma \cup \{\delta_\gamma\}))$.

- $\langle \ell, D \rangle$ *is safe-quiescent, denoted by* $\delta_S(\langle \ell, D \rangle)$, *iff* $\forall d \in D : \langle \ell, D \rangle \xrightarrow{d}$,
- $\langle \ell, D \rangle$ *is enforced-quiescent, denoted by* $\delta_E(\langle \ell, D \rangle)$, *iff* $\forall \mu \in \Sigma_O, \forall d \in D :$
  $\langle \ell, D \rangle \not\xrightarrow{(d, \mu)}$,
- $(\langle \ell, D \rangle \, \mathbf{after}_{\mathcal{Z}} \, \xi) \subseteq \mathcal{Z}$ *is the greatest set satisfying the following rules:*
  - $\langle \ell, D \rangle \in (\langle \ell, D \rangle \, \mathbf{after}_{\mathcal{Z}} \, \epsilon)$ *and*
  - $\langle \ell, D \rangle \in (\langle \ell', D' \rangle \, \mathbf{after}_{\mathcal{Z}}(sp, a) \cdot \xi'')$ *if* $\exists d \in sp : \langle \ell', D' \rangle \xrightarrow{(d, a)} \langle \ell'', D'' \rangle \wedge$
    $\langle \ell, D \rangle \in (\langle \ell'', D'' \rangle \, \mathbf{after}_{\mathcal{Z}} \, \xi'')$,
- *sptraces*$(\langle \ell, D \rangle)$ *is the least set s.t.* $(sp_1, \sigma_1), \ldots, (sp_n, \sigma_n) \in sptraces(\langle \ell, D \rangle)$
  $\Leftrightarrow \langle \ell, D \rangle \xrightarrow{(d_1, \sigma_1) \cdots (d_n, \sigma_n)}$ *where* $d_i \in sp_i, 1 \leq i \leq n$, *and* $\forall \langle \ell', D' \rangle \in \mathcal{Z} :$
  $\langle \ell', D' \rangle \xrightarrow{\delta_\gamma} \langle \ell', D' \rangle$ *iff* $\delta_\gamma(\langle \ell', D' \rangle)$,

- $\mathbf{out}_{\mathcal{Z}}(\langle \ell, D\rangle, \xi) \subseteq (\mathbb{T}_C \times (\mathbb{T}_C \cup \{\infty\}) \times (\Sigma_I \cup \Sigma_O \cup \{\delta_\gamma\}))$ *is the greatest set s.t.*
  $(sp, o) \in \mathbf{out}_{\mathcal{Z}}(\langle \ell, D\rangle, \xi)$ *if* $\langle \ell, D\rangle \overset{\xi}{\rightsquigarrow} \wedge \xi \cdot (sp, o) \in sptraces(z_0) \wedge o \in \Sigma_O \cup \{\delta_\gamma\}$,
- $\widehat{\mathbf{out}}_{\mathcal{Z}}(\mathcal{Z}', \xi)$ *is the least set s.t.* $\forall (sp, o) \in \bigcup_{z \in \mathcal{Z}'} \mathbf{out}_{\mathcal{Z}}(z, \xi) : \exists (sp', o) \in \widehat{\mathbf{out}}_{\mathcal{Z}}(\mathcal{Z}', \xi) : sp \preceq sp'$,
- $im\,\mathbf{ltioco}_{\mathcal{Z}}\ sp :\Leftrightarrow \forall \xi \in sptraces(sp) : \widehat{\mathbf{out}}_{\mathcal{Z}}(im\,\mathbf{after}_{\mathcal{Z}}\ \xi, \xi) \subsetneqq \widehat{\mathbf{out}}_{\mathcal{Z}}(sp\,\mathbf{after}_{\mathcal{Z}}$
  $\xi, \xi)$, *where* $spa \subsetneqq spa' \Leftrightarrow \forall (sp, o) \in spa : \exists (sp', o) \in spa' : sp \preceq sp'$

Example 8. Considering the running example in Figs. 1a and 4, we observe that $[\![\mathcal{A}_1]\!]\,\mathbf{ltioco}_{\mathcal{Z}}\,[\![\mathcal{A}'_1]\!]$ does not hold. Let $\xi = ((20, \infty), \textit{?press}), ((0, 20),$ $\textit{?press})$. Then $((0, \infty), \delta_S) \in \widehat{\mathbf{out}}_{\mathcal{Z}}([\![\mathcal{A}_1]\!]_{\mathcal{Z}}\,\mathbf{after}_{\mathcal{Z}}\ \xi, \xi)$ and $((0, \infty), \delta_S) \notin \widehat{\mathbf{out}}_{\mathcal{Z}}$ $([\![\mathcal{A}'_1]\!]_{\mathcal{Z}}\,\mathbf{after}_{\mathcal{Z}}\ \xi, \xi)$ as it is not safe to wait in *add sugar* of $\mathcal{A}'_1$ due to the invariant $y \leq 15$.

Finally, we prove that for any two TIOA $\mathcal{A}_{im}$ and $\mathcal{A}_{sp}$, checking $\mathbf{ltioco}_{\mathcal{Z}}$ on IOLZG is equivalent to checking $\mathbf{ltioco}_S$ on TIOLTS.

Theorem 4 (Correctness of $\mathbf{ltioco}_{\mathcal{Z}}$). *Let* $\mathcal{A}_{im}$ *and* $\mathcal{A}_{sp}$ *be TIOA.*

$$[\![\mathcal{A}_{im}]\!]_{\mathcal{Z}}\,\mathbf{ltioco}_{\mathcal{Z}}\,[\![\mathcal{A}_{sp}]\!]_{\mathcal{Z}} \Leftrightarrow [\![\mathcal{A}_{im}]\!]_S\,\mathbf{ltioco}_S\,[\![\mathcal{A}_{sp}]\!]_S$$

From Theorems 1 and 4 it also follows that $\mathbf{ltioco}_{\mathcal{Z}}$ is sound with respect to $\mathbf{tioco}_\delta$ and from Theorems 2 and 4 it follows that $\mathbf{ltioco}_{\mathcal{Z}}$ is a preorder on input-enabled IOLZG. Finally, we can likewise conclude compositionality of $\mathbf{ltioco}_{\mathcal{Z}}$.

Corollary 1. *Let* im_1 *and* im_2 *as well as* sp_1 *and* sp_2 *be input-enabled and composable TIOA enabling independent progress. Then* $([\![im_1]\!]_{\mathcal{Z}}\,\mathbf{ltioco}_{\mathcal{Z}}\,[\![sp_1]\!]_{\mathcal{Z}}) \wedge$ $([\![im_2]\!]_{\mathcal{Z}}\,\mathbf{ltioco}_{\mathcal{Z}}\,[\![sp_2]\!]_{\mathcal{Z}}) \Rightarrow [\![im_1 \parallel im_2]\!]_{\mathcal{Z}}\,\mathbf{ltioco}_{\mathcal{Z}}\,[\![sp_1 \parallel sp_2]\!]_{\mathcal{Z}}$.

5   Tool Support

To show practical feasibility of our technique, we implemented a tool based on the concepts of the JTORX tool [6,43], originally being developed for (untimed) **ioco** testing. Similar to JTORX, our tool supports *online white-box testing*: a running implementation is investigated on-the-fly whether it is conforming to a specification both given as TIOA. Our tool supports a generic interface enabling it to be used for checking any kind of implementation (in the current version, the interface is implemented to accept TIOA models as implementation). To check conformance of a given implementation to a specification, the tool checks **ltioco**$_{\mathcal{Z}}$ on the labeled zone-graph representations of both TIOA models. As input TIOA models, our tools supports the exchange format of UPPAAL [29] (a mature model checker for timed systems).

Internally, our tool uses *Difference Bound Matrices* (DBM) being an efficient representation of zones [7,8,18]. In particular, DBM-based representations of zones provide comparison operators $\sim \in \{<, \leq, =, \geq, >\}$. For a consistent representation, a fresh clock 0_C (with constant value zero) is introduced resulting

$$
\begin{array}{c}
\quad\quad 0_C \quad\quad x \quad\quad\quad y \\
\begin{array}{c} 0_C \\ x \\ y \end{array}
\begin{pmatrix}
(0,\leq) & (-1,\leq) & (0,\leq) \\
(2,\leq) & (0,\leq) & \infty \\
(2,\leq) & \infty & (0,\leq)
\end{pmatrix}
\end{array}
$$

Fig. 7. Difference bound matrix for the zone $\langle 1 \leq x \leq 2, y \leq 2 \rangle$

in the set of clocks $\mathcal{C}_0 = \mathcal{C} \cup \{0_C\}$ in which each clock is aligned to 0_C. Based on this construction, atomic clock constraints of the form $x \sim r$ can be represented as $x - y \preceq r$ with $\preceq \in \{<, \leq\}$. Hence, every zone $D \in \mathcal{B}(\mathcal{C}_0)$ can be represented with a maximum of $|\mathcal{C}_0|^2$ atomic clock constraints, and therefore, each zone may be described as a matrix of size $|\mathcal{C}_0| \times |\mathcal{C}_0|$ [8]. Each entry $D_{i,j}$ (row i, column j) thus refers to the atomic clock constraint $x_i - x_j \preceq r$. Hence, entries of the matrix are pairs of difference values $r_{i,j}$ and comparison operators in $\preceq$, being derived as follows. For every entry $D_{i,j}$, we set the value $r_{i,j}$ such that $x_i - x_j \preceq r_{i,j}$ holds. If a difference is unbounded (i.e., x_i and x_j are not related by any constraint), we set the value to $r_{i,j} = \infty$. Additionally, we have to require clocks to have non-negative values (i.e., $0_C - x_i \leq 0$).

Example 9. Figure 7 depicts the DBM for the zone $\langle 1 \leq x \leq 2, y \leq 2 \rangle$. For instance, $D_{x,y} = D_{y,x} = \infty$ as x and y are not related by a comparison. Additionally, $D_{0_C,x} = (-1, \leq)$ as $1 \leq x \leq 2$ such that $0 - x \leq -1$. Furthermore, $D_{x,0_C} = 2$ due to $x \leq 2$.

The tool is available online at https://www.es.tu-darmstadt.de/ltioco.

6   Related Work

Several versions of **tioco** have recently been proposed [14,24,26,27,39], whereas **ltioco** is, to the best of our knowledge, the first approach working on the symbolic thus finite zone-graph representation of TIOA instead of infinitely branching TIOLTS. The only other existing symbolic variant of **tioco** is based on *symbolic* timed automata with data variables, but does neither include quiescence nor ensure finiteness of the state space [41]. In addition, our novel notions of timed quiescence are different from any existing approach, where absence of outputs is either considered only up to a fixed bound M [14,24], or for all possible delays [27,39]. Recent tools implementing variants of **tioco** [12,26,28] also mostly differ in their interpretation of quiescence which can all be simulated in our framework, but not vice versa. Moreover, neither of these approaches distinguishes safe from enforced quiescence as done in our approach.

In addition, compositionality properties have only been considered in [5] so far, where again no notion of quiescence is considered. Furthermore, there are techniques for test-generation from TIOA models. In order to handle infinitely branching state spaces, En-Nouaary and Dssouli [19] derive test cases only for a particular subset of TIOA behaviors, whereas, similar to our approach, Brandán Briones and Röhl [15] use a zone-based representation. However, the latter approach is limited to deterministic TA, which are strictly less expressive than our TIOA. Springintveld et al. [40] propose an algorithm for exhaustive black-box test generation for timed systems, but no notions of quiescence are taken into account.

Besides adopting **ioco**-like conformance notions to timed systems as done by the different variants of **tioco**, the only other timed implementation-relation theory we are aware of uses a refinement-based implementation relation [17]. Moreover, Bornot et al. [13] investigate requirements for ensuring liveness-by-construction of timed systems using trace-based composition operators for TIOLTS, whereas conformance theories are out of scope.

Finally, there are several other **ioco**-based testing theories. Among others, **mioco** [32–34] (i.e., **ioco** for modality-based systems) distinguishes optional transition (which may be implemented) from mandatory transitions (which must be implemented). Furthermore, **featured-ioco** [9] is based on so-called featured transition systems, incorporating feature constraints to to restrict which (pairs of) transitions may be part of the same variant. However, none of these approaches considers real-time constraints.

7   Conclusion

We presented an improved version of a timed input/output conformance testing relation, called **ltioco**, to ensure not only safe but also *live* behaviors of implementations with time-critical behaviors modeled as TIOA. Additionally, we investigated compositionality properties of **ltioco** and we extended the construction of zone graphs to check **ltioco** on a finite semantic representation of TIOA. As a future work, we plan to enrich our framework by further operators including quotienting and conjunction as well as refinement [17] and to extend our tool implementation by automated test-generation and test-execution capabilities. Furthermore, we plan to evaluate our approach by applying our tool to a number of well-known case studies (e.g., [21,25,30]).

References

1. Aceto, L., Burgueño, A., Larsen, K.G.: Model checking via reachability testing for timed automata. In: Steffen, B. (ed.) TACAS 1998. LNCS, vol. 1384, pp. 263–280. Springer, Heidelberg (1998). https://doi.org/10.1007/BFb0054177
2. Alur, R., Dill, D.: Automata for modeling real-time systems. In: Paterson, M.S. (ed.) ICALP 1990. LNCS, vol. 443, pp. 322–335. Springer, Heidelberg (1990). https://doi.org/10.1007/BFb0032042
3. Alur, R., Dill, D.L.: A theory of timed automata. Theor. Comput. Sci. **126**(2), 183–235 (1994). https://doi.org/10.1016/0304-3975(94)90010-8
4. André, É.: What's decidable about parametric timed automata? In: Artho, C., Ölveczky, P.C. (eds.) FTSCS 2015. CCIS, vol. 596, pp. 52–68. Springer, Cham (2016). https://doi.org/10.1007/978-3-319-29510-7_3
5. Bannour, B., Gaston, C., Aiguier, M., Lapitre, A.: Results for compositional timed testing. In: APSEC 2013, pp. 559–564. IEEE (2013). https://doi.org/10.1109/APSEC.2013.81

6. Belinfante, A.: JTorX: a tool for on-line model-driven test derivation and execution. In: Esparza, J., Majumdar, R. (eds.) TACAS 2010. LNCS, vol. 6015, pp. 266–270. Springer, Heidelberg (2010). https://doi.org/10.1007/978-3-642-12002-2_21

7. Bellman, R.: Dynamic Programming. Princeton University Press, Princeton (1957)

8. Bengtsson, J., Yi, W.: Timed automata: semantics, algorithms and tools. In: Desel, J., Reisig, W., Rozenberg, G. (eds.) ACPN 2003. LNCS, vol. 3098, pp. 87–124. Springer, Heidelberg (2004). https://doi.org/10.1007/978-3-540-27755-2_3

9. Beohar, H., Mousavi, M.R.: Input–output conformance testing for software product lines. J. Log. Algebraic Methods Program. **85**(6), 1131–1153 (2016). https://doi.org/10.1016/j.jlamp.2016.09.007

10. Bérard, B., Petit, A., Diekert, V., Gastin, P.: Characterization of the expressive power of silent transitions in timed automata. Fundam. Inform. **36**(2,3), 145–182 (1998). https://doi.org/10.3233/FI-1998-36233

11. Bernot, G.: Testing against formal specifications: a theoretical view. In: Abramsky, S., Maibaum, T.S.E. (eds.) TAPSOFT 1991. LNCS, vol. 494, pp. 99–119. Springer, Heidelberg (1991). https://doi.org/10.1007/3540539816_63

12. Bohnenkamp, H., Belinfante, A.: Timed testing with TorX. In: Fitzgerald, J., Hayes, I.J., Tarlecki, A. (eds.) FM 2005. LNCS, vol. 3582, pp. 173–188. Springer, Heidelberg (2005). https://doi.org/10.1007/11526841_13

13. Bornot, S., Gößler, G., Sifakis, J.: On the construction of live timed systems. In: Graf, S., Schwartzbach, M. (eds.) TACAS 2000. LNCS, vol. 1785, pp. 109–126. Springer, Heidelberg (2000). https://doi.org/10.1007/3-540-46419-0_9

14. Briones, L.B., Brinksma, E.: A test generation framework for *quiescent* real-time systems. In: Grabowski, J., Nielsen, B. (eds.) FATES 2004. LNCS, vol. 3395, pp. 64–78. Springer, Heidelberg (2005). https://doi.org/10.1007/978-3-540-31848-4_5

15. Brandán Briones, L., Röhl, M.: Test derivation from timed automata. In: Broy, M., Jonsson, B., Katoen, J.-P., Leucker, M., Pretschner, A. (eds.) Model-Based Testing of Reactive Systems. LNCS, vol. 3472, pp. 201–231. Springer, Heidelberg (2005). https://doi.org/10.1007/11498490_10

16. Broy, M., Jonsson, B., Katoen, J.-P., Leucker, M., Pretschner, A. (eds.): Model-Based Testing of Reactive Systems. LNCS, vol. 3472. Springer, Heidelberg (2005). https://doi.org/10.1007/b137241

17. David, A., Larsen, K.G., Legay, A., Nyman, U., Wasowski, A.: Timed I/O automata: a complete specification theory for real-time systems. In: HSCC 2010, pp. 91–100. ACM (2010). https://doi.org/10.1145/1755952.1755967

18. Dill, D.L.: Timing assumptions and verification of finite-state concurrent systems. In: Sifakis, J. (ed.) CAV 1989. LNCS, vol. 407, pp. 197–212. Springer, Heidelberg (1990). https://doi.org/10.1007/3-540-52148-8_17

19. En-Nouaary, A., Dssouli, R.: A guided method for testing timed input output automata. In: Hogrefe, D., Wiles, A. (eds.) TestCom 2003. LNCS, vol. 2644, pp. 211–225. Springer, Heidelberg (2003). https://doi.org/10.1007/3-540-44830-6_16

20. Guha, S., Narayan, C., Arun-Kumar, S.: On decidability of prebisimulation for timed automata. In: Madhusudan, P., Seshia, S.A. (eds.) CAV 2012. LNCS, vol. 7358, pp. 444–461. Springer, Heidelberg (2012). https://doi.org/10.1007/978-3-642-31424-7_33

21. Havelund, K., Skou, A., Larsen, K.G., Lund, K.: Formal modeling and analysis of an audio/video protocol: an industrial case study using UPPAAL. In: RTSS 1997, pp. 2–13 (1997). https://doi.org/10.1109/REAL.1997.641264

22. Henzinger, T.A., Manna, Z., Pnueli, A.: Temporal proof methodologies for real-time systems. In: Proceedings of the 18th ACM SIGPLAN-SIGACT Symposium on Principles of Programming Languages, POPL 1991, pp. 353–366. ACM (1991). https://doi.org/10.1145/99583.99629

23. Henzinger, T.A., Nicollin, X., Sifakis, J., Yovine, S.: Symbolic model checking for real-time systems. Inf. Comput. **111**(2), 193–244 (1994). https://doi.org/10.1006/inco.1994.1045

24. Hessel, A., Larsen, K.G., Mikucionis, M., Nielsen, B., Pettersson, P., Skou, A.: Testing real-time systems using UPPAAL. In: Hierons, R.M., Bowen, J.P., Harman, M. (eds.) Formal Methods and Testing. LNCS, vol. 4949, pp. 77–117. Springer, Heidelberg (2008). https://doi.org/10.1007/978-3-540-78917-8_3

25. Jensen, H.E., Larsen, K.G., Skou, A.: Modelling and analysis of a collision avoidance protocol using SPIN and UPPAAL. In: DIMACS 1996 (1996)

26. Krichen, M., Tripakis, S.: Black-box conformance testing for real-time systems. In: Graf, S., Mounier, L. (eds.) SPIN 2004. LNCS, vol. 2989, pp. 109–126. Springer, Heidelberg (2004). https://doi.org/10.1007/978-3-540-24732-6_8

27. Larsen, K.G., Mikucionis, M., Nielsen, B.: Online testing of real-time systems using UPPAAL. In: Grabowski, J., Nielsen, B. (eds.) FATES 2004. LNCS, vol. 3395, pp. 79–94. Springer, Heidelberg (2005). https://doi.org/10.1007/978-3-540-31848-4_6

28. Larsen, K.G., Mikucionis, M., Nielsen, B., Skou, A.: Testing real-time embedded software using UPPAAL-TRON: an industrial case study. In: EMSOFT 2005, pp. 299–306. ACM (2005). https://doi.org/10.1145/1086228.1086283

29. Larsen, K.G., Pettersson, P., Yi, W.: UPPAAL in a nutshell. Int. J. Softw. Tools Technol. Transf. **1**(1), 134–152 (1997). https://doi.org/10.1007/s100090050010

30. Lindahl, M., Pettersson, P., Yi, W.: Formal design and analysis of a gear controller. In: Steffen, B. (ed.) TACAS 1998. LNCS, vol. 1384, pp. 281–297. Springer, Heidelberg (1998). https://doi.org/10.1007/BFb0054178

31. Luthmann, L., Göttmann, H., Lochau, M.: Compositional liveness-preserving conformance testing of timed I/O automata. Technical report, arXiv (2019). arXiv:1909.03703

32. Luthmann, L., Mennicke, S., Lochau, M.: Towards an I/O conformance testing theory for software product lines based on modal interface automata. In: FMSPLE 2015, EPTCS, vol. 182, pp. 1–13. arXiv (2015). https://doi.org/10.4204/EPTCS.182.1

33. Luthmann, L., Mennicke, S., Lochau, M.: Compositionality, decompositionality and refinement in input/output conformance testing. In: Kouchnarenko, O., Khosravi, R. (eds.) FACS 2016. LNCS, vol. 10231, pp. 54–72. Springer, Cham (2017). https://doi.org/10.1007/978-3-319-57666-4_5

34. Luthmann, L., Mennicke, S., Lochau, M.: Unifying modal interface theories and compositional input/output conformance testing. Sci. Comput. Program. **172**, 27–47 (2019). https://doi.org/10.1016/j.scico.2018.09.008

35. Lynch, N.A., Attiya, H.: Using mappings to prove timing properties. Distrib. Comput. **6**(2), 121–139 (1992). https://doi.org/10.1007/BF02252683

36. Merritt, M., Modugno, F., Tuttle, M.R.: Time-constrained automata. In: Baeten, J.C.M., Groote, J.F. (eds.) CONCUR 1991. LNCS, vol. 527, pp. 408–423. Springer, Heidelberg (1991). https://doi.org/10.1007/3-540-54430-5_103

37. Pettersson, P.: Modelling and verification of real-time systems using timed automata: theory and practice. Ph.D. thesis (1999)

38. Rokicki, T.G.: Representing and modeling digital circuits. Ph.D. thesis (1994)

39. Schmaltz, J., Tretmans, J.: On conformance testing for timed systems. In: Cassez, F., Jard, C. (eds.) FORMATS 2008. LNCS, vol. 5215, pp. 250–264. Springer, Heidelberg (2008). https://doi.org/10.1007/978-3-540-85778-5_18
40. Springintveld, J., Vaandrager, F.W., D'Argenio, P.R.: Testing timed automata. Theor. Comput. Sci. **254**(1), 225–257 (2001). https://doi.org/10.1016/S0304-3975(99)00134-6
41. von Styp, S., Bohnenkamp, H., Schmaltz, J.: A conformance testing relation for symbolic timed automata. In: Chatterjee, K., Henzinger, T.A. (eds.) FORMATS 2010. LNCS, vol. 6246, pp. 243–255. Springer, Heidelberg (2010). https://doi.org/10.1007/978-3-642-15297-9_19
42. Tretmans, J.: Test generation with inputs, outputs and repetitive quiescence (1996). http://doc.utwente.nl/65463
43. Tretmans, J., Brinksma, E.: TorX: automated model-based testing, pp. 31–43 (2003)

RecordFlux: Formal Message Specification and Generation of Verifiable Binary Parsers

Tobias Reiher[1]([✉]), Alexander Senier[1], Jeronimo Castrillon[2], and Thorsten Strufe[2]

[1] Componolit, Dresden, Germany
{reiher,senier}@componolit.com
[2] TU Dresden, Dresden, Germany
{jeronimo.castrillon,thorsten.strufe}@tu-dresden.de

Abstract. Various vulnerabilities have been found in message parsers of protocol implementations in the past. Even highly sensitive software components like TLS libraries are affected regularly. Resulting issues range from denial-of-service attacks to the extraction of sensitive information. The complexity of protocols and imprecise specifications in natural language are the core reasons for subtle bugs in implementations, which are hard to find. The lack of precise specifications impedes formal verification.

In this paper, we propose a model and a corresponding domain-specific language to formally specify message formats of existing real-world binary protocols. A unique feature of the model is the capability to define invariants, which specify relations and dependencies between message fields. Furthermore, the model allows defining the relation of messages between different protocol layers and thus ensures correct interpretation of payload data. We present a technique to derive verifiable parsers based on the model, generate efficient code for their implementation, and automatically prove the absence of runtime errors. Examples of parser specifications for Ethernet and TLS demonstrate the applicability of our approach.

1 Introduction

Security issues are common in parsers of communication protocol implementations, and new vulnerabilities are found every day. Vulnerabilities caused by incorrect parsing exist on all protocol layers: from physical and network layer protocols like Bluetooth (BlueBorne [21]) over session-layer protocols like TLS (Heartbleed [19]) to application-layer protocols like SMB (EternalBlue [20]). Communication protocols are an increasingly worthwhile attack target, as more and more devices of our everyday life are connected to the Internet. Their reliability is especially important in business-critical, mission-critical and safety-critical software. Software that suddenly stops working is a potential threat to human

© Springer Nature Switzerland AG 2020
F. Arbab and S.-S. Jongmans (Eds.): FACS 2019, LNCS 12018, pp. 170–190, 2020.
https://doi.org/10.1007/978-3-030-40914-2_9

life, be it in case of patients with an artificial heart or drivers steering their vehicles and braking using x-by-wire. While the problem is quite obvious for highly interconnected cars, even medical devices have at least an interface for software updates, which represent an attack surface for potential compromise by targeted attacks. Therefore, appropriate methods are needed to prevent the introduction of critical errors in protocol implementations.

Message formats of existing real-world protocols are often complex, but rarely formally specified. The simple syntax that is commonly used only defines the basic structure of a message. Additional properties, conditions, and relations between fields are just described in English prose. Such descriptions are imprecise and can easily be misunderstood by developers, which leads to implementation bugs. Lack of formal specification also prevents automatic checks and verification of the implementations.

Manual implementation has yielded 'shotgun parsers' that mix parsing and processing of messages, in the past. The consequence have been various critical vulnerabilities [12]. We assert that generating the parsing code from a formal grammar yields more cleanly separated implementations.

A recurrent cause of vulnerabilities is the widespread use of unsafe programming languages, like C++. Rust and other memory-safe languages have been developed to avoid memory corruptions. Using these languages is a clear progress towards security, but it does not prevent all errors at runtime.

Runtime errors like integer overflows or divisions by zero must still be handled explicitly. Negligence of the matter can have devastating effects, as reported for instance in [16]. Formal verification is the only convincing approach towards this end. Data and control flow analyses can prove their absence, and proving specific properties of software components is the only way to guarantee that unexpected errors do not occur at runtime.

In summary it becomes clear that a suitable process for the secure implementation of message parsers is needed. Concluding from the observations above, we pose the following requirements. At its heart, we need a simple, readable, and expressive domain-specific language (DSL) for a data format specification that is suitable for messages of existing real-world protocols. It shall also cover all invariants of the message parts. It is crucial that the generated code has been verified to be free of runtime errors, to enable its use in security-critical and safety-critical applications. To facilitate application in a wide range of areas, the generated code has to meet the performance requirements and resource limitations, even of embedded systems.

In this paper we introduce a generic approach for the specification of message formats and a methodology for creating verifiable parsers. Our main contributions are:

– We propose a DSL and model for the formal specification of message formats of existing real-world binary protocols, which covers all properties and dependencies of message parts by using invariants.
– We introduce a methodology for the automatic generation of parsers, for which the absence of runtime errors can be shown.

– We show the applicability of our approach on TLS 1.3 and the TLS Heartbeat protocol.

The rest of the paper is organized as follows: Sect. 2 gives an overview of related work. Section 3 introduces the model for the specification of messages. The design and implementation of the RecordFlux toolset is described in Sect. 4. The applicability is shown in Sect. 5 for two case studies. Section 6 gives a conclusion and an outlook for the future.

2 Related Work

In this section we describe related work for interface generators and generic parsers.

Interface Generators. Interface generators like ASN.1 [15], XDR [28], or Protocol Buffers [4] are used for the development of programs which communicate with each other using serialized structured data. Although they are used to describe the message formats in various protocols or applications, they are not compatible to each other and lack the generality to specify messages of already existing protocols. Today's commonly used communication protocols are quite complex. Such protocols have grown historically, and therefore contain ambiguous idioms, like overlapping message fields that need to be parsed before their existence is clear. Their representation hence is impossible with the given interface generators.

Generic Parsers. Generic parsers differ in the way how message formats are specified and which properties the generated code achieves.

One class are parser generators with a declarative description of the message structure. PacketTypes [17] and DataScript [8] use a type-based language to describe the layout of data formats. Binpac [26] is a declarative language for analyzing network protocols. GAPA [11] has a BNF-based specification language which matches the syntax commonly used in RFCs. Kaitai Struct [3] is YAML-based language to specify binary data formats.

Another class are parser combinators. They combine several existing parsers that are represented by functions into a single, new parser. Representatives are Hammer [2], a parsing library for binary formats written in C, attoparsec [25], a parser combinator library for Haskell, and nom [13], a parser combinator written in Rust that leverages Rust's strong type system and memory safety. Parser combinators may contain ambiguities in the grammar which are not reported at compile-time. Consequently, parser combinators do not meet our requirements.

Parser generators in contrast can provide means to prevent ambiguities before generating code. Many parser generators, however, use unsafe programming languages like C or C++: Binpac [26], PADS [14], Nail [10]. Even if the automatic nature of code generation alleviates some risk, these solutions are still prone to low-level bugs. Some parsers like GAPA [11] especially focus on safety or use a

memory-safe language. As many errors still have to be handled correctly at runtime, these approaches are not sufficient for highly critical applications. Lastly,
parsers generated for interpreted languages [1,27], which rely on a complex runtime, have limited use for resource-constraint systems.

Summary. In summary, we observe that no current solution offers expressiveness and legibility, combined with an easy venue towards formal verification and
convincingly efficient generated code. None of the analyzed approaches is expressive enough to parse binary messages of existing real-world protocols including
all its properties, ensures absence of runtime errors, and is suitable for embedded
systems at the same time. Its design and implementation hence remains an open
challenge.

3   Modeling and Processing Message Formats

In this section we introduce a methodology for the specification of message formats and subsequent generation of the corresponding verifiable parsers. Using
Ethernet frames as a running example demonstrates several intricacies that
require consideration. We start with a simple variant of an Ethernet frame and
refine this definition iteratively to reach a complete specification. We set out to
define the specification as a linear list of fields, like several previous approaches,
but turn to a graph-based modeling later, to allow for strict specification of
ambiguities in the standards. We then describe the algorithms to generate the
code of the verifiable parsers. Each parser comprises a number of functions to
validate and access the content of a message. We use Isabelle/HOL to formalize
our model and describe the corresponding algorithms[1].

3.1   Example: Ethernet Frame

Figure 1 depicts the basic structure of an Ethernet frame. Several variants of
Ethernet exist, Ethernet II is most commonly used. It consists of two address
fields of 6 bytes each, a Type field of 2 bytes encoding which protocol is encapsulated in the payload, and a variable-length Payload field which comprises the rest
of the message. Both IEEE 802.3 and Ethernet II frames are used in practice,

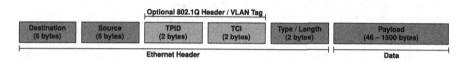

Fig. 1. Ethernet frame structure

---

[1] Isabelle is an automated proof assistant, HOL can be used as a functional programming language that allows proving certain properties. We refer the interested reader
to [24] for an introduction into the matter.

so protocol instances have to distinguish them implicitly on the fly. The payload size is limited to 1500 bytes in Ethernet, so the field following the Source address is interpreted as IEEE 802.3 Length for values below 1500, and Ethernet II Type if the value is 1536 or above. Ethernet also defines extensions, like VLAN tagging (IEEE 802.1Q). It inserts a VLAN tag between the Source address and the Type/Length field, which consists of two fields: the TPID field and the TCI field. To determine if this extension is used, the instance checks the same field for the value 8100_{16}, and interprets the bytes as the TPID, the subsequent two bytes as control information, and only the subsequent bytes as Type or Length.

3.2   Message Representation

To generate a parser and reason about a message format like an Ethernet frame we need a formal specification that captures the message structure and all relevant constraints that must be enforced. The simplest possible representation of a message is a list of fields. For the Ethernet header each field can be represented by an identifier and a fixed length. The protocol may transmit Payload of different sizes, so a variable length value is needed for this part. To make matters worse, the payload length is defined depending on the overall length of the message in Ethernet II, using a mathematical expression. The underlying assumption is that there is a message buffer which comprises a number of bytes and potentially contains the message to parse. We use a deep embedding in our model for the representation (cmp. Appendix A):

datatype $'a$ field $=$ Field $'a$ $'a$ expr

Enumeration types can be used for identifiers, for instance for Ethernet II:

datatype ethernet-v2 $=$ Destination | Source | Type | Payload

The message structure then is described as a list of fields. For the Destination, Source and Type field a fixed number is sufficient to specify the field length, while for the Payload field the length of the header needs to be subtracted from the length of the message. In our deep embedding, the variable *MessageLength* refers to the length of the message buffer. For sake of generality we define all lengths in bits and thereby enable the definition of non-byte-granular fields. An Ethernet II frame is thus defined as follows:

definition ethernet-v2-frame :: ethernet-v2 field list **where**
ethernet-v2-frame $=$ [Field Destination (Num 48), Field Source (Num 48), Field Type (Num 16), Field Payload (Sub MessageLength (Num 112))]

To represent length fields like in the original IEEE 802.3 frame format, we need references to values of other fields in mathematical expressions. Our deep embedding contains the constructor *FieldValue* $'a$ for this purpose. Since our model works on bit granularity, but the Length field is byte-based, the value of the Length field has to be multiplied by 8. The specification for an IEEE 802.3 frame is as follows:

definition *ethernet-frame* :: *ethernet field list* **where**
ethernet-frame = [*Field Destination* (*Num 48*), *Field Source* (*Num 48*), *Field Length* (*Num 16*), *Field Payload* (*Mul* (*FieldValue Length*) (*Num 8*))]

To deal with the parallel use of both Ethernet frames in practice, either both formats could be treated as completely separate and handle their parallel use by other code. We decide to combine both variations in the same model instead. This increases model complexity slightly but prevents the need for manual handling of message variations, which could induce errors. Combining both formats prevents representation as a linear list of fields, as the semantics of the Payload Length depends on the value of the Type/Length field.

We hence have to allow for case distinctions. We described cases by a condition, which refers to the value of Type/Length, and a corresponding length expression for the Payload field. In other words, there are two distinct connections between the Type/Length and the Payload field. When modeling the dependencies between fields such connections can be interpreted as directed edges that define the order of the fields. Such edges contain two attributes: a condition that defines when the target field follows the source field and a length expression that specifies the length of the target field. Consequently, such edges are characterized by a source field, a target field and its attributes. A complete message thus forms a directed-acyclic graph (DAG) where nodes represent fields. Figure 4 shows the graph for the full specification of Ethernet frames.

We define conditions to be handled as boolean expressions, using the same deep embedding as used for mathematical expressions. Expressions must only contain references to field values of preceding nodes. This restriction prevents cyclic dependencies between expressions and ensures that a sequential evaluation of the validity of a message is possible.

Allowing for VLAN tagging further complicates specification. The existence of the 802.1Q header can only be determined by reading the two bytes following the Source address field. To resolve the ambiguity of potential fields we add a virtual node, *Type-Length-TPID* after the *Source*. It is solely used to differentiate the two message formats. It is followed by both *Type-Length* as well as *TPID*. We add a location expression at each edge that defines the position of the first bit of each respective field to be able to deal with the conditional overlay of fields in the model. For the specification of the field location our expressions allow using *FieldFirst* and *FieldLength* to refer to the location and the length of a previous field, respectively. We hence define edges by the following variant type, which is composed of source node, target node, condition, length expression, and location expression.

datatype *'a edge* =
  Edge 'a 'a 'a expr 'a expr 'a expr

Finally, we want to be able to describe the restriction on the payload length, and hence the overall message length, as an invariant. We hence introduce a final node that marks the end of the message, and as it is not followed by any other node it refers to the preceding nodes in the model, only. We introduce an initial node that defines the beginning of a message similarly, to define the length of the first field.

At this point we are able to fully define an Ethernet frame including all its variants. An excerpt of the full specification of Ethernet is shown below. An unabridged version can be found in Appendix B.

> **datatype** *ethernet-node* = *Init* | *Source* | *Destination* | *Type-Length-TPID* | *Type* | *Payload* | *TPID* | *TCI* | *Final*

> **definition** *ethernet-graph* :: *ethernet-node edge list* **where**
> *ethernet-graph* = [
>   *Edge Init Destination True* (*Num 48*) (*Num 0*),
>   *Edge Destination Source True* (*Num 48*) (*Add* (*FieldFirst Destination*) (*FieldLength Destination*)),
>   *Edge Source Type-Length-TPID True* (*Num 16*) (*Add* (*FieldFirst Source*) (*FieldLength Source*)),
>   *Edge Type-Length-TPID Type* (*Ne* (*FieldValue Type-Length-TPID*) (*Num 0x8100*)) (*Num 16*) (*FieldFirst Type-Length-TPID*),
>   *Edge Type-Length-TPID TPID* (*Eq* (*FieldValue Type-Length-TPID*) (*Num 0x8100*)) (*Num 16*) (*FieldFirst Type-Length-TPID*),
>   . . .
>]

We now turn to describing the functions that are generated for the parsers corresponding to our specification.

3.3   Derivation of Validation and Accessor Functions

Parsers are called with a given message as input, and have to extract the content, as specified above. They need to implement validation and accessor functions for each field, which we model as follows. A parser $\mathcal{P}$ consists of a list of validation and accessor functions. The validation function allows checking if all conditions stated in the specification hold for the message field. If this is the case, the corresponding accessor function can safely be used to retrieve the value of the field. For each message field the code for a validation function (*FieldValidFunc*) and the respective accessor function (*FieldAccessFunc*) have to be generated. In this manner all fields of a message can be validated and accessed consecutively.

Several conditions must hold for a bit array to be a valid message. First of all, a message field can only be valid if its first and last bit are within the range of the message buffer. Data out of range indicates an incomplete message. The validity of a message field depends on the specified conditions and the validity of its predecessor. Mapped on the model, this means that one incoming edge must have a valid condition and a valid source (except for the initial node) and, as the conditions of the outgoing edges can constrain the allowed values or the length of the field, the conditions of at least one outgoing edge must be fulfilled. Each path from the initial node to the final node denotes a variant of a message. A whole message is accepted if there is exactly one valid path.

Each node has to be reachable via at least one path from the initial node. The location of a field can vary because of optional or inserted fields and thus depends on a concrete path. Therefore the path has to be known to be able to calculate the location of a field. As conditions can refer to other fields, the path is also needed to evaluate a condition. For this reason, before we can validate or access a field, we have to determine all possible variants of a message, and for each variant the actual conditions and field bounds.

In the following we describe the algorithms for determining path attributes, variant functions, node paths, and field functions. We aim for simplicity in our algorithms. As the parser generation is only done once and the graphs which we use to represent message formats are rather small, the performance of the algorithms is not critical.

Path Attributes. As one of the first steps of the parser generation the *path-attrs* algorithm derives the attributes for all possible paths from the initial node to any node in the graph. All references to other fields in expressions are eliminated during this process. The starting point of the algorithm is a graph definition, where each edge can be uniquely identified by an index number, e.g., by enumerating the edges of the graph definition.

> **type-synonym** $'a\ agraph = (nat \times 'a\ edge)\ list$

The result of the algorithm is a list of tuples. For each path, which is represented by a list of indices, an expression for the condition, the length and the location of the first bit is returned.

The algorithm iterates over all edges of the graph definition. For each edge it determines all paths from the initial node by using the *paths* function. Applying *concat* on the resulting list of lists gives us a list of all paths from the initial node to any other node in the graph. Each path in this list is converted into the corresponding list of edges on the path by the *path-edges* function. From each list of edges the last edge is taken, and for this edge the condition, length and location expression extracted. References to other nodes in these expressions are replaced by the corresponding expression of the referenced node. This is realized by *subs* which recursively looks up the concrete expression in the graph definition.

> **definition** $path\text{-}attrs :: 'a\ agraph \Rightarrow (nat\ list \times 'a\ expr \times 'a\ expr \times 'a\ expr)\ list$ **where**
> $path\text{-}attrs\ graph =$
> $[let\ edges = path\text{-}edges\ graph\ path\ in$
> $(path,\ subs\ edges\ (get\text{-}condition\ (last\ edges)),$
> $\quad subs\ edges\ (get\text{-}length\ (last\ edges)),$
> $\quad subs\ edges\ (get\text{-}first\ (last\ edges)))$
> $.\ path \leftarrow concat\ [paths\ graph\ i\ .\ i \leftarrow map\ fst\ graph]]$

Variant Functions. The parser $\mathcal{P}$ contains a variant validation function *VariantValidFunc* and a variant accessor function *VariantAccessFunc* for each message variant to allow the validation and access of a concrete variant of a field. These variant functions form the building blocks of the field validation functions and the field accessor functions. For each tuple generated by *path-attrs* containing condition, length and location for a specific path, a *VariantValidFunc* and a *VariantAccessFunc* are derived.

The body of a *VariantValidFunc* is based on the condition, a check which ensures that the field is within the bounds of the input buffer, and a call to the validation function of the preceding field, if it is not the first field of the message. Each variant function is identified by a path. As a path is represented by a list of indices, the preceding field can be determined by removing the last element of the current path.[2] Calls to other variant functions are denoted by *VariantValidCall* and *VariantAccessCall*, respectively.

```
fun variant-valid-funcs :: (nat list × 'a expr × 'a expr × 'a expr) list ⇒ 'a func list
where
variant-valid-funcs [] = [] |
variant-valid-funcs ((path, cond, len, first) # xs) =
  VariantValidFunc path (And (if init path ≠ [] then VariantValidCall (init path)
                              else True)
                         (And (Ge BufferLength (Add first len)) cond))
  # variant-valid-funcs xs
```

A *VariantAccessFunc* is defined by the location expression and the length expression of the field.

```
fun variant-access-funcs :: (nat list × 'a expr × 'a expr × 'a expr) list ⇒ 'a func list
where
variant-access-funcs [] = [] |
variant-access-funcs ((path, -, len, first) # xs) =
  VariantAccessFunc path (Value first len) # variant-access-funcs xs
```

The result of *variant-valid-funcs* and *variant-access-funcs* form the variant functions $\mathcal{V}$.

Node Paths. The *node-paths* algorithm determines which paths lead to a field, i.e., which variants of a field exist, and which conditions at outgoing edges a node has. As described in Sect. 3.2 the values of a field can be further restricted by outgoing edges. At least one condition of an outgoing edge has to be fulfilled. Therefore, the corresponding conditions have to be determined as well. Like before, all references to other fields need to be resolved in dependence of a variant.

node-paths iterates over all nodes of the graph. The list of nodes of a graph is provided by *graph-nodes*. For each node all incoming edges are determined by *incoming*. Each incoming edge is used to determine all paths from the initial node by *paths*. For each path it then creates a tuple with two elements: the path represented by a list of indices and a disjunction of all conditions at outgoing edges. The disjunction is created by *any*, which takes a list of conditions from *path-conds*. *path-conds* extracts the conditions of the list of outgoing edges determined by *path-edges* and *outgoing*. Finally, the list of tuples is assigned to the corresponding node identifier.

```
definition node-paths :: 'a agraph ⇒ ('a × (nat list × 'a expr) list) list where
node-paths graph =
  [(node,
    concat [[(path,
              subs (path-edges graph path)
                   (any (path-conds (path-edges graph (outgoing graph node)))))
             . path ← paths graph edge]
            . edge ← incoming graph node])
   . node ← graph-nodes graph {}]
```

---

[2] The *init* function returns a list without its last element.

Field Functions. A *FieldValidFunc* determines if one variant is valid. If a valid variant exists, the *FieldAccessFunc* can be used to return the value of the field. Field functions rely on the functionality provided by variant functions.

The resulting list of *node-paths* is used to generate validation and accessor functions for each field of the message. Each element of this list contains all the necessary information to create a validation and accessor function for one field. A field function is identified by a node identifier.

The algorithm *field-valid-funcs* creates a list of *FieldValidFunc*. In order that a field is valid, a variant of the field and the conditions of one of the outgoing edges must be valid. Hence, the body of a *FieldValidFunc* is a disjunction of calls to all variant validation functions and the corresponding expression which was determined for the conditions at outgoing edges.

> **fun** *field-valid-funcs* :: ($'a \times (nat\ list \times\ 'a\ expr)\ list$) *list* $\Rightarrow\ 'a\ func\ list$ **where**
> *field-valid-funcs* [] = [] |
> *field-valid-funcs* (($path$, $path\text{-}cond$) # xs) =
>   *FieldValidFunc path* (*valid-calls path-cond*) # *field-valid-funcs xs*

The body of each function is determined by *valid-calls*. *valid-calls* iterates over all path-condition tuples which it receives as arguments. For each path it creates a call to a *VariantValidFunc* and combines this call with the corresponding expression derived from the outgoing edges by a conjunction, as a variant is only valid if one of the conditions at the outgoing edges is valid. All created conjunctions are connected by a disjunction, as only one variant has to be valid.

> **fun** *valid-calls* :: ($nat\ list \times\ 'a\ expr$) *list* $\Rightarrow\ 'a\ expr$ **where**
> *valid-calls* [] = *True* |
> *valid-calls* (($path$, $out\text{-}cond$) # []) = *And* (*VariantValidCall path*) *out-cond* |
> *valid-calls* (($path$, $out\text{-}cond$) # xs) = *Or* (*And* (*VariantValidCall path*) *out-cond*)
>   (*valid-calls xs*)

The list of field accessor functions is created by *field-access-funcs*. A *FieldAccessFunc* checks subsequently which variant of a field is valid and calls the corresponding *VariantAccessFunc*.

> **fun** *field-access-funcs* :: ($'a \times (nat\ list \times\ 'a\ expr)\ list$) *list* $\Rightarrow\ 'a\ func\ list$ **where**
> *field-access-funcs* [] = [] |
> *field-access-funcs* (($path$, $path\text{-}cond$) # xs) =
>   *FieldAccessFunc path* (*access-calls path-cond*) # *field-access-funcs xs*

The body of such a function is created by *access-calls*. It iterates over the list of paths and creates a nested if-expression, where the else-branch is created recursively. Each if-expression has a call to a *VariantValidFunc* as condition and a call to the corresponding *VariantAccessFunc* as body.

> **fun** *access-calls* :: ($nat\ list \times\ 'a\ expr$) *list* $\Rightarrow\ 'a\ expr$ **where**
> *access-calls* [] = *Null* |
> *access-calls* (($path$, -) # xs) =
>   *IfThenElse* (*VariantValidCall path*) (*VariantAccessCall path*) (*access-calls xs*)

The result of *field-valid-funcs* and *field-access-funcs* form the field functions $\mathcal{F}$. The parser $\mathcal{P}$ comprises all variant functions $\mathcal{V}$ and field functions $\mathcal{F}$.

3.4   Message Refinement

Communication protocols are typically structured in layers. A protocol message contains a message of a higher layer protocol as its payload. We model this relation by message refinements. A message refinement is a tuple consisting of an identifier of the message, the name of the payload field, an identifier for the contained message and an expression. The expression describes under which conditions a message is contained in the payload field of another message. In the case of Ethernet the expression could specify that an IPv4 packet is contained in the Ethernet frame's payload field, if the Type/Length field has the value 0800_{16}. For each message multiple message refinements can be defined.

4   Implementation

The RecordFlux toolset[3] comprises multiple parts (Fig. 2). The specification language allows describing message formats and the relation of a message field to messages of higher protocol layers. The specification parser transforms this textual description into the model introduced in Sect. 3, which is used by the code generator. We chose SPARK [7] as the target language for code generation, as it already provides simple verification including all required tools. It is supported by the standard GCC toolchain and suitable for resource constrained systems.[4] We hence generate SPARK code, including all necessary function contracts. We then use the SPARK verification toolset to ensure the absence of runtime errors and the functional correctness of the generated code.

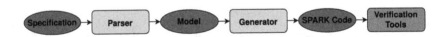

Fig. 2. Architecture

4.1   Specification Language

To specify message formats in a simple and readable manner, we have designed a specification language that allows expressing all properties of a message in accordance to our model. The specification language describes messages based on types. A type definition has the form: **type** NAME **is** DEFINITION;

The language supports two integer types to represent numbers: modular and range integers. A modular type represents the values from zero to one less than

---

[3] RecordFlux is available as open source [5].

[4] We refer the interested reader [18] for an introduction into the language including all of its beneficial properties.

the modulus. The bit size of a modular type is determined by calculating the binary logarithm of the modulus. The destination and source address fields of Ethernet is represented by the following modular integer:

```
type Address is mod 2**48;
```

A range integer allows restricting the range of numbers by bounds. The set of values of a range type consists of all numbers from the lower bound to the upper bound. For a range type the bit size has to be specified explicitly. A range integer can be used for the Type/Length field, and allows incorporating the minimum length restriction of the payload field into the type definition:

```
type Type_Length is range 46 .. 2**16 - 1 with Size => 16;
```

This defines a type with a size of 16 bit which comprises all numbers from 46 to $2^{16} - 1$.

A message format is specified by a message type. A message type is a collection of components. Each component corresponds to one field in a message and is of form: FIELD_NAME : FIELD_TYPE. A simplified specification of an Ethernet II frame is as follows:

```
type Simplified_Frame is
   message
      Destination : Address;
      Source : Address;
      Type_Length : Type_Length;
      Payload : Payload;
   end message;
```

But as argued in Sect. 3 such a simple specification is not sufficient for Ethernet in general.

A then clause following a component allows defining which field follows. If no then clause is given, it is assumed that always the next component of the message follows. If no further component follows, it is assumed that the message ends with this field. A then clause can contain a condition under which the corresponding field follows and aspects which allow defining the length of the next field and the location of its first bit. The condition can refer to previous fields (including the component containing the then clause). In case of Ethernet two then clauses can be added to the Type/Length field to differentiate the two different meanings of this field:

```
Type_Length : Type_Length
   then Payload
      with Length => Type_Length * 8
      if Type_Length <= 1500,
   then Payload
      with Length => Message'Last - Type_Length'Last
      if Type_Length >= 1536;
```

The full specification of an Ethernet frame including VLAN tags is shown in Fig. 3. Figure 4 depicts the corresponding graph representation. The package Ethernet consists of multiple integer types and a message type Frame. Packages are used to structure a specification and thus make the specification modular.

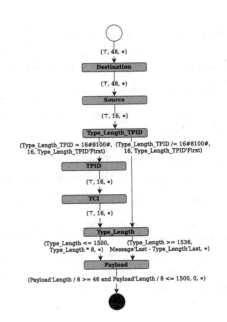

Fig. 3. Full specification of an Ethernet frame covering Ethernet II, IEEE 802.3, and IEEE 802.1Q

Fig. 4. Graph representation of Ethernet frame specification (Notation: For an edge $e = (s, t, c, l, f)$: $*$ denotes $f = s'\text{First} + s'\text{Length}$, $\top$ denotes $c = \text{True}$)

A type refinement describes the relation of a component in a message type to another message type. It states under which condition a specific protocol message is expected inside of a payload field. Only components of the built-in type `Payload` can be refined. Types defined in other packages are referenced by a qualified name in the form `package_name.message_type_name`. The condition can refer to components of the refined type.

```
type IPv4_In_Ethernet is new Ethernet.Frame (Payload => IPv4.Packet)
    if Type_Length = 16#0800#;
```

In this example the relation between an Ethernet frame and an IPv4 packet is specified. The message type `Frame` in package `Ethernet` contains a `Packet` defined in package `IPv4` if the `Type_Length` field of the Ethernet frame equals to `0x0800`.

4.2   Code Generation

The basis for the code generation is the model described in Sect. 3. The generated code takes a plain byte array as input and allows validating and accessing the message data in a structured way. For each specified message a number of functions is generated. The user of the generated code finds a validation function and accessor function for each field of the message. The validity of a field must be checked before accessing its value. This is realized by preconditions. By this means it is ensured that the value of a field is only accessible if all previous fields and the value of the field is valid.

Applying a function to a wrong input buffer or to an incorrect part of the buffer could lead to unexpected results. To prevent this a buffer has to be labeled correctly. This is realized by a predicate used as precondition of all validation and accessor functions. A label is added automatically if the relation between a payload field and a contained message is specified by a type refinement, and a contains function is used to check if the corresponding conditions are fulfilled for the input buffer in question. A contains function is the representation of a type refinement in the generated code. If the input data is received from an external source, the input buffer must be labeled explicitly.

The structure of the specification is reflected in the generated code. As a result it is possible to keep the code as well as the specification modular and extendable. For example type refinements can be defined in a separate specification. This allows adding further higher layer protocols transmitted in an already specified protocol without changing existing code.

4.3   Verification

The SPARK programming language allows the detailed specification of the behavior of software components by the use of contracts. This specification is used by the SPARK verification tools to formally proof that the stated properties of the program hold. The achievable assurance ranges from showing that no runtime exceptions occur to ensuring functional correctness based on a formal specification. This is realized by analyzing the source code and generating verification conditions, which are then passed to multiple theorem provers to formally verify the correctness of the code.

The use of SPARK allows us proving the absence of runtime errors and the correct use of the generated code. All of the generated code is valid SPARK code and will be analyzed by the verification tools. The incorrect use of the generated code, e.g. accessing a field value without prior verification, is prevented by adding appropriate contracts to these functions.

A key benefit of using SPARK is that the code generator need not to be trusted with regard to the absence of runtime errors in the generated code, as this property is proved by the verification tools. Furthermore, the SPARK verification tools assist ensuring the correctness of the specified message format. For example the tools will find potential integer overflows in expressions, which could indicate a missing restriction of the value range of a field.

5   Case Studies

We demonstrate the applicability of RecordFlux using the example of TLS in two case studies. In the first study we replaced the whole parsing code of an existing TLS library by an implementation specified and generated by RecordFlux and analyzed its impact. In the second study we used RecordFlux to specify and parse messages of the TLS Heartbeat protocol, an optional extension of the TLS standard, which is not supported by the library used in the first study.

5.1 Verified TLS Parser

Fizz [22] is a TLS 1.3 implementation developed and used by Facebook, written in C++. As a proof of concept we have replaced the C++ parser by verified SPARK code. Therefore, we used our specification language (see Sect. 4.1) to specify the messages of TLS 1.3, as standardized in RFC8446. Based on this specification, RecordFlux generated SPARK code for TLS Record messages, TLS Handshake messages and TLS extensions. We integrated the existing C++ code with the generated SPARK code manually. The glue code mainly performs the conversion between C++-specific structures like vectors and SPARK-compatible data formats.

Security. Parsing of protocol messages is a sensitive part of a TLS implementation. [9] reports an integer overflow in Fizz. An exploit could have left the application using Fizz in an infinite loop, just by sending a short sequence of messages with a well-chosen value in the length field of a TLS Record message. Facebook fixed the bug by choosing a bigger integer type (size_t instead of uint16). RecordFlux checks the length field for allowed values before continuing the parsing, which we argue is a better solution to the problem. The SPARK verification tools can then prove the absence of unexpected integer overflows.

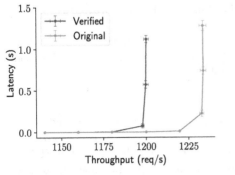

Fig. 5. Performance at Handshake layer; separate TLS handshake for each request

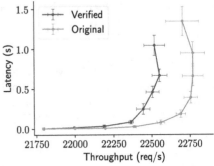

Fig. 6. Performance at Record layer; one TLS handshake for multiple requests

Performance. Performance is considered at least of equal importance as security in practice. We hence evaluated the performance impact of replacing the original message parser with the code generated by RecordFlux. For that purpose we used a modified version of wrk2 [6], where we added the possibility to run in two different modes. To measure the impact of RecordFlux on the TLS handshake, the first mode of wrk2 creates a TLS connection for each HTTP request. The impact of RecordFlux during data transmission is measured by the second mode of wrk2 that only creates one TLS connection before sending requests. wrk2 sends requests in a constant rate and measures the resulting throughput and latency of the responses. The sending rate of wrk2 is increased iteratively until the throughput is not improving any more. For each sending rate the mean values of 40 measurements with a duration of 60 s each are calculated. To minimize the impact of network hardware on the results we run Fizz and wrk2 on the same machine.

We expected to see some performance impact due to the additional validation checks in the generated code and the conversions between C++ and SPARK structures. The diagrams in Figs. 5 and 6 show the resulting mean values of throughput and latency and the corresponding 95 % confidence intervals. The maximum throughput is around 2.7 % lower in the Handshake layer and 1.1 % lower in the Record layer compared to the original parser. An analysis of the CPU cycles used by both variants with Valgrind [23] showed that the majority of additional cycles are spent on memory allocations and processing of data conversions in the glue code.

The results show that there is no significant performance degradation. We conclude that the approach is generally applicable, although mixing existing C++ code with SPARK code is not ideal from the point of view of performance.

5.2   TLS Heartbeat

The Heartbeat extension adds keep-alive functionality to TLS. It gained inglorious prominence by Heartbleed [19], a security vulnerability in the OpenSSL library that affected millions of devices. Heartbleed allowed to extract sensitive data from a TLS endpoint due to an improper input validation.

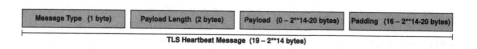

Fig. 7. Message format of a TLS Heartbeat

Both sides of a TLS connection can request the use of the Heartbeat protocol during the TLS handshake. If accepted by the other side, the initiator is allowed to periodically send Heartbeat requests during the lifetime of the TLS connection. Each request contains payload of arbitrary length and content. The receiver of a Heartbeat request must send a response back which contains the same payload as the request. The format of a TLS Heartbeat message is shown in

Fig. 7. The corresponding RecordFlux specification of a TLS Heartbeat message is as follows:

```
package TLS_Heartbeat is

   type Message_Type is (HEARTBEAT_REQUEST => 1, HEARTBEAT_RESPONSE => 2) with
      Size => 8;
   type Length is range 0 .. 2**14 - 20 with Size => 16;

   type Heartbeat_Message is
      message
         Message_Type : Message_Type;
         Payload_Length : Length
            then Payload with Length = Payload_Length * 8;
         Payload : Payload
            then Padding with Length = Message'Last - Payload'Last;
         Padding : Payload
            then null if Message'Length <= 2**14 * 8 and Padding'Length >= 16 * 8;
      end message;

end TLS_Heartbeat;
```

Heartbeat_Message represents a TLS Heartbeat Message. Such a message consists of four fields: The Message_Type field specifies the type of the message. It is represented by a enumeration type with a size of 1 byte and comprises two valid values: 1 for a request and 2 for a response. All other values are considered invalid. Payload_Length defines the length of the following Payload field. The Payload field contains the content of the message, and Padding comprises the rest of the message.

The lengths of the Payload and the Padding field is explicitly defined by length expressions. The whole message is restricted to a length of 2^{14} bytes. The Padding field must be at least 16 bytes long.

The following excerpt shows some of the generated SPARK subprogram declarations for the a Heartbeat_Message:

```
function Is_Contained (Buffer : Bytes) return Boolean with Ghost, Import;

procedure Label (Buffer : Bytes) with Ghost, Post => Is_Contained (Buffer);

function Valid_Message_Type (Buffer : Bytes) return Boolean
   with Pre => Is_Contained (Buffer);

function Get_Message_Type (Buffer : Bytes) return Message_Type
   with Pre => (Is_Contained (Buffer) and then Valid_Message_Type (Buffer));

function Valid_Payload (Buffer : Bytes) return Boolean
   with Pre => Is_Contained (Buffer);

procedure Get_Payload (Buffer : Bytes; First : out Natural; Last : out Natural)
   with Pre => (Is_Contained (Buffer) and then Valid_Payload (Buffer)),
        Post => (First = Get_Payload_First (Buffer) and then
                 Last = Get_Payload_Last (Buffer));

function Is_Valid (Buffer : Bytes) return Boolean
   with Pre => Is_Contained (Buffer);
```

The Is_Contained function which is a precondition for all validation and accessor functions ensures that always the correct message buffer is used. The Is_Contained predicate is set automatically for all defined message refinements. If a message buffer is received from an external source it can be explicitly labeled with the `Label` function. For each message field a validation function (prefixed with `Valid`) and an accessor function (prefixed with `Get`) is created. Each accessor function has the corresponding validation function as a precondition. This ensures that the validity is always checked before accessing a field value. The `Is_Valid` checks the validity of the whole message. It returns `True` if the input buffer contains one valid message variant.

The following code example shows how the generated code is used:

```
with IO;
with TLS.Heartbeat_Message; use TLS.Heartbeat_Message;

procedure Main is
   Buffer : Bytes := IO.Read;
   Tag    : Message_Type;
   First  : Natural;
   Last   : Natural;
begin
   Label (Buffer);
   if Is_Valid (Buffer) then
      Tag := Get_Message_Type (Buffer);
      Get_Payload (Buffer, First, Last);
      Process_Payload (Buffer (First .. Last));
   end if;
end Main;
```

The message buffer is read from an external source. In this example the message buffer is explicitly labeled as a buffer which should contain a TLS Heartbeat message. By also specifying the Record layer with RecordFlux and defining a message refinement between Record message and Heartbeat message, no labeling would be needed. The validity of content of `Buffer` is checked by `Is_Valid`. Alternatively a user could also check the validity of each field on his own. As the `Padding` must be ignored and the relation between the `Payload_Length` field and the `Payload` field is internally known, a user only needs to access the `Message_Type` and the `Payload`.

The SPARK verification tools ensure the correct use of the generated code. If a user would not check the validity of the input buffer, the tools will find this mistake when proving the correctness of the code. `Get_Message_Type` and `Get_Payload` will be flagged with `precondition might fail`.

While the Heartbeat protocol appears quite simple, a flawed implementation can have serious implications. Heartbleed allowed to send a request with a high length value while sending just a short payload and padding. On the receiver side the length value was not checked against the actual received payload. This led to a buffer overflow, so that not only the payload of the request was sent back, but also data following the message buffer.

RecordFlux enforces that the length of a payload field is always defined by a length expression. The code generator adds checks to ensure that the value of a length field is in the allowed range and the message to parse is long enough, and so prevents the issue seen in Heartbleed. Even if RecordFlux would erroneously miss adding a necessary check, the SPARK verification tools will find the potential buffer-overflow before faulty code is used involuntarily.

6 Conclusion and Outlook

We have created a methodology for specification of message formats of communication protocols and the automatic generation of a parser. Based on this methodology we have created a practical implementation, which comprises a DSL for describing message formats and code generator that creates SPARK code, for which the absence of runtime errors can be shown. The generated code is applicable in real-world applications as demonstrated for TLS 1.3, which proved to suffer only from minor performance penalties despite its proven security.

So far we only handled the parsing of messages, but this is only one part of a protocol. In the future we will also look into the protocol logic. We aim to extend the current methodology to get a full formal specification of a protocol, to generate provable secure code.

Acknowledgements. This work is partially funded by the European Union (EU), the European Social Fund (ESF), tax money on the basis of the budget approved by the members of the parliament of Saxony, and the Cluster of Excellence EXC 2050 "Centre for Tactile Internet with Human-in-the-Loop" (CeTI).

A   Deep Embedding

```
datatype 'a expr =
Num int |
Add 'a expr 'a expr |
Mul 'a expr 'a expr |
Sub 'a expr 'a expr |
Div 'a expr 'a expr |
Value 'a expr 'a expr |
FieldValue 'a |
FieldLength 'a |
FieldFirst 'a |
MessageLength |
MessageFirst |
MessageLast |
BufferLength |
VariantValidCall nat list |
VariantAccessCall nat list |
Null |
True |
And 'a expr 'a expr |
Or 'a expr 'a expr |
Eq 'a expr 'a expr |
Ne 'a expr 'a expr |
Lt 'a expr 'a expr |
Le 'a expr 'a expr |
Gt 'a expr 'a expr |
Ge 'a expr 'a expr |
IfThenElse 'a expr 'a expr 'a expr
```

B   Formal Specification of Ethernet Frame

```
definition ethernet-graph :: ethernet-node edge list where
ethernet-graph = [
  Edge Init Destination True (Num 48) (Num 0),
    Edge Destination Source True (Num 48) (Add (FieldFirst Destination) (FieldLength
Destination)),
```

Edge Source Type-Length-TPID True (Num 16) (Add (FieldFirst Source) (FieldLength Source)),
    Edge Type-Length-TPID Type (Ne (FieldValue Type-Length-TPID) (Num 0x8100)) (Num 16) (FieldFirst Type-Length-TPID),
    Edge Type-Length-TPID TPID (Eq (FieldValue Type-Length-TPID) (Num 0x8100)) (Num 16) (FieldFirst Type-Length-TPID),
    Edge TPID TCI True (Num 16) (Add (FieldFirst TPID) (FieldLength TPID)),
    Edge TCI Type True (Num 16) (Add (FieldFirst TCI) (FieldLength TCI)),
    Edge Type Payload (Le (FieldValue Type) (Num 1500)) (Mul (FieldValue Type) (Num 8)) (Add (FieldFirst Type) (FieldLength Type)),
    Edge Type Payload (Ge (FieldValue Type) (Num 1536)) (Add (Sub MessageLast (Add (FieldFirst Type) (FieldLength Type))) (Num 1)) (Add (FieldFirst Type) (FieldLength Type)),
    Edge Payload Final (And (Ge (FieldLength Payload) (Num 368)) (Le (FieldLength Payload) (Num 12000))) (Num 0) (Add (FieldFirst Payload) (FieldLength Payload))
]

References

1. Construct: Declarative data structures for python that allow symmetric parsing and building (2019). https://github.com/construct/construct
2. Hammer (2019). https://github.com/UpstandingHackers/hammer
3. Kaitai Struct: declarative language to generate binary data parsers (2019). https://github.com/kaitai-io/kaitai_struct
4. Protocol Buffers (2019). https://developers.google.com/protocol-buffers/
5. RecordFlux (2019). https://github.com/Componolit/RecordFlux
6. wrk2: A constant throughput, correct latency recording variant of wrk (2019). https://github.com/treiher/wrk2
7. SPARK (2019). https://www.adacore.com/about-spark
8. Back, G.: DataScript - a specification and scripting language for binary data. In: Batory, D., Consel, C., Taha, W. (eds.) GPCE 2002. LNCS, vol. 2487, pp. 66–77. Springer, Heidelberg (2002). https://doi.org/10.1007/3-540-45821-2_4
9. Backhouse, K.: Facebook fizz integer overflow vulnerability (CVE-2019-3560) (2019). https://blog.semmle.com/facebook-fizz-CVE-2019-3560/
10. Bangert, J., Zeldovich, N.: Nail: a practical tool for parsing and generating data formats. In: 11th USENIX Symposium on Operating Systems Design and Implementation (OSDI 14), pp. 615–628. USENIX Association, Broomfield (2014). https://www.usenix.org/conference/osdi14/technical-sessions/presentation/bangert
11. Borisov, N., Brumley, D., Wang, H.J., Dunagan, J., Joshi, P., Guo, C.: Generic application-level protocol analyzer and its language. In: NDSS (2007)
12. Bratus, S., Patterson, M.L., Hirsch, D.: From shotgun parsers to more secure stacks. Shmoocon, Nov (2013)
13. Couprie, G.: Nom, a byte oriented, streaming, zero copy, parser combinators library in rust. In: 2015 IEEE Security and Privacy Workshops, pp. 142–148. IEEE (2015)
14. Daly, M., et al.: PADS: an end-to-end system for processing ad hoc data. In: Proceedings of the 2006 ACM SIGMOD International Conference on Management of Data - SIGMOD 2006, p. 727. ACM Press, Chicago (2006). https://doi.org/10.1145/1142473.1142568. http://portal.acm.org/citation.cfm?doid=1142473.1142568
15. ITU-T: Recommendation X.680 - Abstract Syntax Notation One (ASN.1): Specification of basic notation (2015)
16. Lions, J.L., et al.: Flight 501 failure. Report by the Inquiry Board (1996)

17. McCann, P.J., Chandra, S.: Packet types: abstract specification of network protocol messages. In: Proceedings of the Conference on Applications, Technologies, Architectures, and Protocols for Computer Communication, SIGCOMM 2000, pp. 321–333. ACM, New York (2000). https://doi.org/10.1145/347059.347563. http://doi.acm.org/10.1145/347059.347563

18. McCormick, J.W., Chapin, P.C.: Building High Integrity Applications with SPARK. Cambridge University Press, Cambridge (2015)

19. MITRE: CVE-2014-0160. https://cve.mitre.org/cgi-bin/cvename.cgi?name=CVE-2014-0160

20. MITRE: CVE-2017-0144. https://cve.mitre.org/cgi-bin/cvename.cgi?name=CVE-2017-0144

21. MITRE: CVE-2017-14315. https://cve.mitre.org/cgi-bin/cvename.cgi?name=CVE-2017-14315

22. Nekritz, K., Iyengar, S., Guzman, A.: Deploying TLS 1.3 at scale with Fizz, a performant open source TLS library (2018). https://code.fb.com/security/fizz/

23. Nethercote, N., Seward, J.: Valgrind: a framework for heavyweight dynamic binary instrumentation. In: ACM SIGPLAN Notices, vol. 42, pp. 89–100. ACM (2007)

24. Nipkow, T., Paulson, L.C., Wenzel, M.: Isabelle/HOL: A Proof Assistant for Higher-Order Logic, vol. 2283. Springer, Heidelberg (2002). https://doi.org/10.1007/3-540-45949-9

25. O'Sullivan, B.: Attoparsec (2019). https://github.com/bos/attoparsec

26. Pang, R., Paxson, V., Sommer, R., Peterson, L.: Binpac: a Yacc for writing application protocol parsers. In: Proceedings of the 6th ACM SIGCOMM Conference on Internet Measurement, IMC 2006, pp. 289–300. ACM, New York (2006). https://doi.org/10.1145/1177080.1177119. http://doi.acm.org/10.1145/1177080.1177119

27. Rodriguez, A., McGrath, R.E.: Daffodil: A New DFDL Parser (2010)

28. Srinivasan, R.: XDR: external data representation standard. Technical report (1995)

State Identification for Labeled Transition Systems with Inputs and Outputs

Petra van den Bos[✉] and Frits Vaandrager

Institute for Computing and Information Sciences, Radboud University,
Nijmegen, The Netherlands
{petra,f.vaandrager}@cs.ru.nl

Abstract. For Finite State Machines (FSMs) a rich testing theory has been developed to discover aspects of their behavior and ensure their correct functioning. Although this theory is widely used, e.g., to check conformance of protocol implementations, its applicability is limited by restrictions of the FSM framework: the fact that inputs and outputs alternate in an FSM, and outputs are fully determined by the previous input and state. Labeled Transition Systems with inputs and outputs (LTSs), as studied in ioco testing theory, provide a richer framework for testing component oriented systems, but lack the algorithms for test generation from FSM theory.

In this article, we propose an algorithm for the fundamental problem of *state identification* during testing of LTSs. Our algorithm is a direct generalization of the well-known algorithm for computing adaptive distinguishing sequences for FSMs proposed by Lee & Yannakakis. Our algorithm has to deal with so-called *compatible* states, states that cannot be distinguished in case of an adversarial system-under-test. Analogous to the result of Lee & Yannakakis, we prove that if an (adaptive) test exists that distinguishes all pairs of incompatible states of an LTS, our algorithm will find one. In practice, such adaptive tests typically do not exist. However, in experiments with an implementation of our algorithm on an industrial benchmark, we find that tests produced by our algorithm still distinguish more than 99% of the incompatible state pairs.

1 Introduction

Starting with Moore's famous 1956 paper [17], a rich theory of testing finite-state machines (FSMs) has been developed to discover aspects of their behavior and ensure their correct functioning; see e.g. [13] for a survey. One of the classical testing problems is *state identification*: given some FSM, determine in which state it was initialized, by providing inputs and observing outputs.

Various forms of *distinguishing sequences* were proposed, ranging from sets of sequences to single sequences solving the problem. Moreover, when combined

---

Funded by the Netherlands Organisation of Scientific Research (NWO) under project 13859: Supersizing Model-Based testing (SUMBAT).

F. Arbab and S.-S. Jongmans (Eds.): FACS 2019, LNCS 12018, pp. 191–212, 2020.
https://doi.org/10.1007/978-3-030-40914-2_10

with state access sequences, so called n-complete test suites can be constructed [9]. The challenge in using n-complete test suites is to keep their size as small as possible. Using a single (adaptive) sequence for state identification [12], helps to reach this objective. If such a single sequence does not exist, then a distinguishing sequence distinguishing most states may be supplemented with some additional distinguishing sequences that distinguish the remaining states [16].

Although state identification algorithms for FSMs have been widely used, e.g., to check conformance of protocol implementations, their applicability is limited by the expressivity of the FSM framework. In FSMs, inputs and outputs strictly alternate, outputs are fully determined by the previous input and state, and inputs must be enabled in every state. Labeled Transition Systems with inputs and outputs (LTSs), as studied in ioco testing theory [25], provide a richer framework for testing component oriented systems: transitions are labeled by either an input or an output, allowing any combination of inputs and outputs, multiple outputs may be starting from the same state, allowing (observable) output nondeterminism, and states do not need to have transitions for all inputs, allowing partiality. However, LTSs lack the algorithms for test generation from FSM theory. Although progress has been made in defining and constructing n-complete test suites for LTSs [5], an algorithm to solve the state identification problem as in [12], and hence to provide slim n-complete test suites, is missing.

Therefore we generalize the construction algorithms for adaptive distinguishing sequences, as given in [12]. As in [5], we have to face the problem of compatible states [19,21], which does not occur for FSMs. States are *compatible* when they cannot be distinguished in case of an adversarial system-under-test, e.g. when two states have a transition for the same output to the same state. As it is easy to construct LTSs with compatible states, we made sure our algorithms can deal with such LTSs: they accept LTSs with compatible states, but they 'work around' them, dealing with all incompatible states.

The outline of the paper is as follows. We first introduce graphs, LTSs, and some syntax for denoting trees. Then we elaborate on compatibility and the related concept of validity. Furthermore, we introduce test cases, and define when they distinguish states of an LTS. After that we define a data structure called *splitting graph*, present an algorithm that constructs a splitting graph for a given LTS, and another algorithm that extracts a test case from a splitting graph. We show that, unlike for FSMs, the splitting graph may have an exponential number of nodes. However, this is worst case behaviour, as our experiments on an industrial case study will show. Analogous to FSMs, it may not be possible to distinguish all states of an LTS with a single test case. Our experiments show that this is typically the case in practice, but nevertheless more than 99% of the incompatible state pairs are distinguished by the constructed test case. Following [12], we show that our algorithms constructs a test case distinguishing all incompatible state pairs, if it exists.

Related Work. There are (at least) three orthogonal ways in which the classical FSM (or Mealy machine) model can be generalized.

A first generalization is to add nondeterminism. Whereas an FSM has exactly one outgoing transition for each state q and input i, a *nondeterministic FSM* allows for more than one transition. Alur, Courcoubetis and Yannakakis [1] propose an algorithm to generate adaptive distinguishing sequences for nondeterministic FSMs, using (overlapping) subsets of states, similar to our algorithm. However, their sequences only distinguish pairs of states, and are not designed to distinguish more states at the same time. In between FSMs and nondeterministic FSMs we find the *observable* FSMs, which have at most one outgoing transition for each state q, input i and output o; one may use a determinization construction to convert any nondeterministic FSM into an observable one. The LTSs that we consider have observable nondeterminism.

A second generalization of FSMs is to relax the requirement that each input is enabled in each state. In a *partial FSM*, states do not necessarily have outgoing transitions for every state and every input. Petrenko and Yevtushenko [18] derive complete test suites for partial, observable FSMs, which is the closest to the automata model that we study in this paper. Their test generation is based on (adaptive) state counting [11], which is a trace search-based method which recognizes when states are distinguished, but does not provide a constructive way to build a test that distinguishes (many) states at once. Yannakakis and Lee [27] present a randomized algorithm which generates, with high probability, checking sequences, i.e., n-complete test suites consisting of a single sequence. This approach is also applicable to partial FSMs, as opposed to the adaptive distinguishing sequence construction algorithms of [12], which apply to plain FSMs.

A third generalization of FSMs is to relax the requirement that inputs and outputs alternate. In our LTS, inputs and outputs may occur in arbitrary order. Bensalem, Krichen and Tripakis [20] give an algorithm for extracting adaptive distinguishing sequences for all states of a given LTS, by translating back and forth between a corresponding Mealy machine. This translation is only possible, if all states of the LTS have at most one outgoing output transition. Van den Bos, Janssen and Moerman [5] do not need such a restriction. They propose an algorithm that generates an adaptive distinguishing sequence for all pairs of incompatible states. In this paper, we generalize the result of [5] to distinguish more states at the same time.

Due to page limits, all proofs have been omitted. They can be found in the full version [4].

2   Preliminaries

We write $f : X \rightharpoonup Y$ to denote that f is a partial function from X to Y. We write $f(x) \downarrow$ to mean $\exists y : f(x) = y$, i.e., the result is defined, and $f(x) \uparrow$ if the result is undefined. We often identify a partial function f with the set of pairs $\{(x, y) \in X \times Y \mid f(x) = y\}$.

If Σ is a set of symbols then Σ^* denotes the set of all finite words over Σ. The empty word is denoted by ϵ, the word consisting of symbol $a \in \Sigma$ is denoted a, and concatenation of words is denoted by juxtaposition.

Throughout this article, we use standard notations and terminology related to finite directed graphs (digraphs) and finite directed acyclic graphs (DAGs), as for instance defined in [2,7]. If $G = (V, E)$ is a digraph and $v \in V$, then we let $Post_G(v)$, or briefly $Post(v)$, denote the set of direct successors of v, that is, $Post(v) = \{w \in V \mid (v, w) \in E\}$. Similarly, $Pre_G(v)$, or briefly $Pre(v)$, denotes the set of direct predecessors of v, that is, $Pre(v) = \{w \in V \mid (w, v) \in E\}$. Vertex v is called a *root* if $Pre(v) = \emptyset$, a *leaf* if $Post(v) = \emptyset$, and *internal* if $Post(v) \neq \emptyset$. We write $leaves(G) = \{v \in V \mid Post(v) = \emptyset\}$, and $internal(G) = V \backslash leaves(G)$.

The automata considered this paper are deterministic, finite labeled transition systems with transitions that are labeled by inputs or outputs. Since a single state may have outgoing transitions labeled with different outputs, and since outputs are not controllable, the behavior of our automata is nondeterministic: in general, for a given sequence of inputs, the resulting sequence of outputs is not uniquely determined. Nevertheless, our automata are deterministic in the sense of classical automata theory: for any observed sequence of inputs and outputs the resulting state is uniquely determined. We say that our automata have *observable nondeterminism*.

Because the inputs and outputs will be fixed throughout this article, we fix I and O as nonempty, disjoint, finite sets of input and output labels, respectively, and write $L = I \cup O$. We will use a, b to denote input labels, x, y, z to denote output labels, and μ for labels that are either inputs or outputs.

Definition 1. *An* automaton (with inputs and outputs) *is a triple* $A = (Q, T, q_0)$ *with* Q *a finite set of* states, $T : Q \times L \rightharpoonup Q$ *a transition function, and* $q_0 \in Q$ *the* initial state. *We associate a digraph to A as follows*

$$digraph(A) = (Q, \{(q, q') \mid \exists \mu \in L : T(q, \mu) = q'\}).$$

Concepts and notations for $digraph(A)$ extend to A. Thus we say, for instance, that automaton A is acyclic when $digraph(A)$ is acyclic, and we write $Post(q)$ for the set of direct successors of a state q. For $A = (Q, T, q_0)$ and $q \in Q$ we write A/q for (Q, T, q), that is, the automaton obtained from A by replacing the initial state by q.

Figure 1 shows an example automaton. Below, we recall the definitions of some basic operations on (sets of) automata states. Operations *in*, *out* and *init* retrieve all the inputs, outputs, or labels enabled in a state, respectively. To every set of states P and every sequence of labels σ we can associate three sets of states: P *after* σ, P *before* σ, and *enabled*(P, σ). The set P *after* σ comprises all states that can be reached starting from a state of P via a path with trace σ, whereas the set P *before* σ consists of all the states from where it is possible to reach a state in P via a trace in σ, and *enabled*(P, σ) consists of all states in P from where

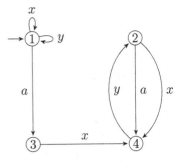

Fig. 1. Running example

a path with trace σ is possible. The *traces* operation provides the sequences of labels that can be observed from one or more of the states. We use a subscript if confusion may arise due to the use of several automata in the same context, e.g. $out_A(q)$ denotes the enabled outputs of q in automaton A.

Definition 2. *Let* $A = (Q, T, q_0)$ *be an automaton,* $q \in Q$, $\mu \in L$ *and* $\sigma \in L^*$. *Then we define:*

$$in(q) = \{a \in I \mid T(q, a) \downarrow\}$$
$$out(q) = \{x \in O \mid T(q, x) \downarrow\}$$
$$q \text{ after } \epsilon = \{q\}$$
$$q \text{ after } \mu\sigma = \begin{cases} T(q, \mu) \text{ after } \sigma & \text{if } T(q, \mu) \downarrow \\ \emptyset & \text{otherwise} \end{cases}$$

$$enabled(q, \sigma) = \begin{cases} \emptyset & \text{if } q \text{ after } \sigma = \emptyset \\ \{q\} & \text{otherwise} \end{cases}$$
$$q \text{ before } \sigma = \{q' \in Q \mid q \in q' \text{ after } \sigma\}$$
$$A \text{ after } \sigma = q_0 \text{ after } \sigma$$
$$traces(q) = \{\rho \in L^* \mid q \text{ after } \rho \neq \emptyset\}$$

Definitions are lifted to sets of states by pointwise extension. Thus, for $P \subseteq Q$, $in(P) = \bigcup_{p \in P} in(p)$, $P \text{ after } \sigma = \bigcup_{p \in P} p \text{ after } \sigma$, *etc. We sometimes write the automaton, instead of the singleton set containing the initial state.*

We find it convenient to use a fragment of Milner's Calculus of Communicating Systems [15] as syntax for denoting acyclic automata. In particular, its recursive definition will allow us to incrementally construct test cases in Sects. 5 and 6.

Definition 3. *The set of expressions* E_{CCS} *is defined by the BNF grammar*

$$F := \mathbf{0} \mid F + F \mid \mu.F$$

The set $T_{CCS} \subseteq E_{CCS} \times L \times E_{CCS}$ *is the smallest set of triples such that, for all* $\mu \in L$ *and* $F, F', G \in E_{CCS}$,

1. $(\mu.F, \mu, F) \in T_{CCS}$
2. *If* $(F, \mu, G) \in T_{CCS}$ *then* $(F + F', \mu, G) \in T_{CCS}$
3. *If* $(F, \mu, G) \in T_{CCS}$ *then* $(F' + F, \mu, G) \in T_{CCS}$

An expression $F \in E_{CCS}$ *is deterministic iff, for all subexpressions* G *of* F,

$$(G, \mu, G') \in T_{CCS} \wedge (G, \mu, G'') \in T_{CCS} \Rightarrow G' = G''$$

To each deterministic expression $F \in E_{CCS}$ *we associate an automaton* $A_F = (Q, T, F)$, *where* Q *is the set of subexpressions of* F, *and transition function* T *is defined by*

$$T(G, \mu) = \begin{cases} G' & \text{if } (G, \mu, G') \in T_{CCS} \\ undefined & otherwise \end{cases}$$

Example 4. The CCS expression $a.(x.\mathbf{0} + y.\mathbf{0})$ has subexpressions $a.(x.\mathbf{0} + y.\mathbf{0})$, $x.\mathbf{0} + y.\mathbf{0}$, $x.\mathbf{0}$, $y.\mathbf{0}$, and $\mathbf{0}$. These are the states of its associated automaton. The automaton's transition relation is: $\{(a.(x.\mathbf{0} + y.\mathbf{0}), a, x.\mathbf{0} + y.\mathbf{0}), (x.\mathbf{0} + y.\mathbf{0}, x, \mathbf{0}), (x.\mathbf{0} + y.\mathbf{0}, y, \mathbf{0})(x.\mathbf{0}, x, \mathbf{0}), (y.\mathbf{0}, y, \mathbf{0})\}$. Note that states $x.\mathbf{0}$ and $y.\mathbf{0}$ are not reachable from initial state $a.(x.\mathbf{0} + y.\mathbf{0})$.

Suspension automata are automata with the additional property that in each state at least one output label is enabled. We note that this requirement can be easily enforced by adding a self-loop for an additional output label, that denotes 'no-output' or *quiescence* [25], in each state that has no output transition. We note that our definition of suspension automata, which is taken from [5], is more general than the one from [25, 26], since we only require states to be non-blocking, while suspension automata from [25, 26] adhere to some additional properties associated to this special quiescence output.

Definition 5. *Let* $A = (Q, T, q_0)$ *be an automaton. We call a state* $q \in Q$ *blocking if* $out(q) = \emptyset$, *and call* A *non-blocking if none of its states is blocking. A non-blocking automaton is also called a* suspension automaton.

We will use suspension automata as the specifications to derive test cases from. Figure 1 shows a suspension automaton. Plain automata are sometimes used as an intermediate structure to do computations, and test cases will be acyclic automata adhering to some additional properties.

3   Validity and Compatibility

In this section, we recall the definitions of the related notions of validity and compatibility [5]. We first give an efficient algorithm for computing valid states. After that, we show how the relation between validity and compatibility can be used to efficiently compute all pairs of compatible states occurring in a suspension automaton. We will need this last relation when constructing test cases to distinguish incompatible states.

3.1   Validity

We consider the following 2-player concurrent game, which is a minor variant of reachability games studied e.g., in [6, 14]. Two players, the tester and the System Under Test (SUT), play on a state space consisting of an automaton $A = (Q, T, q_0)$. At any point during the game there is a *current state*, which is q_0 initially. To advance the game, both the tester and the SUT choose an action from the current state q:

- The tester chooses either an input from $in(q)$, or the special action $\theta \notin L$. By choosing θ, the tester indicates that she performs no input and allows the SUT to execute any output he wishes.
- The SUT chooses an output from $out(q)$, or θ if no output is possible.

The game moves to a next state according to the following rule (this is the input-eager assumption from [6]): If the tester chooses an enabled input a this will be executed, i.e., the current state changes to $T(q, a)$; if the SUT chooses an enabled output x this will only be executed when the tester has chosen θ, in this case the current state changes to $T(q, x)$; when both players choose θ, the game terminates. The tester wins the game if she reaches a blocking state, and the SUT wins if he has a strategy that ensures that the tester will never win. A (memoryless) strategy for the tester is a function $move : Q \rightarrow I \cup \{\theta\}$. We say a strategy is *winning* if the tester will always win the game (within a finite number of moves) when selecting actions according to this strategy, no matter which actions the SUT takes. Following Beneš et al. [3] and Van den Bos et al. [5], we call states for which the tester has a winning strategy *invalid*, and the remaining states in Q *valid*. The sets of valid and invalid states are characterized by the following lemma (cf Proposition 2.18 of [14]):

Lemma 6. *Let* $A = (Q, T, q_0)$ *be an automaton.*

1. *The set of invalid states of A is the smallest set $P \subseteq Q$ such that $q \in P$ if*

$$\exists a \in in(q) : T(q, a) \in P \text{ or } \forall x \in out(q) : T(q, x) \in P.$$

2. *The set of valid states of A is the largest set $P \subseteq Q$ such that $q \in P$ implies*

$$\forall a \in in(q) : T(q, a) \in P \text{ and } \exists x \in out(q) : T(q, x) \in P.$$

Based on Lemma 6(1), Algorithm 1 computes the set of invalid states of an automaton A and, for each invalid state q, the first move $move(q)$ of a winning strategy for the tester, as well as the maximum number $level(q)$ of moves required to win the game. Algorithm 1 is a minor variation of the classical algorithm for computing attractor sets and traps in 2-player concurrent games [14] and the procedure described by Beneš et al. [3], which takes as input an automaton, of which each state q has $in(q) = L_I$, and prunes away invalid states. Key invariants of the while-loop of lines 13–33 are that states in $W \cup P$ are invalid, and for $q \in Q \backslash (P \cup W)$, $count(q)$ gives the number of output transitions to states in $Q \backslash P$.

Let n be the number of states in Q, and m the number of transitions in T. We assume, for convenience, that $m \geq n$. If we use an adjacency-list representation of A and represent the set of incoming transitions using a linked list, then the time complexity of the initialization part (lines 2–11) is $\mathcal{O}(m)$. The while-loop (lines 13–33) visits each transition of A at most twice (in lines 15 and 26) and performs a constant amount of work. Thus the time complexity of the while loop is $\mathcal{O}(m)$. This means that the time complexity of Algorithm 1 is also $\mathcal{O}(m)$.

3.2 Compatibility

Two states of a suspension automaton are *compatible* [19,21] if a tester may not be able to distinguish them in the presence of an adversarial SUT. For example, if the tester wants to determine whether the SUT behaves according to state 2 or 3 of the suspension automaton of Fig. 1, taking output transition x will result in reaching state 4, from both states, but after reaching state 4, it cannot be determined, from which of the two states the x transition was taken. Hence, states 2 and 3 are compatible.

Input: An automaton $A = (Q, T, q_0)$.

Output: The subset $P \subseteq Q$ of invalid states and, for each state $q \in P$, the first move $move(q)$ from a winning stragegy for the tester and the maximum number $level(q)$ of moves required to win.

```
1  Function ComputeWinningTester (Q, T, q0):
2      W := ∅ ;            // winning states for tester that need processing
3      foreach q ∈ Q do
4          count(q) :=  | out(q) |;
5          incomingtransitions(q) := set of incoming transitions of q;
6          if count(q) = 0 then
7              W := W ∪ {q} ;                        // state q is invalid
8              move(q) := θ;
9              level(q) := 0
10         end
11     end
12     P := ∅ ;        // winning states for tester that have been processed
13     while W ≠ ∅ do
14         p := any element from W;
15         foreach (q, μ, p) ∈ incomingtransitions(p) do
16             if q ∉ P ∪ W then
17                 if μ ∈ I then
18                     W := W ∪ {q} ; // state q has input to winning state
19                     move(q) := μ;
20                     level(q) := level(p) + 1
21                 else
22                     count(q) := count(q) − 1;
23                     if count(q) = 0 then
24                         W := W ∪ {q} ; // all outputs q to winning states
25                         move(q) := θ;
26                         level(q) := 1 + max_{x∈out(q)} level(T(q, x))
27                     end
28                 end
29             end
30         end
31         W := W \ {p};
32         P := P ∪ {p}
33     end
34     return set P, function move, and function level;
```

Algorithm 1. Computing the invalid states.

Definition 7. *Let (Q, T, q_0) be a suspension automaton. A relation $R \subseteq Q \times Q$ is a* compatibility relation *if for all $(q, q') \in R$ we have*

$$\forall a \in in(q) \cap in(q') : (T(q, a), T(q', a)) \in R, and$$
$$\exists x \in out(q) \cap out(q') : (T(q, x), T(q', x)) \in R$$

Two states $q, q' \in Q$ are compatible, *denoted $q \Diamond q'$, if there exists a compatibility relation R relating q and q'. Otherwise, the states are* incompatible, *denoted by $q \,\not\!\!\Diamond\, q'$. For $P \subseteq Q$ a set of states, we write $\Diamond(P)$ to denote that all states in P are pairwise compatible, i.e., $\forall q, q' \in P : q \Diamond q'$.*

We note that the compatibility relation is symmetric and reflexive, but not transitive. For an elaborate discussion of compatibility, we refer the reader to [5]. The notions of compatibility and validity can be related using the following synchronous composition operator:

Definition 8. *Let $A_1 = (Q_1, T_1, q_0^1)$ and $A_2 = (Q_2, T_2, q_0^2)$ be automata. The* synchronous composition *of A_1 and A_2, notation $A_1 \| A_2$, is the automaton $A = (Q_1 \times Q_2, T, (q_0^1, q_0^2))$, where transition function T is given by:*

$$T((q_1, q_2), \mu) = \begin{cases} (T_1(q_1, \mu), T_2(q_2, \mu)) & \text{if } T(q_1, \mu) \downarrow \text{ and } T(q_2, \mu) \downarrow \\ undefined & otherwise \end{cases}$$

The next lemma asserts that states q and q' are compatible precisely when the pair (q, q') is a valid state of S composed with itself.[1]

Lemma 9. *Let $S = (Q, T, q_0)$ be a suspension automaton with $q, q' \in Q$. Then $q \Diamond q'$ if and only if (q, q') is a valid state of $S \| S$.*

Example 10. Figure 2 shows the synchronization of the suspension automaton of Fig. 1. It has 6 valid states, and in particular it shows that $2 \Diamond 3$.

Lemma 9 suggests an efficient algorithm for computing compatibility of states. Suppose S is a suspension automaton with n states and m transitions, with $m \geq n$. Then we may compute composition $S \| S$ in time $\mathcal{O}(m(n + \log m))$. The idea is that we first sort the list of transitions on the value of their action label, which takes $\mathcal{O}(m \log m)$ time. Next we check for each transition $t = (q, \mu, q')$ what are the possible transitions that may synchronize with t. Since t may only synchronize with μ-transitions, and since there are at most n μ-transitions (as S is deterministic), we may compute the list of transitions of the composition in $\mathcal{O}(mn)$ time. Thus, the overall time complexity of computing $S \| S$ is $\mathcal{O}(m(n + \log m))$. The composition $S \| S$ has n^2 states and $\mathcal{O}(mn)$ transitions. Next we use Algorithm 1 to compute the set of invalid states of $S \| S$, which requires $\mathcal{O}(mn)$ time. Two states q and q' of S are compatible iff (q, q') is not in this set. Altogether, we need $\mathcal{O}(m(n + \log m))$ time to compute the compatible state pairs.

4 Test Cases

In this section, we introduce a simple notion of *test cases*. The goal of these test cases is *state identification*, i.e., to explore whether a state of the SUT,

---

[1] This is a variation of Lemma 22 from [5], which is stated for a slightly different composition operator that involves demonic completions. Adding demonic completions is useful in the setting of [5], but not needed for our purposes.

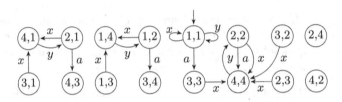

Fig. 2. Synchronous composition of the suspension automaton from Fig. 1.

that is reached after some initial interactions, has the same traces as the state where it should be according to a given suspension automaton. Our test cases are adaptive in the sense that inputs that are sent to the SUT may depend on previous outputs generated by the SUT. They are similar to the adaptive distinguishing sequences of Lee and Yannakakis [12], except that inputs and outputs do not necessarily alternate, and the graph structure is a DAG rather than a tree.

Definition 11. *A test case is an acyclic automaton $A = (Q, T, q_0)$ such that each state $q \in Q$ enables either a single input action, or zero or more output actions. We refer to states that enable a single input as* input states, *and states that enable at least one output as* output states. *Thus each state from a test case is either an input state, an output state, or a leaf.*

To each test case A we associate a set of *observations*: maximal traces that we may observe during a run of A.

Definition 12. *For each test case A, $Obs(A)$ is the set of traces that reach a leaf of A: $Obs(A) = \{\sigma \in traces(A) \mid A \text{ after } \sigma \subseteq leaves(A)\}$.*

Given a suspension automaton S, we only want to consider test cases A that are consistent with S in the sense that each input that is provided by A is also specified by S, and conversely each output that is allowed by S also occurs in A.

Definition 13. *Let $A = (Q, T, q_0)$ be a test case and $S = (Q', T', q'_0)$ a suspension automaton. We say that A is a test case for S if, for each state (q, q') of $A\|S$ reachable from the initial state (q_0, q'_0):*

- *if q is an input state then $in(q) \subseteq in(q')$,*
- *if q is an output state then $out(q') \subseteq out(q)$.*

We say A is a test case for state $q' \in Q'$ *if A is a test case for S/q'. Furthermore, A is a test case for* a set of states $P \subseteq Q'$ *if A is a test case for all $q' \in P$.*

If A is a test case for a suspension automaton S then the composition $A\|S$ is also a test case. We can view $A\|S$ as the subautomaton of A in which all outputs that are not enabled in S have been pruned away. A test case distinguishes two states, if the states enable different observable traces of the test case.

Lemma 14. *If A is a test case for a suspension automaton S, then the composition $A\|S$ is also a test case for S, satisfying $Obs(A\|S) \subseteq Obs(A)$.*

Definition 15. *Let A be a test case for states q and q' of suspension automaton S. Then A distinguishes q and q' if $Obs(A\|(S/q)) \cap Obs(A\|(S/q')) = \emptyset$.*

Example 16. The associated automaton of the CCS expression $a.(x.\mathbf{0}+y.\mathbf{0})$ (see Example 4) is a test case for states 1 and 2 of the suspension automaton from Fig. 1. Its observable traces are $\{ax, ay\}$, and it distinguishes states 1 and 2.

Lemma 17. *Let $S = (Q, T, q_0)$ be a suspension automaton with $q, q' \in Q$. Then $q \not\sim q'$ iff there exists a test case that distinguishes q and q'.*

The following definition generalizes the notion of adaptive distinguishing sequence for FSM's [10,12] to the setting of suspension automata.

Definition 18. *Let $S = (Q, T, q_0)$ be a suspension automaton, $P \subseteq Q$, and A a test case for P. We say that A is an adaptive distinguishing graph for P if, for all $q, q' \in P$ with $q \not\sim q'$, A distinguishes q and q'. Test case A is an adaptive distinguishing graph for S if it is an adaptive distinguishing graph for the set Q of states of S.*

Just like there are FSMs without an adaptive distinguishing sequence, there are suspension automata for which no adaptive distinguishing graph exists. This is the case for the suspension automaton from Fig. 3. We cannot construct an adaptive distinguishing graph by choosing the root node to be an output state, since states 1 and 3 cannot be distinguished, as they both go to state 2 with their single output transition y. The root also cannot be an input state for either of all inputs a or b. After a, states 1 and 2 both reach state 1, and after b, states 2 and 3 both reach state 3.

In the remainder of this paper, we present algorithms for constructing an adaptive distinguishing graph for S from a suspension automaton S, if it exists.

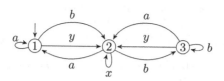

Fig. 3. A suspension automaton without adaptive distinguishing graph.

5   Splitting Graphs

In this section, we present the concept of a splitting graph, as well as an algorithm for constructing such a graph. Our algorithm generalizes the algorithm of Lee and Yannakakis [12] for computing a splitting tree for an FSM. In the next section, we will construct an adaptive distinguishing graph by extracting its parts from the splitting graph. An adaptive distinguishing graph that distinguishes all incompatible state pairs, is only guaranteed to be found, if some additional requirements on the splitting graph construction are satisfied. We will delay the discussion of adaptive distinguishing graphs to the next section, and focus on splitting graphs first.

We will first give the definition of a splitting graph, and the outer loop of our algorithm for constructing it. Then we define when a leaf node of a splitting graph is splittable (i.e., when child nodes can be added), and show that a splittable leaf exists whenever some leaf contains incompatible states. After that, we explain how to construct the child nodes for splittable leaves.

5.1   Splitting Graph Definition

A splitting graph for suspension automaton $S = (Q, T, q_0)$ is a directed graph in which the vertices are subsets of states of S; there is a single root Q, and an internal node is the union of its children. We require that, for each edge (v, c) of the splitting graph, c is a proper subset of v; this implies that a splitting graph is a DAG. We associate a test case $W(v)$ to each internal node v and require a tight link between the observations of $W(v)$ and the children of v: each observation σ has one child c that contains all states enabling σ. As we have $|c| < |v|$, this means that, after following any trace σ from test case $W(v)$, the states $v \backslash c$ have been distinguished from states from the states c.

Definition 19. *A splitting graph for suspension automaton $S = (Q, T, q_0)$ is a triple $Y = (V, E, W)$ with*

- *$Q \in V \subseteq \mathcal{P}(Q) \backslash \emptyset$*
- *$E \subseteq V \times V$ such that*
  1. *Q is the only root of Y,*
  2. *$(v, w) \in E \implies v \supset w$, and*
  3. *$v \in internal(Y) \implies v = \bigcup Post(v)$.*
- *$W : internal(Y) \to E_{CCS}$ is a witness function such that, for all internal vertices v, $A_{W(v)}$ is a test case such that:*
  $\forall \sigma \in Obs(A_{W(v)}), \exists c \in Post(v) : enabled(c, \sigma) = enabled(v, \sigma)$.

Splitting graph Y is complete if, for each leaf l, the states contained in l are pairwise compatible, i.e., $\Diamond(l)$.

Algorithm 2 shows the main loop for constructing a splitting graph for a given suspension automaton. The idea is to start with the trivial splitting graph with just a single node, and then repeatedly split leaf nodes, i.e., add child nodes, until all leaves only contain pairwise compatible states. This means that incompatible states are in different leaves when the algorithm terminates. Since nodes in a splitting graph are finite sets of states, and children are strict subsets of their parents, Algorithm 2 terminates after a finite number of refinements. With $\perp$, we denote the empty function.

Input: A suspension automaton $S = (Q, T, q_0)$
Output: A complete splitting graph for S

1 $Y := (\{Q\}, \emptyset, \bot)$;
2 **while** $\exists l \in leaves(Y) : \neg\Diamond(l)$ **do**
3      $Y :=$**splitnode** (S, Y);
4 **end**
5 **return** Y;

Algorithm 2. Constructing a splitting graph.

5.2   Splitting Conditions

Before we elaborate on the algorithm for the method `splitnode`, we first explore what conditions should hold for a leaf l to be splittable. The formal definition of these conditions is given below in Definition 21.

If we are lucky we can find, for each output $x \in out(l)$, a state $q \in l$ that does not enable x. In this case, observing an output allows us to distinguish at least one state from some other states. Otherwise, we may check whether, for certain enabled inputs, or all outputs, the states of l have a transition to the states of an internal node, i.e., a node that has already been split, because l then may be split as well when these labels occur. In particular, the states of l can be split for some label μ if the reached node is a *least common ancestor* of l *after* μ. An internal node v is least common ancestor for a set of states P if it contains P but none of its children does.

Definition 20. *Let Y be a splitting graph for suspension automaton S and let P be a set of states of S. An internal node v of Y is a* least common ancestor *of P if $P \subseteq v$ and, for all $c \in Post(v)$, $P \not\subseteq c$. We write $LCA(Y, P)$ for the set of least common ancestors of P contained in Y.*

Note that we can compute the set of least common ancestors for any set P in a time that is linear in the size of the splitting graph.

Definition 21. *Let Y be a splitting graph for suspension automaton S.*

1. *A leaf l of Y is* splittable on output *if*

$$\forall x \in out(l) : (\exists q \in l : x \notin out(q)) \vee LCA(Y, l \text{ after } x) \neq \emptyset$$

2. *A leaf l of Y is* splittable on input *if*

$$\exists a \in in(l) : LCA(Y, l \text{ after } a) \neq \emptyset$$

A leaf l of Y is splittable *if it is splittable on output or splittable on input.*

Lemma 22. *Each incomplete splitting graph has a splittable leaf.*

5.3   Splitting Graph Construction

Based on the condition of Definition 21 that holds, we assign children to split-table leaf nodes, and update the witness function. This is worked out in the method `splitnode` of Algorithm 3. The algorithm may choose nondeterministically between a split on output or a split on input, if both are possible. Such a choice is denoted with the syntax for guarded commands [8], i.e., as the guards on lines 5 and 16, and their respective statements on lines 6–15, and 17–20.

If a leaf l is split on output, then children are added for each output $x \in out(l)$. If $enabled(l, x) \neq l$, then we add $enabled(l, x)$ as a child, as those are the only states from which x can be observed. We also add (i.e., by using $+$) the term $x.\mathbf{0}$ to the witness of l, as observing x distinguishes states in $enabled(l, x)$ from states in $l \backslash enabled(l, x)$.

If $enabled(l, x) = l$, observing x will not distinguish any states. We then use that there is a $v \in LCA(Y, l\ after\ x)$, which means that some states of $l\ after\ x$ are distinguished by the witness $W(v)$. Hence, by taking output x, followed by $W(v)$, some states of l are distinguished. Therefore, we add $x.W(v)$ to the witness of l, and split l in the same way v was split, i.e., if $d \subseteq l$ are all the states with $d\ after\ x \subseteq c$ for some child $c \in Post(v)$, then d is a child of l. We call such a split an *induced split*.

For splitting on some input a, we also use an induced split to obtain the children for l. Since there exists some $v \in LCA(Y, l\ after\ a)$, at least two states of l may be distinguished by the witness constructed for v, after taking input a. To each element of the induced split, we add all the states not enabling a. If we would not do this, Algorithm 3 may assign the empty set as children to a splittable leaf, such that it remains a leaf. As a consequence, Lemma 25 and also Corollary 26 then do not hold. Corollary 26 shows termination of our splitting graph construction algorithm. It follows from the consecutive application of Lemma 25.

Definition 23. *Let Y be a splitting graph for suspension automaton S. Let v be an internal node of Y, P a set of states of S, and $\mu \in L$, such that $P\ after\ \mu \subseteq v$. Then the induced split of P with μ to v is:*

$$\Pi(P, \mu, v) = \{(c\ before\ \mu) \cap P \mid c \in Post_Y(v)\} \backslash \emptyset.$$

Example 24. We compute the splitting graph of the suspension automaton from Fig. 1, using Algorithm 2, and show the result in Fig. 4(left).

For the root node $\{1, 2, 3, 4\}$, we observe that state 4 does not enable x, while states 2 and 3 do not enable y. Hence, the root is split on output, gets children $\{1, 2, 3\}$ and $\{1, 4\}$, and witness $x.\mathbf{0} + y.\mathbf{0}$.

Node $\{1, 2, 3\}$ can be split on input a, as states 1 and 2 enable a, and since the root node is an LCA of $\{1, 2, 3\}$ *after* a: from $T(1, a) = 3$ and $T(2, a) = 4$, we obtain that the root node is an LCA, since $\{3, 4\} \subseteq \{1, 2, 3, 4\}$, but $\{3, 4\} \not\subseteq \{1, 2, 3\}$, and $\{3, 4\} \not\subseteq \{1, 4\}$. The induced split is $\{\{1\}, \{2\}\}$. We then need to add state 3 to both sets, because state 3 does not enable a, so node $\{1,2,3\}$ gets

Input: A suspension automaton $S = (Q, T, q_0)$
Input: An incomplete splitting graph $Y = (V, E, W)$ for S
Output: A splitting graph Y' that extends Y with additional leaf nodes

```
1  Function splitnode (S, Y):
2      l := a splittable leaf of Y;
3      C := ∅;
4      F := 0;
5      if l splittable on output →
6          foreach x ∈ out(l) do
7              if ∃q ∈ l : x ∉ out(q) then
8                  C := C ∪ {enabled(l, x)};
9                  F := F + x.0;
10             else
11                 Let v ∈ LCA(Y, l after x);
12                 C := C ∪ Π(l, x, v);
13                 F := F + x.W(v);
14             end
15         end
16     [] l splittable on input →
17         Let a ∈ in(l) with LCA(Y, l after a) ≠ ∅;
18         Let v ∈ LCA(Y, l after a);
19         C := {d ∪ (l \ enabled(l, a)) | d ∈ Π(l, a, v)};
20         F := a.W(v);
21     fi
22     return (V ∪ C, E ∪ {(l, c) | c ∈ C}, W ∪ {l ↦ F});
```

Algorithm 3. Splitting a leaf node of a splitting graph.

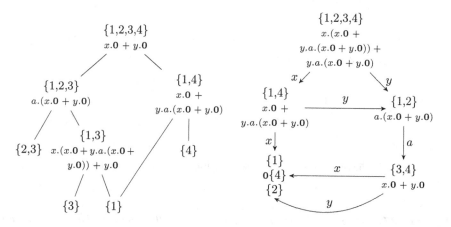

Fig. 4. Splitting graph for the suspension automaton of Fig. 1 (left), where we shorten all CCS expressions $0 + F$ to F, and an adaptive distinguishing graph for the suspension automaton of Fig. 1 (right), annotated with current state sets P, as used in Algorithm 4.

children $\{1,3\}$ and $\{2,3\}$. Prepending a to the witness of the root node gives us witness $a.(x.0 + y.0)$ for $\{1,2,3\}$.

Node $\{1,4\}$ can be split on output. As state 4 does not enable x, we only need to find an LCA for $\{1,4\}$ *after* $y = \{1,2\}$, which is the previously split node $\{1,2,3\}$. For x we have witness $x.0$, and for y we use the witness of $\{1,2,3\}$, so the witness for $\{1,4\}$ is $x.0 + y.a.(x.0 + y.0)$.

Next, node $\{1,3\}$ can be split on output using $\{1,4\}$ as LCA for x. Node $\{2,3\}$ does not need to be split, as we have $2\lozenge 3$. All other leaves are singletons, so we have obtained a complete splitting graph.

Lemma 25. *Algorithm 3 returns a splitting graph Y' for S, when given some splitting graph Y, such that one leaf l of Y, has become an internal node in Y'.*

Corollary 26. *Algorithm 2 returns a complete splitting graph for S.*

The algorithm of [12] constructs a splitting tree in polynomial time, because leaves of a node form a partition of that node. Our splitting graphs do not have this property. Clearly, a splitting graph for a suspension automaton with n states cannot have more than 2^n nodes, as the set of nodes is a subset of $\mathcal{P}(Q)\backslash \emptyset$ by Definition 19. For $n \in \mathbb{N}$ with $n \geq 3$, consider suspension automaton $S_n = (\{1,\ldots,n\}, T_n, 1)$, where T_n consists of the following output transitions:

$$T_n = \{(n,n,1)\} \cup \{(s,x,s+1) \mid s \in \{1,\ldots,n-1\}, x \in \{1,\ldots,n-1\}, s \neq x\}.$$

Figure 5 depicts suspension automata S_n for $n = 3, 4, 5$. We can prove Lemma 27 by showing that S_n has a splitting graph with 2^{n-1} nodes.

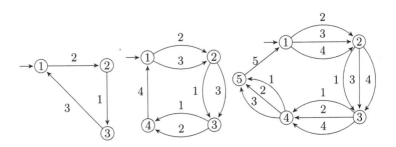

Fig. 5. Suspension automaton S_n for $n = 3$, $n = 4$, and $n = 5$

Lemma 27. *Let S be a suspension automaton with n states. Then a splitting graph returned by Algorithm 2 has $\mathcal{O}(2^n)$ nodes. This bound is tight.*

6   Extracting Test Cases from a Splitting Graph

Algorithm 4 retrieves CCS terms, of which the associated automata are test cases that distinguish states. The algorithm "concatenates" several CCS terms while keeping track of the current set of states. Each CCS term ensures that one state is distinguished from the rest because it lacks some output. We compute the current states for the leaves of the CCS term, and attach another CCS term to this leaf, if the current set of states consists of some incompatible pair of states. Hence in total, the automaton of the resulting CCS term distinguishes multiple pairs of states.

Input: A suspension automaton $S = (Q, T, q_0)$
Input: A complete splitting graph $Y = (V, E, W)$ for S
1   compDG $(S, Y, Q, \mathbf{0})$;
2   where
3   **Function** compDG (S, Y, P, F):
4   **if** $\Diamond(P)$ **then**
5   **return** F
6   **else**
7   **if** $F = \mathbf{0}$ **then**
8   Let $v \in LCA(Y, P)$;
9   **return** compDG $(S, Y, P, W(v))$
10   **else if** $F = \mu.F_1$ for some CCS term F_1 **then**
11   **return** $\mu.$compDG $(S, Y, P \text{ after } \mu, F_1)$
12   **else if** $F = F_1 + F_2$ for some CCS terms F_1, F_2 **then**
13   **return** compDG (S, Y, P, F_1) + compDG (S, Y, P, F_2)
14   **end**
15   **end**
Output: A CCS term F such that, for each $\sigma \in Obs(A_F)$, $\Diamond(Q \text{ after } \sigma)$.

Algorithm 4. Retrieving a test case from a splitting graph

Example 28. We construct the adaptive distinguishing graph for the suspension automaton from Fig. 1, using the splitting graph from Fig. 4, which also shows the result of this example. Algorithm 4 starts with $P = \{1, 2, 3, 4\}$ and $F = \mathbf{0}$. Hence, we search for a least common ancestor for Q. This will be the root node of the splitting graph, with witness $x.\mathbf{0} + y.\mathbf{0}$.

The function is then called with $F = x.\mathbf{0} + y.\mathbf{0}$, and will result in two recursive calls of the function on line 13 for $P = \{1, 2, 3, 4\}$ and $F = x.\mathbf{0}$, and $P = \{1, 2, 3, 4\}$ and $F = y.\mathbf{0}$ respectively. In the first case, the condition of line 10 holds, and we the function is called for $P = \{1, 2, 3, 4\}$ *after* $x = \{1, 4\}$ and $F = \mathbf{0}$, which means that lines 8–9 are executed next, using the only LCA for $\{1,4\}$, namely $\{1,4\}$.

The algorithm will then do some more recursive calls, checking whether the witness of $\{1,4\}$ must be extended further to distinguish more states. This will not be the case, because only singleton sets are reached at the leaves of the witness, and $\Diamond\{q\}$ holds for any state q, since $\Diamond$ is reflexive. Hence, we need to prepend x to the witness of $\{1,4\}$ to obtain the left term of the $+$ operator of the resulting CCS term of the algorithm: $x.(x.\mathbf{0} + y.a.(x.\mathbf{0} + y.\mathbf{0}))$.

As $P = \{1, 2, 3, 4\}$ *after* $y = \{1, 2\}$, its LCA $\{1,2,3\}$ will be used to complete the construction of the right term of the $+$ operator of the result.

The associated automaton of the resulting CCS term is an adaptive distinguishing graph for the suspension automaton, as it distinguishes all incompatible state pairs.

Lemma 29. *Algorithm 4 terminates and outputs a CCS term F that denotes a test case satisfying, for each $\sigma \in Obs(A_F)$, $\Diamond(Q$ after $\sigma)$.*

Algorithm 4 does not always construct an adaptive distinguishing graph for all incompatible state pairs. To ensure this, it must be able to select an "injective" splitting node as LCA on line 8. This will guarantee that a transition never maps two incompatible states to two compatible states (which cannot be distinguished any more), or that an input is used that is not enabled in some states.

Definition 30. *Let $S = (Q, T, q_0)$ be a suspension automaton, $P \subseteq Q$ a set of states, and $\mu \in L$ a label. Then μ is* injective *for P if*

$$\forall q, q' \in P : q \not\wedge q' \implies T(q, \mu) \downarrow \wedge T(q', \mu) \downarrow \wedge T(q, \mu) \not\wedge T(q', \mu)$$
$$\vee \mu \in O \backslash (out(q) \cap out(q'))$$

Analogous to the result of [12], Theorem 31 asserts that if an adaptive distinguishing graph exists our algorithms will find it, provided there are no compatible states. This last assumption is motivated in Example 32.

Theorem 31. *Let S be a suspension automaton such that all pairs of distinct states are incompatible. Then S has an adaptive distinguishing graph if and only if, during construction of a splitting graph Y for S, Algorithm 3 can and does only perform injective splits, that is, whenever Algorithm 3 splits a leaf l on output, then x is injective for l, for all $x \in out(l)$, and whenever it splits a leaf l on input a, then a is injective for l. Moreover, in this case Algorithm 4 constructs an adaptive distinguishing graph for S, when Y is given as input.*

Example 32. Without the assumption that there are no compatible state pairs, Theorem 31 does not hold. The suspension automaton S of Fig. 6 has an adaptive distinguishing graph, but our algorithm does not find it. Note that states 2 and 3 are compatible, and also states 6 and 7 are compatible. An adaptive distinguishing graph for S is denoted by CCS term $x.a.b.(z.\mathbf{0} + t.\mathbf{0}) + y.a.b.(z.\mathbf{0} + t.\mathbf{0}) + z.\mathbf{0} + t.\mathbf{0}$. When we construct a splitting graph for S, the set of all states $\{1, 2, 3, 4, 5, 6, 7, 8\}$ will be split on output, resulting in children $\{1\}$, $\{2, 3, 4\}$, $\{5\}$ and $\{6, 7, 8\}$. Now a split of $\{2, 3, 4\}$ on input b is not injective and a split on input a is not possible since the set of LCAs is empty. Similarly, there is no injective split of $\{6, 7, 8\}$.

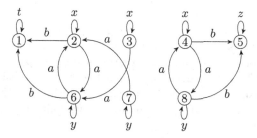

Fig. 6. Theorem 31 fails in presence of compatible state pairs.

7   Experimental Results on a Case Study

In [23], an FSM model, with over 10.000 states, was learned of an industrial piece of software, called the Engine Status Manager (ESM). During the learning process, testing against the ESM posed a significant challenge: it turned out to be extremely difficult to find counterexamples for hypothesis models. Initially, existing conformance testing algorithms were used to find counterexamples for hypothesis models (random walk, W-method, Wp-method, etc), but for larger hypothesis models these methods were unsuccessful. However, adaptive distinguishing sequences as in [12], augmented with additional pairwise distinguishing sequences for states not distinguished by the adaptive sequence, were able to find the required counterexamples. Therefore, the ESM models are good candidates to show the strength of the adaptive distinguishing graphs of this paper too.

Of course, applying our adaptive distinguishing graphs directly on the Mealy machine models, would not show our capability to handle the more expressive suspension automata. We therefore transformed the FSM models in such a way that they exhibit output nondeterminism. We first split all Mealy i/o transitions in two consecutive transitions i and o, and added a self-loop output transition 'quiescence' (denoting absence of response) to all states only having input transitions, to make it non-blocking. To ensure determinism, information about data parameters from the ESM was added to the labels of the Mealy machine in [23]. For our experiments, we removed this information again, resulting in suspension automata with states with multiple outgoing output transitions.

For performance reasons, we reduced the Mealy machine model with a subalphabet, before applying the transformation steps described above, i.e., we removed all i/o transitions with i not in the subalphabet. We obtained these subalphabets from [22], which contains a figure displaying interesting subalphabets based on domain knowledge. Table 1 shows that the resulting suspension automata still have a significant size.

We applied the algorithms of this paper to obtain a splitting graph and an adaptive distinguishing graph. The splitting graph was constructed as in Algorithm 3, so without requiring injectivity of the used labels. However, in the construction of the adaptive distinguishing graph (Algorithm 4) we chose on line 8 an LCA which was injective for the most pairs of states.

Table 1 shows that there are many pairs of incompatible states to distinguish. However, the number of nodes of the splitting graph are in the order of magnitude of the number of states of the suspension automaton, and the longest observable trace (i.e., the depth) of the adaptive distinguishing graphs is not long at all. Moreover, over 99% of the pairs of incompatible states are distinguished by the adaptive distinguishing graph. This indicates that the adaptive distinguishing graphs, although constructed from a non-injective splitting graph, can be very effective in testing.

Table 1. Computation statistics

Subalphabet	Number of states	Pairs of compatible states	Nodes in splitting graph	Depth distinguishing graph	Incompatible pairs not distinguished
InitIdleSleep	1616	16638 (0.64%)	1121	33	1145 (0.044%)
InitIdleStandbyRunning	2855	14171 (0.17%)	2082	33	2183 (0.027%)
InitIdleStandbySleep	3168	25974 (0.26%)	2226	33	3826 (0.038%)
InitIdleStandbyLowPower	2614	13834 (0.20%)	1809	33	2920 (0.043%)
InitError	2649	373427 (5.3%)	3097	35	17972 (0.27%)

8   Conclusions and Future Work

We studied the state identification problem for suspension automata, generalizing results from [12]. We presented algorithms to construct test cases that distinguish all incompatible state pairs, if possible, or many, if not. Experiments suggest that this approach is quite effective.

We see several directions for future research. First, though we did apply our algorithms to instances of an industrial benchmark, we would like to apply it to different case studies as well, to further explore the applicability of our approach. We note however that there are not that many (large) LTS benchmarks available.

An open problem is to give a bound on the depth of the distinguishing graph that our algorithms constructs. For FSMs, a quadratic bound is known [12], with examples to show it is tight [12,24]. These examples extend to our setting, as we generalize from the FSM setting, but the proof for the quadratic bound on adaptive distinguishing sequences from [12] does not.

If our algorithm returns an adaptive distinguishing graph that does not distinguish all incompatible state pairs, the question remains how to efficiently distinguish these remaining states. Graphs distinguishing pairs of states can be obtained directly from our splitting graph, or by computing them as in [5], but distinguishing all remaining pairs results in a large overhead compared to the small size of the distinguishing graph we obtained in our experiments. On the one hand, we can optimize the obtained distinguishing graph by improving the splitting graph's quality by applying heuristics that optimize the choice of labels for splitting leaves. On the other hand, we can use causes for states not being distinguished to construct a distinguishing graph that distinguishes all or at least many of the not distinguished states.

Though our distinguishing graphs significantly improve the size of an n-complete test suite, the problem to compute good access sequences for such a test suite requires further research as well [5]. Due to the output nondeterminism of suspension automata, we need an input-fairness assumption, to ensure that all outputs enabled from a state may eventually be observed. However, for access sequences we rather have a more adaptive strategy, in the spirit of [6], that reacts on the outputs as produced by the tested system rightaway. Adaptively choosing access sequences means that for reaching the same state, different access sequences may be used. However, the proof of n-completeness of a test suite depends on using one unique access sequence for accessing the same state. It remains an open problem whether using different access sequences breaks n-completeness or not.

References

1. Alur, R., Courcoubetis, C., Yannakakis, M.: Distinguishing tests for nondeterministic and probabilistic machines. In: STOC, vol. 95, pp. 363–372. Citeseer (1995)
2. Baier, C., Katoen, J.P.: Principles of Model Checking. The MIT Press, Cambridge (2008)
3. Beneš, N., Daca, P., Henzinger, T.A., Křetínský, J., Ničković, D.: Complete composition operators for IOCO-testing theory. In: Proceedings of the 18th International ACM SIGSOFT Symposium on Component-Based Software Engineering. CBSE 2015, pp. 101–110. ACM, New York (2015). https://doi.org/10.1145/2737166.2737175
4. van den Bos, P., Vaandrager, F.W.: State identification for labeled transition systems with inputs and outputs. CoRR abs/1907.11034 (2019). http://arxiv.org/abs/1907.11034
5. van den Bos, P., Janssen, R., Moerman, J.: n-complete test suites for IOCO. Softw. Qual. J. **27**(2), 563–588 (2019). https://doi.org/10.1007/s11219-018-9422-x
6. van den Bos, P., Stoelinga, M.: Tester versus bug: a generic framework for model-based testing via games. In: Orlandini, A., Zimmermann, M. (eds.) Proceedings Ninth International Symposium on Games, Automata, Logics, and Formal Verification. Electronic Proceedings in Theoretical Computer Science, Saarbrücken, Germany, 26–28th September 2018, vol. 277, pp. 118–132. Open Publishing Association (2018). https://doi.org/10.4204/EPTCS.277.9
7. Cormen, T.H., Leiserson, C.E., Rivest, R.L., Stein, C.: Introduction to Algorithms, 3rd edn. The MIT Press, Cambridge (2009)
8. Dijkstra, E.W.: Guarded commands, nondeterminacy, and formal derivation of programs. In: Gries, D. (ed.) Programming Methodology: A Collection of Articles by Members of IFIP WG2.3. Texts and Monographs in Computer Science, pp. 166–175. Springer, New York (1978). https://doi.org/10.1007/978-1-4612-6315-9_14
9. Dorofeeva, R., El-Fakih, K., Maag, S., Cavalli, A.R., Yevtushenko, N.: FSM-based conformance testing methods: a survey annotated with experimental evaluation. Inf. Softw. Technol. **52**(12), 1286–1297 (2010)
10. Gill, A.: Introduction to the Theory of Finite-State Machines. McGraw-Hill, New York (1962)
11. Hierons, R.M.: Testing from a nondeterministic finite state machine using adaptive state counting. IEEE Trans. Comput. **53**(10), 1330–1342 (2004). https://doi.org/10.1109/TC.2004.85

12. Lee, D., Yannakakis, M.: Testing finite-state machines: state identification and verification. IEEE Trans. Comput. **43**(3), 306–320 (1994). https://doi.org/10.1109/12.272431
13. Lee, D., Yannakakis, M.: Principles and methods of testing finite state machines – a survey. Proc. IEEE **84**(8), 1090–1123 (1996)
14. Mazala, R.: Infinite games. In: Grädel, E., Thomas, W., Wilke, T. (eds.) Automata Logics, and Infinite Games. LNCS, vol. 2500, pp. 23–38. Springer, Heidelberg (2002). https://doi.org/10.1007/3-540-36387-4_2
15. Milner, R.: Communication and Concurrency. Prentice-Hall Inc., Upper Saddle River (1989)
16. Moerman, J.: Nominal techniques and black box testing for automata learning. Ph.D. thesis, Radboud University Nijmegen, July 2019
17. Moore, E.F.: Gedanken-experiments on sequential machines. In: Annals of Mathematics Studies, vol. 34, pp. 129–153. Princeton University Press, Princeton (1956)
18. Petrenko, A., Yevtushenko, N.: Conformance tests as checking experiments for partial nondeterministic FSM. In: Grieskamp, W., Weise, C. (eds.) FATES 2005. LNCS, vol. 3997, pp. 118–133. Springer, Heidelberg (2006). https://doi.org/10.1007/11759744_9
19. Petrenko, A., Yevtushenko, N.: Adaptive testing of deterministic implementations specified by nondeterministic FSMs. In: Wolff, B., Zaïdi, F. (eds.) ICTSS 2011. LNCS, vol. 7019, pp. 162–178. Springer, Heidelberg (2011). https://doi.org/10.1007/978-3-642-24580-0_12
20. Bensalem, S., Krichen, M., Tripakis, S.: State identification problems for input/output transition systems. In: 2008 9th International Workshop on Discrete Event Systems, pp. 225–230, May 2008. https://doi.org/10.1109/WODES.2008.4605949
21. Simão, A., Petrenko, A.: Generating complete and finite test suite for ioco: is it possible? In: Proceedings Ninth Workshop on Model-Based Testing, MBT 2014, Grenoble, France, 6 April 2014, pp. 56–70 (2014). https://doi.org/10.4204/EPTCS.141.5
22. Smeenk, W.: Applying automata learning to complex industrial software. Master's thesis, Radboud University Nijmegen (2012)
23. Smeenk, W., Moerman, J., Vaandrager, F., Jansen, D.N.: Applying automata learning to embedded control software. In: Butler, M., Conchon, S., Zaïdi, F. (eds.) ICFEM 2015. LNCS, vol. 9407, pp. 67–83. Springer, Cham (2015). https://doi.org/10.1007/978-3-319-25423-4_5
24. Sokolovskii, M.N.: Diagnostic experiments with automata. Cybernetics **7**(6), 988–994 (1971). https://doi.org/10.1007/BF01068822
25. Tretmans, J.: Model based testing with labelled transition systems. In: Hierons, R.M., Bowen, J.P., Harman, M. (eds.) Formal Methods and Testing. LNCS, vol. 4949, pp. 1–38. Springer, Heidelberg (2008). https://doi.org/10.1007/978-3-540-78917-8_1
26. Willemse, T.A.C.: Heuristics for ioco-based test-based modelling. In: Brim, L., Haverkort, B., Leucker, M., van de Pol, J. (eds.) FMICS 2006. LNCS, vol. 4346, pp. 132–147. Springer, Heidelberg (2007). https://doi.org/10.1007/978-3-540-70952-7_9
27. Yannakakis, M., Lee, D.: Testing finite state machines: fault detection. J. Comput. Syst. Sci. **50**(2), 209–227 (1995). https://doi.org/10.1006/jcss.1995.1019. http://www.sciencedirect.com/science/article/pii/S0022000085710197

Combining State- and Event-Based Semantics to Verify Highly Available Programs

Peter Zeller[✉], Annette Bieniusa, and Arnd Poetzsch-Heffter

University of Kaiserslautern, Kaiserslautern, Germany
{p_zeller,bieniusa,poetzsch}@cs.uni-kl.de

Abstract. Replicated databases are attractive for managing the data of distributed applications that require high availability, low latency, and high throughput. However, these benefits entail weak consistency which comes at a price: it becomes harder to reason about application correctness. We address this difficulty with a verification technique for highly available programs. We augment an existing sequential programming language with primitives for interacting concurrently with a highly available database and extend the state-based operational semantics of that language accordingly. To this end we make use of existing event-based database semantics.

We then present a reduction of the extended semantics to a simpler one, which is again sequential and therefore easier to handle in verification tools. Our verification tool *Repliss* uses this technique and demonstrates its feasibility.

1 Introduction

Replication is an essential mechanism for implementing highly available and scalable information systems like social networks, e-government, or e-commerce applications. One popular approach for building highly available apps is to delegate all synchronization aspects to a distributed database while the application processes are typically stateless. To achieve high availability even in the presence of network failures, the underlying databases have to weaken their consistency guarantees. This trade-off has been formalized in the CAP [9] and PACELC [1] theorems.

When prioritizing high availability over strong consistency, reasoning about the correctness of an application that utilizes a replicated distributed database becomes significantly more difficult as developers have to consider the consequences of concurrent updates of shared data and its implication for the guarantees when reading data. This complexity extends also to the verification effort to prove the correctness of such programs. Concurrent requests on the database give rise not only to interleaved executions within a replica, but also temporal divergence of replicas. These aspects are hard to tackle in a direct way such as induction over all possible executions. Another challenge is in the different

© Springer Nature Switzerland AG 2020
F. Arbab and S.-S. Jongmans (Eds.): FACS 2019, LNCS 12018, pp. 213–232, 2020.
https://doi.org/10.1007/978-3-030-40914-2_11

approaches to formalize database semantics and programming languages. Consistency models of highly available databases are often formalized axiomatically using event graphs [7]. Programming language semantics are on the other hand often formalized using operational semantics.

In this paper, we present a novel semantics and verification technique for programs utilizing weakly-consistent databases. We combine axiomatic consistency models with operational semantics, by extending a sequential programming language with constructs for concurrent requests and primitives for accessing the database. This allows us to reduce the verification problem for distribution and replication to a sequential program with nondeterministic steps for simulating the possible effects of concurrent requests (Sect. 3). In Sect. 4, we present our soundness proof for this reduction which we have formally proven using Isabelle/HOL [19]. Our database semantics targets recent trends emerging in highly available databases, like convergent and commutative replicated data types (CRDTs) [20] and transactional causal$^+$ consistency (TCC$^+$) [2,16].

Finally, we have built a verification tool named *Repliss* (Sect. 5) which employs this reduction for partially automating the verification of highly available programs. The tool takes a program and a specification of functional properties as input. The functional properties may use history invariants, which can describe causal relations between several requests or the effects that requests have on the database state. Further, the user must also provide additional invariants to guide the prover. The tool then uses random testing to find counterexamples and symbolic execution to verify the absence of errors. As a case study, we demonstrate how Repliss can be used to verify a highly available chat application.

Before we present the details of our work, we show why existing verification approaches do not solve our problem.

Related Work. The challenge of weak consistency in verification is well known and has been approached with a number of different techniques. Weak memory models have been studied in depth in the context of concurrent programming for multi-core machines [8]. However, the techniques in this area usually target linearizability as a correctness criterion and employ hardware-supported synchronization mechanisms such as memory fences or CAS-operations. In distributed systems, it is neither feasible to consider linearizability as consistency notion nor to implement the same concurrency control mechanisms as in weak memory system. This precludes the direct applicability of these techniques to our scenario. In the following, we therefore focus on related work that shares our application domain.

Composite Replicated Data Types [10] allow to compose basic data types into application-specific data representations that are synchronized atomically. Their area of application is similar to our setting, though our approach is more widely applicable as we model procedures involving several transactions on arbitrary combinations of objects. More importantly, their approach is axiomatic and based on a denotational semantics, which is more difficult to adapt in a tool implementing the technique.

CISE [11,18] is a tool, which can automatically determine the procedures in an application, which require stronger consistency guarantees for correctness. This line of work focuses on combining weak consistency with strong synchronization for some operations, whereas our work only considers weak consistency. CISE does not consider features like transactions or replicated data types directly. Instead, application procedures are assumed to have a single atomic effect which is applied on every replica asynchronously. This is similar to the implementation technique of operation-based CRDTs, where effects have to be commutative to ensure convergence. Soteria [17] is a similar tool which is based on state-based implementations of CRDTs instead. In contrast, our model handles data types as components with a high-level (axiomatic) specification and not their concrete implementation.

QUELEA [21] is another tool supporting the development of applications on top of weakly consistent databases. Unlike our approach and the previously discussed approaches, the specifications in QUELEA are not given as invariants. Instead, the user specifies constraints on the order between operations and the tool automatically chooses the necessary consistency level.

Q9 [12] is a symbolic execution engine for finding bugs in programs written on top of weakly consistent databases. The tool only supports bounded verification, where the number of concurrent effects is limited, so unlike Repliss, it cannot be used to prove the absence of errors in the general case. Weak consistency is modeled using commutative effects, which works well for symbolic execution, but is less suitable when working with invariants as we do.

Chapar [14] is a framework for verifying causally consistent, replicated databases and applications employing such databases. The development is formally verified using Coq and the goal of verifying application is similar to ours. Their approach is different, though. They have implemented a model checker for applications, thus providing automation. However, the kind of applications which can be analyzed is restricted, since the model checker can check all possible reorderings of one concrete execution where all parameters have fixed values.

None of the work discussed so far handles the integration of transactions into a technique to reason about programs. This aspect has been tackled in work in different contexts, for example in a program logic for handling Java Card's transaction mechanism [4]. Transactions in Java Card provide atomicity, but do not handle concurrency. We are not aware of other work integrating weakly consistent transactions into a verification technique.

2   A Formal Semantics for Highly Available Programs

In this section, we extend the operational semantics of an imperative core calculus with primitives for concurrent procedure invocations and database interactions. Procedures define the external interface (API) of a program. Clients invoke the procedures of a program which are then processed concurrently.

In highly available systems, operations must be able to progress even if only the local replica is available. This is reflected in our system model, where database operations are executed locally first and propagated to other replicas asynchronously.

Figure 1 shows the definitions regarding system state that we use in our formalization. In the initial state, all fields are empty maps or sets. We grouped the fields by the different aspects of our semantics, which we explain in detail below. To support database operations, we model the database state using event graphs including information about all database calls, the happens-before relation between them and the respective transactions. For procedure invocations and the sequential semantics we keep the local state and the currently active procedure per invocation. Furthermore, we record the history of procedure invocations to make them available in specifications. Finally, we add direct support for generating and using unique identifiers. Besides these aspects, we also explain how we handle partial failures and invariants when we explain the rules below (these do not require fields in the system state).

Database operations:

$\quad$ *call* : *callId* $\hookrightarrow$ *callInfo*

$\quad$ *happensBefore* : (*callId* $\times$ *callId*) *set*

$\quad$ *visibleCalls* : *invocId* $\hookrightarrow$ *callId set*

$\quad$ *currentTransaction* : *invocId* $\hookrightarrow$ *txid*

$\quad$ *callOrigin* : *callId* $\hookrightarrow$ *txid*

$\quad$ *txStatus* : *txid* $\hookrightarrow$ *txStatus*

$\quad$ *transactionOrigin* : *txid* $\hookrightarrow$ *invocId*

Procedure invocations:

$\quad$ *localState* : *invocId* $\hookrightarrow$ *localState*

$\quad$ *currentProc* : *invocId* $\hookrightarrow$ *procedureImpl*

History Recording:

$\quad$ *invocationOp* : *invocId* $\hookrightarrow$ (*procName* $\times$ *any list*)

$\quad$ *invocationRes* : *invocId* $\hookrightarrow$ *any*

Unique Identifiers:

$\quad$ *generatedIds* : *any* $\hookrightarrow$ *invocId*

$\quad$ *knownIds* : *any set*

Programs:

$\quad$ *querySpec* : (*operationContext* $\times$ *operation* $\times$ *any list* $\times$ *res*) $\to$ *bool*

$\quad$ *procedure* : (*procName* $\times$ *any list*) $\hookrightarrow$ (*localState* $\times$ *procedureImpl*)

$\quad$ *invariant* : *invariantContext* $\to$ *bool*

Fig. 1. Fields of the system state. We use $\times$ for product types, $\hookrightarrow$ for map types (functions returning an option type), τ *set* for sets containing elements of type τ, and τ *list* for lists. The type *any* is the type we use for arbitrary values used in the program. Other relevant types are explained in the text.

The rules of our semantics are shown in Fig. 2. We write $S \xrightarrow{i,a} S'$ to denote that the system makes a step from state S to S' by executing action a in procedure invocation i. In every step a different invocation can progress, resulting in a fine-grained interleaving semantics. $S \xrightarrow{tr}{}^* S'$ denotes the reflexive, transitive closure with the trace tr. A trace is a sequence of (*invocId*, *action*) pairs.

Each rule describes the complete effect of a single action which includes some orthogonal aspects of our semantics. In the following we therefore describe the different aspects and how they manifest in the rules.

$$\frac{}{S \xrightarrow{\epsilon}{}^* S} \text{(steps-empty)} \qquad \frac{S_1 \xrightarrow{tr}{}^* S_2 \quad S_2 \xrightarrow{i,a} S_3}{S_1 \xrightarrow{tr \cdot (i,a)}{}^* S_3} \text{(steps)}$$

$$\frac{\begin{array}{c} procedure_{prog}(procName, args) \triangleq (initialState, impl) \\ uniqueIdsInList(args) \subseteq knownIds(S) \qquad invocationOp(S,i) = \bot \end{array}}{S \xrightarrow{i,invoc(procName,args)} S \begin{bmatrix} currentProc(S,i) := impl \\ localState(S,i) := initialState \\ invocationOp(S,i) := (procName, args) \end{bmatrix}} \text{(invocation)}$$

$$\frac{\begin{array}{c} localState(S,i) \triangleq ls \\ currentProc(S,i) \triangleq f \quad f(ls) = return(res) \quad currentTransaction(S,i) = \bot \end{array}}{S \xrightarrow{i,return(res)} S \begin{bmatrix} localState(S,i) := \bot \\ invocationRes(S,i) := res \\ knownIds(S) := knownIds(S) \cup uniqueIds(res) \end{bmatrix}} \text{(return)}$$

$$\frac{localState(S,i) \triangleq ls \quad currentProc(S,i) \triangleq f \quad f(ls) = localStep(ls')}{S \xrightarrow{i,local} S \left[localState(S,i) := ls'\right]} \text{(local)}$$

$$\frac{\begin{array}{c} localState(S,i) \triangleq ls \\ currentProc(S,i) \triangleq f \quad f(ls) = beginAtomic(ls') \quad currentTransaction(S,i) = \bot \\ txStatus(S,t) = \bot \quad visibleCalls(S,i) \triangleq vis \quad newTxns \subseteq committedTransactions(S) \\ newCalls = callsInTransaction(S, newTxns) \downarrow_{happensBefore(S)} \quad snapshot = vis \cup newCalls \end{array}}{S \xrightarrow{i,beginAtomic(t,newTxns)} S \begin{bmatrix} localState(S,i) := ls' \\ currentTransaction(S,i) := t \\ txStatus(S,t) := uncommitted \\ transactionOrigin(S,t) := i \\ visibleCalls(S,i) := snapshot \end{bmatrix}} \text{(atomic)}$$

$$\frac{\begin{array}{c} localState(S,i) \triangleq ls \\ currentProc(S,i) \triangleq f \quad f(ls) = endAtomic(ls') \quad currentTransaction(S,i) \triangleq t \end{array}}{S \xrightarrow{i,endAtomic} S \begin{bmatrix} localState(S,i) := ls' \\ currentTransaction(S,i) := \bot \\ txStatus(S,t) := committed \end{bmatrix}} \text{(commit)}$$

$$\frac{\begin{array}{c} localState(S,i) \triangleq ls \\ currentProc(S,i) \triangleq f \quad f(ls) = dbOperation(ls', op, args) \quad currentTransaction(S,i) \triangleq t \\ call(S,c) = \bot \quad querySpec_{prog}(operationContext(S,i), op, args, res) \quad visibleCalls(S,i) \triangleq vis \end{array}}{S \xrightarrow{i,dbOp(c,op,args,res)} S \begin{bmatrix} localState(S,i) := ls'(res) \\ call(S,c) := (op, args, res) \\ callOrigin(S,c) := t \\ visibleCalls(S,i) := vis \cup \{c\} \\ happensBefore(S) := happensBefore(S) \cup (vis \times \{c\}) \end{bmatrix}} \text{(DB-operation)}$$

$$\frac{\begin{array}{c} localState(S,i) \triangleq ls \quad currentProc(S,i) \triangleq f \\ f(ls) = newId(ls') \quad generatedIds(S, uid) = \bot \quad uniqueIds(uid) = \{uid\} \quad ls'(uid) \triangleq ls'' \end{array}}{S \xrightarrow{i,newId(uid)} S \begin{bmatrix} localState(S,i) := ls'' \\ generatedIds(S, uid) := i \end{bmatrix}} \text{(new-id)}$$

$$\frac{}{S \xrightarrow{i,crash} S \left[localState(S,i) := \bot\right]} \text{(crash)} \qquad \frac{res = invariant_{prog}(invContext(S))}{S \xrightarrow{i,invCheck(txns,res)} S} \text{(inv)}$$

Fig. 2. Interleaving semantics. We use $x \triangleq y$ as an abbreviation for $x = Some(y)$ for option types.

Procedure Invocations. In our semantics, a procedure invocation is triggered by an application request from some client. Clients may invoke procedures concurrently, but each single invocation executes sequentially.

Programs are modeled with a partial function (field *procedure*) that takes the procedure name and the arguments of the invocation and, if the procedure is defined for the given arguments, returns the initial local state of the procedure and its implementation. The implementation (*procedureImpl*) is given by a function which calculates the next action based on the current invocation state. Possible actions are local evaluations (*local*), generating unique identifiers (*newId*), database related actions (*beginAtomic, endAtomic, dbOp*) and returning from an invocation (*return*). In the system state, we use the fields *currentProc* to store the implementation and the field *localState* to store the local invocation state.

The rule *invocation* describes the start of a procedure invocation. The precondition of the rule enforces that the procedure is defined for the given arguments. The remaining aspects of this rule are related to tracking the history and handling of unique identifiers (see below). The *return* rule is similar.

For local actions in a procedure invocation we have the rule (*local*), which subsumes the standard sequential semantics of the core calculus.

Database Operations. Instead of modeling the database state explicitly and thus assuming a concrete implementation of the database, we represent the current state using event graphs of the database calls [6]. This is a common practice for specifying the semantics of highly available databases and of replicated data types [7,10,23].

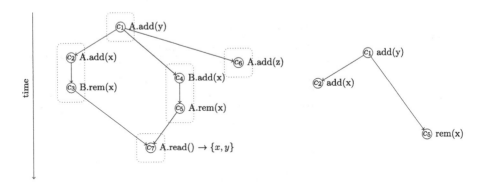

Fig. 3. Illustration of an event graph (left) and the extracted operation context for query c_7 (right).

Figure 3 shows an event graph of an execution involving two replicated sets A and B. Each call to the database is represented by a node. Several calls can be bundled together in a transaction (boxes with dashed line). We add an edge from a call c_1 to a call c_2 if call c_1 happened before call c_2. Calls that are not reachable from each other, such as c_2 and c_5, are concurrent. The result of a read operation, like c_7 in the example, depends only on the calls that happened before. Also, each call operates only on one object. The subgraph of calls that happened before an operation is called the operation context. For query c_7, the operation context is visualized on the right part of Fig. 3. Given an operation context and the parameters of the database operation, the specification (*querySpec* in the formal model) of the corresponding data type yields the result of the operation. For example, the specification of an add-wins replicated data type is defined as:

$$spec_{\text{CrdtAWSet}}(\mathit{call}, \mathit{happensBefore}, \mathit{read}()) =$$
$$\{x.\ \exists a \in \mathit{call}.\ a.op = add(x)$$
$$\wedge \neg(\exists r \in \mathit{call}.\ r.op = rem(y) \wedge (a,r) \in \mathit{happensBefore})\}$$

Since no remove happened after the $add(x)$ in c_2, the read operation c_7 from our example returns a set containing x. Though we do not model replicas explicitly, they are represented as concurrent events in the event graph.

In the formal model, each database call is identified by a *callId*; the partial function *call* in the system state stores information about each call. The *callInfo* consists of the arguments to the operation and its return value. The *happensBefore* relation records the partial order between calls and *callOrigin* stores the transaction each call originated from. Transactions are identified by a *txid*, and its *txStatus* can be "running" or "committed". The procedure invocation that started a transaction is stored in *transactionOrigin*. Together, this information represents the history of database calls. Additionally, the field *visibleCalls* keeps the set of database calls that are visible at an procedure invocation and *currentTransaction* the currently running transaction (if any) for an invocation.

The rule *atomic* describes what happens on the database when starting a new transaction. Recall that we here consider databases in which transactions work on causally consistent snapshots, modeled as a set of visible updates on the database. To obtain the snapshot at the start of a transaction, we choose an arbitrary set of committed transactions *newTxns* which are to become visible. We derive the new set of visible calls by first taking all calls to update operations from these transactions: $callsInTransaction(S, newTxns) = \{c.\ \exists t \in newTxns.\ callOrigin(S, t) \triangleq t\}$. Next, we calculate the downwards closure of this set with respect to the happens-before relation on calls. The downwards closure of a set S and relation R is defined as $S \downarrow_R = S \cup \{x.\ \exists y.\ (x,y) \in R \wedge y \in S\}$. In our model, the happens-before relation is by construction transitive, so that this definition is sufficient to include all causal dependencies in the snapshot, therefore providing causal consistency (we have verified this in Isabelle and omit the proofs here).

Finally, we add the visible calls from the current invocation to the snapshot. We also set the transaction status of the new transaction id t to *uncommitted* and store the current transaction for the invocation.

When ending the current transaction (rule *commit*), its *txStatus* is set to *committed*. This allows it to be included in new snapshots, which eventually makes the database calls in the transaction visible to others. As the atomic rule can only pick committed transactions and no new calls can be added to it after committing, transactions are atomic.

When executing a database operation (rule *DB-operation*), we extract the *operationContext* from the current state. As stated earlier, the operation context consists of the currently visible calls (formally: $\lambda x.$ **if** $x \in visibleCalls(S, i)$ **then** $call(S, x)$ **else** $\bot$) and the happens-before relation restricted to the visible calls (formally: $happensBefore(S) \cap (visibleCalls(S, i) \times visibleCalls(S, i))$). We then nondeterministically pick a result *res*, which satisfies the query specification of the program in the operation context. The database call is then recorded in the state by adding the operation with its arguments and result to the existing calls. We also record the current transaction as the originating transaction for the new call. The happens-before relation is also updated by making the new call causally depend on all currently visible calls. The new call is then added to the set of visible calls, such that following operations depend on it.

History Recording. As for the database, we also store a history of invocations of API-procedures including the respective arguments (*invocationOp*) and the result for completed invocations (*invocationRes*). By using these in specifications, we can relate different procedure invocations and link procedure invocation with their corresponding changes in the database state. The rules *invocation* and *return* update this information accordingly.

Unique Identifiers. In practice, unique identifiers are often generated using UUIDs or using a replica-specific identifier together with a locally unique identifier. Since identifiers for database entries appear in most applications, we include a builtin action, which lets applications generate globally unique identifiers (see rule *new-id*). With this extension, we avoid proving the correctness of an identifier generator for every application. Moreover, it allows us to handle generated identifiers as special values, which cannot be forged by clients.

To model unique identifiers, we require that the type *any* comes with a function *uniqueIds: any $\rightarrow$ any set* extracting the unique identifiers of a value. The *new-id* rule ensures that the generated value includes exactly one unique identifier and that the generated value is in the domain of the *ls'* function. This allows us to include a kind of type-check in the action to generate a unique identifier of a specific form.

To describe the semantics, we keep track of all generated unique identifiers in *generatedIds*. The set *knownIds* represents the identifiers which could be known to clients, i.e. identifiers which have been returned from an invocation of the application API (see rule *return*). In the *invocation* rule, we enforce that clients can only invoke the API with known identifiers.

Partial Failures. Since we are considering a distributed application, it is necessary to handle partial failures. This is captured in the semantics with the rule *crash*, which models a crash of a single procedure invocation and loses all locally stored information. Afterwards, the invocation cannot continue, since there is no local state.

Invariants. We use invariants to specify the application. Invariants can refer to the database state as well as the history of procedure invocations, which makes the specification language more expressive than related work where invariants can only refer to the database state.

Formally, rule *inv* is always enabled so the invariant can be checked (and therefore must hold) at all times. However, the invariant cannot involve arbitrary aspects of the current state. The function *invContext* provides a special view on the state for checking invariants, which only contains part of the system state. All information local to a particular procedure invocation is not included (i.e. *txStatus, generatedIds, localState, currentProc, currentTransaction,* and *visibleCalls*). Moreover, the function *invContext* only includes database calls from committed transactions. These restrictions are essential for simplifying the verification efforts, which we discuss in the next section.

Using the invariant checks in our transition relation $S \xrightarrow{tr}{}^* S'$ we can define program correctness as follows:

$$traces(program) := \{tr \mid \exists S'.\ initialState(program) \xrightarrow{tr}{}^* S'\}$$
$$traceCorrect(trace) := \forall i.\ (i, invCheck(false)) \notin trace$$
$$programCorrect(program) := \forall trace \in traces(program).\ traceCorrect(trace)$$

The set *traces* includes all traces admitted by a program; the predicate *traceCorrect* defines that a trace is correct if it does not contain an invariant check on any procedure invocation i that evaluates to *False*; and the last definition states that a program is correct iff all its traces are correct.

Using these definitions to reason about correctness is however not very practical. We have to consider all possible traces, which includes the interleavings of several concurrent procedure invocations that are allowed by the transition relation.

3   Single-Invocation Semantics

To address the challenge of handling concurrency in our setting, we have developed a proof technique, which reduces the formal verification problem to a simpler problem, where we can reason about only one procedure invocation at a time.

Our proof technique is based on invariants and reduces the proof obligations to checking that the initial system state satisfies the invariant and that each procedure invocation maintains the invariant. When verifying a single procedure invocation, the effects of other invocations only need to be considered at specific program points, namely at the procedure invocation and before the start of transactions. We use the invariant and generic properties of executions to reason about possible state changes at these program points. For the procedure to be verified, we must then guarantee that the invariant is maintained at the end of transactions, right after the start of a procedure invocation and after returning from a procedure invocation. The latter two are necessary because at these program points the information stored in the history of procedure invocations is updated.

Technically, we formalize the reduction using a second operational semantics, the *single-invocation semantics*, which we present in this section. In Sect. 4, we then prove that reduction from the interleaving semantics to the single-invocation semantics is sound: If a program is correct with respect to the single-invocation semantics, it is also correct in the interleaving semantics.

The main difference between the two semantics is that the single-invocation semantics only allows steps in a single invocation. Effects from different invocations are treated with nondeterministic steps in the rules for starting a procedure invocation and beginning a transaction. In these cases, the rules of the single-invocation semantics assume an arbitrary state change, assuming that the invariant is maintained, the new state is well-formed, and the history of the new state is an extension of the former history.

Moreover, the single-invocation semantics does not include a dedicated step to check the invariant. The invariant has to be checked in the following three steps: directly after a procedure invocation (*S-invocation*), after the end of a transaction (*S-commit*), and after a procedure invocation returns to the client (*S-return*).

Correspondingly, we adapted the transition relation to be $S \xrightarrow{i,(a,v)} S'$ for a single step. The value v is *true* if the step fulfills the necessary invariant checks. A program is *correct* with respect to the single-invocation semantics if for all possible executions the trace contains only actions for which v is true.

Figure 4 shows the rules of the single-invocation semantics that differ from the interleaving semantics. The rule *S-steps* enforces that only steps on one invocation can be taken in a trace. The rules *local*, *DB-operation*, and *new-id* are not included in Fig. 4, because they do not involve the invariant – the parameter v is always *true* in these transitions and the rules are otherwise equivalent to the interleaving semantics. The interesting aspects are in the handling of transactions and invocations. Here, we add the nondeterministic state changes and check the invariant.

Rule S-atomic. At the beginning of a transaction a new snapshot is determined, which means that this is a place where changes from concurrent invocations might become visible to the current invocation. We model this with a nondeterministic state change from the current state S to a new state S'. The predicate $growing(i, S, S')$ requires that S' must contain at least the invocations and database operations from S. However, it might have grown with further events:

$$growing(i, S, S') :=$$
$$wellformed(S) \land \left(\exists tr.\ S \xrightarrow{tr}{}^* S' \land (\forall (i', a) \in tr.\ i' \neq i) \land \left(\forall i'.\ (i', crash) \notin tr \right) \right)$$

$$\frac{}{S \xrightarrow{i,\epsilon}{}^{*} S}\;(S\text{-}steps\text{-}empty) \qquad \frac{S_1 \xrightarrow{i,tr}{}^{*} S_2 \quad S_2 \xrightarrow{i,t} S_3}{S_1 \xrightarrow{i,tr\cdot t}{}^{*} S_3}\;(S\text{-}steps)$$

$$\frac{\begin{array}{c} invocationOp(S,i) = \bot \\ procedure_{prog}(procName, args) \triangleq (initialState, impl) \quad uniqueIdsInList(args) \subseteq knownIds(S') \\ wellformed(S') \qquad \forall tx.\ txStatus(S', tx) \not\cong uncommitted \\ invariant\_all(S') \qquad invocationOp(S', i) = \bot \qquad v = invariant\_all(S'') \\ \forall tx.\ transactionOrigin(S'', tx) \not\cong i \qquad S'' = S' \begin{bmatrix} visibleCalls(S,i) := \emptyset \\ currentProc(S,i) := impl \\ localState(S,i) := initialState \\ invocationOp(S,i) := (procName, args) \end{bmatrix} \end{array}}{S \xrightarrow{i,(invoc(procName,args),v)} S''}\;(S\text{-}invocation)$$

$$\frac{\begin{array}{c} localState(S,i) \triangleq ls \\ currentProc(S,i) \triangleq f \quad f(ls) = return(res) \quad currentTransaction(S,i) = \bot \\ v = invariant\_all(S') \qquad S' = S \begin{bmatrix} visibleCalls(S,i) := \bot \\ currentProc(S,i) := \bot \\ localState(S,i) := \bot \\ invocationRes(S,i) := res \\ knownIds(S) := knownIds(S) \cup uniqueIds(res) \end{bmatrix} \end{array}}{S \xrightarrow{i,(return(res),v)} S'}\;(S\text{-}return)$$

$$\frac{\begin{array}{c} localState(S,i) \triangleq ls \quad currentProc(S,i) \triangleq f \quad f(ls) = beginAtomic(ls') \\ currentTransaction(S,i) = \bot \quad txStatus(S,t) = \bot \quad growing(i,S,S') \\ \forall t.\ transactionOrigin(S,t) \triangleq i \leftrightarrow transactionOrigin(S',t) \triangleq i \quad invariant\_all(S') \\ \forall tx.\ txStatus(S', tx) \not\cong uncommitted \quad wellformed(S') \quad wellformed(S'') \\ localState(S',i) \triangleq ls \quad currentProc(S',i) \triangleq f \quad currentTransaction(S',i) = \bot \\ visibleCalls(S,i) = vis \quad visibleCalls(S',i) = vis \quad newTxns \subseteq committedTransactions(S) \\ newCalls = callsInTransaction(S, newTxns) \downarrow_{happensBefore(S)} \quad vis' = vis \cup newCalls \\ consistentSnapshot(S', vis') \quad txStatus(S',t) = \bot \quad \forall c.\ callOrigin(S', c) \not\cong t \\ transactionOrigin(S',t) = \bot \quad S'' = S' \begin{bmatrix} txStatus(t) := uncommitted \\ transactionOrigin(t) := i \\ currentTransaction(i) := t \\ localState(i) := ls' \\ visibleCalls(i) := vis' \end{bmatrix} \end{array}}{S \xrightarrow{i,(beginAtomic(t,newTxns),true)} S''}\;(S\text{-}atomic)$$

$$\frac{\begin{array}{c} localState(S,i) \triangleq ls \\ currentProc(S,i) \triangleq f \quad f(ls) = endAtomic(ls') \quad currentTransaction(S,i) \triangleq t \\ v = invariant\_all(S') \quad wellformed(S') \quad S' = S \begin{bmatrix} localState(S,i) := ls' \\ currentTransaction(S,i) := \bot \\ txStatus(S,t) := committed \end{bmatrix} \end{array}}{S \xrightarrow{i,(endAtomic,v)} S'}\;(S\text{-}commit)$$

Fig. 4. Single invocation semantics.

This definition allows us to use any general property we can prove about steps taken in other invocations. Additionally, the rule allows us to assume the invariant for the new state S'.

The remaining aspects of the rule describe the other state changes and are equivalent to the interleaving semantics.

Rule S-commit. When a transaction is committed, we check the invariant in the state after the commit and record the result of the invariant check in the trace. This ensures that an execution is considered incorrect, if a transaction breaks the invariant.

Rule S-invocation. An invocation can only be executed at the beginning of the trace, since the rule demands that the invocation i is not yet used in the current state S. The rule then nondeterministically chooses a state S' which satisfies the invariant and starts the procedure invocation, which yields state S''. The rule also allows us to assume that there are no uncommitted transactions at the start of an invocation and that we start from a well-formed state. A state is defined to be well-formed if it is reachable from the initial state.

We then check whether the invariant holds in S'' and record the result in the trace. This is necessary, because invariants can refer to unfinished procedure invocations, so starting an invocation can cause an invariant violation.

Rule S-return. For a return statement, we check the invariant in the state after completing the invocation.

4 Soundness of the Reduction

We now show that it is in fact sufficient to prove a program correct with respect to the single-invocation semantics in order to ensure correctness in all possible concurrent executions according to the interleaving semantics. Figure 5 illustrates the main steps in our proof with an example. The full proof is formalized in Isabelle/HOL[1]. Below we give the corresponding definitions and lemmas.

Definition 4.1 (Packed traces). A trace tr is **packed** for a procedure invocation i if it only switches to invocation i at the start of a procedure invocation (*invoc*) or at the start of a transaction (*beginAtomic*) action:
$\forall i', a.\ i' \neq i \wedge [(i', \_), (i, a)] \in tr \rightarrow (is\text{-}invoc(a) \vee is\text{-}beginAtomic(a))$. We say a trace is **packed** if it is packed for all procedure invocations.

Lemma 4.2 (Reduction to packed traces). A program is correct if all its traces that are packed and do not contain *crash*-steps are correct.

Proof Sketch. The proof is essentially a reduction argument [15] which uses commutativity of actions performed on different invocations.

Let tr be a failing trace of the program, i.e. a trace containing a failing invariant check. We show that we can then construct a packed trace which is also failing. We consider the prefix of tr up to the first failing invariant check. We then can reorder the actions in this prefix to get a packed trace.

We prove this by induction over the number of procedure invocations, which are not yet packed. To show that a single invocation i can be packed without

---

[1] The Isabelle proofs are available at https://github.com/peterzeller/repliss-isabelle/.

Trace actions: ●– begin a procedure invocation, –● return from an invocation, ↙ start a transaction, ↘ commit a transaction, ◖ database operation, ✳ create a new unique identifier, ◠ local steps, and ✗ a failing invariant. We annotate each start of a transaction with a transaction id and the set of transactions that are visible to this transaction.

Step 1: Assume we have a failing trace in the interleaving semantics, for example:

Step 2: In Lemma 4.2, we show that in this case there is an equivalent packed trace that is also failing. We construct this packed trace by moving actions to the front. This reordering does not invalidate any snapshots and since snapshots are fixed in the trace, the effect of the trace is preserved. For the example above, we can construct the following packed trace which ends with a failing invariant:

Step 3: In Lemma 4.3, we show that there is a corresponding trace in the single invocation semantics, where we only consider procedure invocation i_1 while actions from other invocations are summarized using invariants (visualized by lightcones). The proof obligation for the single invocation semantics is to check the invariant after an invocation return and after each transaction commit. In the picture below, the proof obligation is always to check that the actions between a light source and a check mark preserve the invariant.

Thus, if we prove that no invariant check ever fails in the single invocation semantics, we have shown the correctness of the application.

Fig. 5. Overview of our soundness proof applied to an exemplary trace.

unpacking any other invocation, we use another induction over the minimal index k with a not-allowed invocation switch. This cannot be the first action for invocation i, since *invoc* is an allowed invocation switch. Thus, let k' be the last action on invocation i before k. We then split the trace into $tr = tr_{[..k'-1]} \cdot tr_{k'} \cdot tr_{[k'+1..k-1]} \cdot tr_k \cdot tr_{[k+1..]}$ and reorder it by moving tr_k to the front to get $tr' = tr_{[..k'-1]} \cdot tr_{k'} \cdot tr_k \cdot tr_{[k'+1..k-1]} \cdot tr_{[k+1..]}$. By changing the order of the trace, we have eliminated all unwanted invocation switches up to index k. We can show that the action of tr_k commutes with the actions it is swapped with, which are all from a different invocation.

The commutativity proof involves a case distinction over all possible actions in the system. An interesting case here is moving an *endAtomic* action to the front. In principle, this could affect other transactions, but since we do not change the transaction snapshots when reordering the trace, we are guaranteed to get the same results. □

Lemma 4.3 (Simulation). Let tr be a packed trace of an execution starting in state S and ending in state S' where S is well-formed (i.e. reachable from the initial state) and satisfies the invariant and S' does not satisfy the invariant. Moreover, assume that tr is packed and does not contain any crashes. Then, there is an execution in the single-invocation semantics starting with state S and the trace of this execution is not correct.

Proof Sketch. We show that the single-invocation semantics can simulate the distributed (interleaving) semantics. To this end, we define a coupling relation between a state S_d of the distributed execution and a state S_i of a single-invocation execution with invocation i. The coupling invariant distinguishes two cases: (1) When the last step in the distributed execution was in invocation i, then the states must be equal. (2) When the last step was on an invocation different from i, it must hold that S_i is greater than S_d with respect to the *growing* relation and that the local state of invocation i is equivalent in S_i and S_d.

In the simulation proof, the case of switching between invocations can only occur at the beginning of invocations or at the start of a transaction (because we assumed the trace is packed). In both cases, the single-invocation semantics allows nondeterministic state-transitions, which allow the single-invocation execution to catch up with the distributed invocation. The remaining cases are straight-forward. □

Theorem 4.4 (Soundness of verification technique). When a program is correct with respect to the single-invocation semantics and the initial state satisfies the invariant, then the program is correct with respect to the distributed (interleaving) semantics.

Proof Sketch. We show that all executions are correct. Because of Lemma 4.2, it is sufficient to consider executions with packed traces without crashes. Let tr be a trace for such an execution.

For the sake of a contradiction assume tr is not a correct trace, i.e. there is a failing invariant check in the trace. We now consider the first time when an invariant-violating state is reached and the prefix of tr leading to this state. As the state does not satisfy the invariant, though the initial state does (by assumption), we can apply Lemma 4.3 and obtain a failing trace in the single-invocation semantics.However, this is a contradiction to the assumption that the program is correct in the single-invocation semantics. □

5   The Repliss Tool

We have developed the Repliss verification tool, which uses our technique to partially automate the verification of highly available applications. In principle, our reduction to the sequential single-invocation semantics could be combined with any technique for verifying sequential programs. For Repliss we chose to implement a symbolic execution engine with the CVC4 [3] automated theorem solver as a backend.

Let us illustrate how Repliss can be used to verify a chat application. This example is inspired by an experience report from Discord [22], who migrated their chat service from a single centralized database to the replicated and weakly consistent database Cassandra [13]. Although the code had been well tested prior to deployment, when the new solution was first used in production, some messages ended up with missing metadata. This problem occurred when a user edited a message while another user concurrently deleted the message. When modeling the chat application in Repliss, a small counter example is generated illustrating this problematic case. Below, we present a model where the Bug is fixed, which can then be verified using Repliss.

5.1 Implementation

An essential aspect of developing a highly available program is to find a suitable data model for storing the persistent state. To this end, Repliss provides a library of built-in replicated data types (CRDTs [20]). There are different variants of the same data type, which differ in how concurrent updates are handled. It is important to choose the appropriate variant to get the desired application behavior.

Our implementation of the chat application is shown in Fig. 6. The data model for the chat application is defined in lines 1–4 of the code. Repliss uses a suffix, such as rw, to denote how conflicts between concurrent updates are resolved. In the example, we use a set named chat of type Set_rw to store the set of messages that belong to the chat. We further use a Map_dw to store the data for each message. Each message has an author and a message content. For the author, we use a simple register with last-writer-wins semantics, since it is only written when creating a new message. The message content is stored in a multi-value register, which keeps multiple versions in case of concurrent assignments.

The choice of data types determines the *querySpec* in the formal semantics. Repliss includes the specifications for all supported CRDTs and knows how to compose the semantics in the case of nested CRDTs (e.g. a CRDT-set used as the value in a CRDT-map). In the example, we use the delete-wins (dw) variant of maps. This is important for the message map to ensure that a message is deleted in the case of concurrent invocations of deleteMessage and editMessage. This choice fixes the bug in the original application. For the set of messages we use the remove-wins variant (rw) to have semantics compatible with the delete-wins behavior of the map. Choosing a different semantics would make it hard to maintain consistency between the set of messages and the related information stored in the message map.

Starting from line 6 in Fig. 6 the procedures of the program are implemented. The **atomic**-blocks correspond to the *beginAtomic* and *endAtomic* actions in the formal semantics, the references to the chat- and message-CRDTs correspond to a database call, and the expression **new** MessageId triggers the generation of a new unique identifier.

```
1    crdt chat: Set_rw[MessageId]
2    crdt message: Map_dw[MessageId, {
3        author: Register[UserId],
4        content: MultiValueRegister[String]}]
5
6    def sendMessage(from: UserId, content: String): MessageId {
7        var m: MessageId
8        atomic {
9            m = new MessageId
10           call message_author_assign(m, from)
11           call message_content_assign(m, content)
12           call chat_add(m) }
13       return m }
14   def editMessage(id: MessageId, newContent: String) {
15       atomic {
16           if (message_exists(id)) {
17               call message_content_assign(id, newContent) }}}
18   def deleteMessage(message_id: MessageId) {
19       atomic {
20           if (message_exists(message_id)) {
21               call chat_remove(message_id)
22               call message_delete(message_id) }}}
23   def getMessage(m: MessageId): getMessageResult {
24       atomic {
25           if (message_exists(m)) {
26               return found(message_author_get(m), message_content_getFirst(m))
27           } else {
28               return notFound() }}}
```

Fig. 6. Model of Chat application in Repliss.

5.2 Specification

To prove the correctness of our application, we next specify its invariants. To address the database state, we use queries (as in Property 1 below). Further, invariants can address a history of procedure invocations and database calls. This enables us to express some temporal properties and relate effects of procedure invocations, as demonstrated in Property 2.

Property 1. For the chat application, referential integrity is important: every MessageId occurring in the message set should have a corresponding entry in the message map. Formally, this relation can be expressed as a first order logical formula using the queries defined on the data types:

invariant forall m: MessageId :: chat_contains(m) ==> message_exists(m)

When queries are used in an invariant, we have to define in which operation context the query should be evaluated. Remember that invariants cannot access the current database snapshot of a specific procedure invocation (field *visibleCalls*). We therefore specify that this invariant has to hold in all valid database snapshots. Repliss automatically adds this quantification to the invariant if free queries are used.

Property 2. We use the history of procedure invocations to express properties at the external interface of the application. To this end, we use the *invocationOp* and *invocationRes* maps, which contain the operation (input) and result of each procedure invocation. In the invariant below, we relate the results of the getMessage procedure with invocations of sendMessage: If getMessage returns a certain user u as part of a message, then there must be an invocation of sendMessage with that user.

invariant forall g: invocationId, m: MessageId, u: UserId, c: String ::
 g.info == getMessage(m) && g.result == getMessage_res(found(u, c))
 ==> (**exists** s: invocationId, c2: String :: s.info == sendMessage(u, c2))

The original implementation of the chat application did not satisfy this property – it returned null as the author value, even though the user null never sent a message.

```
1   // For every author assignment, there is a corresponding invocation of sendMessage:
2   invariant forall c: callId, m: MessageId, u: UserId ::
3      c.op == message_author_assign(m, u)
4      ==> (exists i: invocationId, s: String :: i.info == sendMessage(u, s))
5   // For assignments of the content field, there is a prior assignment to the author field:
6   invariant forall c1: callId, m: MessageId, s: String ::
7      c1.op == message_content_assign(m, s)
8      ==> (exists c2: callId, u: UserId ::
9         c2.op == message_author_assign(m, u) && c2 happened before c1)
10  // There is no update after a delete:
11  invariant !(exists write: callId, delete: callId, m: MessageId ::
12     ((exists u: UserId :: write.op == message_author_assign(m, u))
13     || (exists s: String :: write.op == message_content_assign(m, s)))
14     && delete.op == message_delete(m) && delete happened before write)
```

Fig. 7. Further invariants for Chat application.

5.3 Correctness

Repliss can verify the referential integrity constraint from Property 1 automatically. For the verification to succeed it is important that we used transactions for sendMessage and deleteMessage such that no client can observe a database state where the chat-set is updated and the message-map is not. Moreover, it was necessary that we chose data types with compatible semantics such that concurrent updates are merged into concurrent states.

For the history invariant in Property 2, the verification is more involved. As explained before, our proof approach is based on an invariant and fully compositional, i.e. each procedure is checked individually. Therefore, the invariant needs to be strong enough, such that assuming the invariant in a pre-state gives us enough information to prove the invariant in the post-state. We thus define additional invariants (Fig. 7) which enable Repliss to verify Property 2. The verification by Repliss takes approx. 40 s when verifying both properties together.

6   Conclusion and Future Work

We have presented a novel proof technique for verifying applications built on top of weakly consistent databases. Our proof technique is designed to be used with partially automated tools like the Repliss verification tool. The soundness of our technique is formally verified using Isabelle/HOL with a proof based on a formal small-step semantics. The semantics comprises only eight different steps to facilitate the validation of the assumptions made for the soundness proof.

While we restricted our consistency model to causally consistent transactions, we are confident that the central ideas of our approach can be transferred to other consistency models. To support stronger consistency models, we could extend the predicate *growing* used in the *atomic* rule in Fig. 4. Here, a model based on tokens as used by CISE [11] could restrict the changes that may be done by concurrent invocations. For weaker consistency models, it suffices to drop assumptions from our proof rules. None of these changes requires adaptations of the proof technique. Essential for our approach is simply that transactions are isolated (i.e. concurrent transactions cannot see each others updates) and that application replicas only communicate via the replicated database.

In this paper, we have demonstrated the applicability of our proof technique with the chat application. In current work, we are improving the automation of the approach to facilitate the invariant preservation proofs. Application on larger case studies is work-in-progress, but we expect it to scale well in the number of procedures, as each procedure can be verified individually against the invariants. In that sense, our approach is composable. However, the number or size of invariants is likely to grow with the number of procedures, which could restrict the scalability of the technique. We expect that further techniques to partition invariants according to components will be necessary to handle more complex applications.

Acknowledgement. This research is supported in part by the EU H2020 project "LightKone" (732505) https://www.lightkone.eu/.

References

1. Abadi, D.: Consistency tradeoffs in modern distributed database system design: CAP is only part of the story. IEEE Comput. **45**(2), 37–42 (2012). https://doi.org/10.1109/MC.2012.33. http://dx.doi.org/10.1109/MC.2012.33
2. Ahamad, M., Neiger, G., Burns, J.E., Kohli, P., Hutto, P.W.: Causal memory: definitions, implementation, and programming. Distrib. Comput. **9**(1), 37–49 (1995). https://doi.org/10.1007/BF01784241
3. Barrett, C., et al.: CVC4. In: Gopalakrishnan, G., Qadeer, S. (eds.) CAV 2011. LNCS, vol. 6806, pp. 171–177. Springer, Heidelberg (2011). https://doi.org/10.1007/978-3-642-22110-1_14
4. Beckert, B., Mostowski, W.: A program logic for handling JAVA CARD'S transaction mechanism. In: Pezzè, M. (ed.) FASE 2003. LNCS, vol. 2621, pp. 246–260. Springer, Heidelberg (2003). https://doi.org/10.1007/3-540-36578-8_18

5. Bodík, R., Majumdar, R. (eds.): Proceedings of the 43rd Annual ACM SIGPLAN-SIGACT Symposium on Principles of Programming Languages, POPL 2016, St. Petersburg, FL, USA, 20–22 January 2016. ACM (2016). http://dl.acm.org/citation.cfm?id=2837614
6. Burckhardt, S.: Principles of eventual consistency. Found. Trends Program. Lang. 1(1–2), 1–150 (2014). https://doi.org/10.1561/2500000011
7. Burckhardt, S., Gotsman, A., Yang, H., Zawirski, M.: Replicated data types: specification, verification, optimality. In: Jagannathan, S., Sewell, P. (eds.) The 41st Annual ACM SIGPLAN-SIGACT Symposium on Principles of Programming Languages, POPL 2014, San Diego, CA, USA, 20–21 January 2014, pp. 271–284. ACM (2014). https://doi.org/10.1145/2535838.2535848. http://doi.acm.org/10.1145/2535838.2535848
8. Dongol, B., Derrick, J.: Verifying linearisability: a comparative survey. ACM Comput. Surv. 48(2), 19:1–19:43 (2015). https://doi.org/10.1145/2796550
9. Gilbert, S., Lynch, N.A.: Brewer's conjecture and the feasibility of consistent, available, partition-tolerant web services. SIGACT News 33(2), 51–59 (2002)
10. Gotsman, A., Yang, H.: Composite replicated data types. In: Vitek, J. (ed.) ESOP 2015. LNCS, vol. 9032, pp. 585–609. Springer, Heidelberg (2015). https://doi.org/10.1007/978-3-662-46669-8_24
11. Gotsman, A., Yang, H., Ferreira, C., Najafzadeh, M., Shapiro, M.: 'Cause I'm strong enough: reasoning about consistency choices in distributed systems. In: Bodík, Majumdar [5], pp. 371–384. https://doi.org/10.1145/2837614.2837625. http://doi.acm.org/10.1145/2837614.2837625
12. Kaki, G., Earanky, K., Sivaramakrishnan, K., Jagannathan, S.: Safe replication through bounded concurrency verification. In: 32nd ACM SIGPLAN International Conference on Object-Oriented Programming, Systems, Languages and Applications (OOPSLA) (2018)
13. Lakshman, A., Malik, P.: Cassandra: a decentralized structured storage system. SIGOPS Oper. Syst. Rev. 44(2), 35–40 (2010). https://doi.org/10.1145/1773912.1773922. http://doi.acm.org/10.1145/1773912.1773922
14. Lesani, M., Bell, C.J., Chlipala, A.: Chapar: certified causally consistent distributed key-value stores. In: Bodík, Majumdar [5], pp. 357–370. https://doi.org/10.1145/2837614.2837622. http://doi.acm.org/10.1145/2837614.2837622
15. Lipton, R.J.: Reduction: a method of proving properties of parallel programs. Commun. ACM 18(12), 717–721 (1975). https://doi.org/10.1145/361227.361234
16. Lloyd, W., Freedman, M.J., Kaminsky, M., Andersen, D.G.: Don't settle for eventual: scalable causal consistency for wide-area storage with COPS. In: Wobber, T., Druschel, P. (eds.) Proceedings of the 23rd ACM Symposium on Operating Systems Principles 2011, SOSP 2011, Cascais, Portugal, 23–26 October 2011, pp. 401–416. ACM (2011). https://doi.org/10.1145/2043556.2043593. http://doi.acm.org/10.1145/2043556.2043593
17. Nair, S., Petri, G., Shapiro, M.: Invariant safety for distributed applications. CoRR abs/1903.02759 (2019). http://arxiv.org/abs/1903.02759
18. Najafzadeh, M., Gotsman, A., Yang, H., Ferreira, C., Shapiro, M.: The CISE tool: proving weakly-consistent applications correct. In: Alvaro, P., Bessani, A. (eds.) Proceedings of the 2nd Workshop on the Principles and Practice of Consistency for Distributed Data, PaPoC@EuroSys 2016, London, United Kingdom, 18 April 2016, pp. 2:1–2:3. ACM (2016). https://doi.org/10.1145/2911151.2911160. http://doi.acm.org/10.1145/2911151.2911160

19. Nipkow, T., Paulson, L.C., Wenzel, M.: Isabelle/HOL—A Proof Assistant for Higher-Order Logic. LNCS, vol. 2283. Springer, Heidelberg (2002). https://doi.org/10.1007/3-540-45949-9
20. Shapiro, M., Preguiça, N., Baquero, C., Zawirski, M.: A comprehensive study of convergent and commutative replicated data types. Rapport de recherche RR-7506, INRIA, January 2011. http://hal.inria.fr/inria-00555588
21. Sivaramakrishnan, K.C., Kaki, G., Jagannathan, S.: Declarative programming over eventually consistent data stores. In: Grove, D., Blackburn, S. (eds.) Proceedings of the 36th ACM SIGPLAN Conference on Programming Language Design and Implementation, Portland, OR, USA, 15–17 June 2015, pp. 413–424. ACM (2015). https://doi.org/10.1145/2737924.2737981. http://doi.acm.org/10.1145/2737924.2737981
22. Vishnevskiy, S.: How discord stores billions of messages. https://blog.discordapp.com/how-discord-stores-billions-of-messages-7fa6ec7ee4c7. Accessed 16 Nov 2018
23. Zeller, P., Bieniusa, A., Poetzsch-Heffter, A.: Formal specification and verification of CRDTs. In: Ábrahám, E., Palamidessi, C. (eds.) FORTE 2014. LNCS, vol. 8461, pp. 33–48. Springer, Heidelberg (2014). https://doi.org/10.1007/978-3-662-43613-4_3

Short Papers

Reowolf: Synchronous Multi-party Communication over the Internet

Christopher A. Esterhuyse and Hans-Dieter A. Hiep$^{(\boxtimes)}$ (ID)

Centrum Wiskunde & Informatica,
Science Park 123, 1098 XG Amsterdam, The Netherlands
{esterhuy,hdh}@cwi.nl

1 Introduction

In this position paper we introduce Reowolf: an on-going project that aims to replace the decades-old application programming interface, BSD sockets, for communication on the Internet. A novel programming interface is being implemented at the systems level that is inter-operable with existing Internet applications. It should provide support for middleware to further improve quality of service without having to give up on privacy, and makes programming of decentralized Internet applications simpler: we give arguments as to why we hold this position.

The main idea in Reowolf is to offer a high-level abstraction for communication, called *connectors*. Connectors are complexes of synchronization primitives among multiple data streams, generalizing end-to-end sockets. Programmers create a connector and configure it using a protocol description language (PDL), that allows for the declarative specification of what synchronization and data exchange primitives applications require to communicate on an abstract level. These connectors are to sockets what high-level programming languages are to machine code: we compile high-level *application-defined protocols* to low-level operational code that realizes the actual communication.

To see how Reowolf is used, we draw a parallel between GPU programming and network programming with Reowolf. In GPU programming, one writes a high-level program (e.g., GLSL) that describes graphical manipulations that take place on dedicated hardware (e.g., graphics cards). This alleviates programmers from having to write hardware-specific programs that use low-level, device-specific assembly instructions. Each vendor of graphics hardware supports the high-level programming language and compiles it down to their own target. The novelty of Reowolf is to approach Internet application programming in the same way: one writes a high-level 'program' that *configures* the possible communications between multiple parties on a network. The Reowolf implementation compiles into code that targets available communication mechanisms (e.g., TCP) that performs synchronization and exchange of data.

Received support from NLnet: NGI Zero PET fund from the European Union's Horizon 2020 research and innovation programme under grant agreement No. 825310.

F. Arbab and S.-S. Jongmans (Eds.): FACS 2019, LNCS 12018, pp. 235–242, 2020.
https://doi.org/10.1007/978-3-030-40914-2_12

The main benefits of programming Internet applications in this way are:

1. Raising the level of **abstraction** relieves the programmer from controlling a quadratic number of data streams, while still being able to communicate in a decentralized manner. In realizing the abstraction there is ample room for optimization and innovation for the implementer: different optimization techniques apply to different networking circumstances[1], and applications are agnostic to the underlying network protocols that realize communication.
2. A clean **separation** of application-defined protocols from pure computations results in application programs that are more isolated, making them simpler to reason about, validate, and verify. Moreover, separation of application content from its signaling information allows encryption of content only, while granting middleware the insight in the nature of traffic.
3. Working with application-defined protocols as **explicit** objects allows them to be collected in a standard library, that facilitates protocol reuse. Moreover, application-defined protocols are publicly visible within the network, allowing middleware to perform informed traffic monitoring such as deviation detection without having to guess application's intent.

These main benefits do not fall out of the sky: in the Reowolf project, we turn theoretical and prototypical results of the past two decades [4] into practical, low-level systems work. The Protocol Description Language of Reowolf is largely based on Reo, an exogenous coordination language for synchronous communication [2]. Although Reo has seen recent practical applications, such as the distributed implementation Dreams [13], and compilers for shared-memory synchronization [11], Reo was never integrated deeper into operating systems.

Reowolf is an on-going project with the ultimate goal of replacing the *socket* with the *connector* for programming internet applications. In the rest of this paper, we discuss the main ideas underpinning Reowolf, and highlight some of the interesting challenges of implementing Reowolf that remain to be solved.

2   Background

Programming Internet applications has essentially remained unchanged since the 1980s. The Berkeley Software Distribution (BSD) implementation allows applications to create *sockets* for communication over the Internet, e.g. Transmission Control Protocol over Internet Protocol (TCP/IP), that either listen for incoming connections or are connected outward, resulting in a bi-directional, reliable channel between two peers on the outer edges of the network. Applications consequently control the stream of data into the sending side of a channel, to be received in order by the other side of the channel. Although applications of the 1980s were process-driven, send and receive operations were performed

---

[1] E.g. local host (virtualized) networking, wireless sensor networks, Beowulf class cluster computing, high-speed local area networks in datacenters, wide-area networks spanning the globe, and satellite networks.

by cooperating with an operating system responsible for scheduling application processes. More recently, applications have become event-driven [12], allowing for more fine-grained scheduling decisions by applications themselves. However, the essence of programming Internet applications remains based on controlling a channel between two peers in a network.

In turn, this programming discipline seems to naturally support application architectures that favor centralization. With sockets, each channel between two peers requires a program controlling the stream of data on *both* sides of the socket connection. In a fully connected graph, where each peer maintains a connection to each other peer, one controls a *quadratic* number of data streams. However, a graph with a single central peer that connects to all other peers, requires controlling only a linear number of data streams at the central peer! In practice, clients use sockets to connect to a single centralized server and that server provides a shared service to its clients. Keeping the client simple solves the control problem: only the server-side has to control most of the data streams.

Recently, a drawback to centralized architecture became apparent by the increasing tension between central service providers (e.g. video streaming, search engines, content delivery) and network operators (providing Internet connectivity). The Workshop on Internet Economy (WIE2017) has discussed that service providers, "deliberately obfuscate both content and signaling information from network operators providing transit for the traffic." This leads to difficulties "to improve traffic engineering, police (or secure) usage, and improve their own services. Increasingly, the access and transit network operators have less insight as to the nature of the traffic, and fewer effective traffic management tools" [5].

That service providers "deliberately obfuscate" their data seems reasonable: they namely employ data encryption to increase user privacy, and this is even required by EU regulation [14]. It also seems reasonable that more traffic will be sent encrypted in the future, as this and other similar regulations are implemented over time. Moreover, service providers may use non-standard protocols, for example, to gain a competitive edge against competing service providers.

Network operators typically deploy *middleware* to increase the quality of their service: to optimize Internet traffic to improve latency and throughput, and to monitor traffic to detect intrusions and abuse such as Denial of Service (DOS). Middleware uses, among other techniques, *deep packet inspection*: scanning further down the packet than the (TCP/)IP headers to take action. On high bandwidth networks, middleware can be implemented close to the metal to keep throughput high, for example, using field programmable gate arrays [6,15].

A drawback of deep packet inspection is that it is non-standardized: middleware tries to *guess* an application's intent by scanning its traffic. As a simplified example, consider middleware that caches HTTP resources. As more Internet traffic is sent encrypted, deep packet inspection becomes less effective, e.g. HTTPS prohibits caching by a man-in-the-middle. Currently deployed middleware may therefore be wasting effort and may be hard to adapt due to its closeness to the hardware level. Moreover, the opportunity to optimize and monitor may not apply to certain traffic, e.g. obscure or innovative protocols.

3   Comparing Reo and Reowolf

From the perspective of history of computing, Reowolf is the logical next step for programming Internet applications. Compare this to computing before the Internet age, where early programming languages such as Algol or Fortran lacked the ability to deal with data abstractions. A data structure was treated implicitly, and search and manipulation algorithms were littered all over the program. With abstract data types, programs can be written against an abstract interface that describes the possible operations. Abstract data types can be specified *explicitly*, for example, by algebraic data structures of a given signature and a (e.g., equational, first-order) specification.

A similar case can be made for today's programs, that lack the ability to deal with concurrency and communication abstractions. A program implicitly deals with protocols, and low-level concurrency code for synchronization is all over the place. Similar to abstract data types is the idea of abstract behavior types [1,9]; programs can be written against an abstract interface that describes all possible behaviors a program could expect. Abstract behavior types can be specified explicitly by a protocol description language such as Reo.

Reo is a language for specifying exogenous communication protocols by constraining the possible interactions available to participating components: a protocol coordinates its components. Reo structurally separates computation from coordination, unlike other exogenous coordination languages [7]. A primitive *component* is treated as an indivisible black-box that publicly exposes a set of named *ports*. Components are linked together by complex connectors: connectors are graph-like structures of nodes and primitive channels between nodes. A set of primitive channels is provided: e.g. synchronous channels, lossy channels, and asynchronously buffered channels. Multiple connectors (between possibly the same set or different sets of components) can be composed into larger connectors: Reo's compositional semantics ensures that properties of a composite connector are derived from properties of its constituent connectors.

Where Reo remains abstract, Reowolf becomes concrete. Reo components can be implemented as follows: each component is identified by an IP address, or by a domain name that resolves to an IP address. Components do not correspond 1-to-1 to (physical) machines, as one component can be implemented by multiple machines behind a router with network address translation, and a single machine can host multiple components. Components have several ports, each identified by a number. Later, *port name systems* may be used to resolve ports by name.

Components are responsible for setting up connectors. Each component provides a local view of the connector, by configuring it with a protocol description as it sees fit. The protocol description includes references to the ports of other components to connect to, and it includes which local ports are left open. Multiple components can provide different local views of a connector, and the established connector is the composition of all local views of its components. Components that dynamically allocate new ports over time, and dynamically change their view of the connector by reconfiguring with a new protocol description, are interesting research challenges.

4   Programming with Connectors

Among existing environments for Internet applications are UNIX-like operating systems, where applications are implemented by one or more processes that perform system calls. There are system calls to create sockets, sending and receiving payloads, and handling exceptional network conditions such as timeouts. The implementation of those socket system calls is provided as part of the operating system. Reowolf connectors have a programming interface similar to sockets: creating connectors, configuring the expected behavior of a connector, putting and getting payloads, and handling error conditions such as non-conformance.

Sockets establish a channel between *two* peers and communication is fundamentally *asynchronous*, connectors establish a protocol between *multiple* peers and communication is fundamentally *synchronous*:

Multi-party Communication

Two socket connections are assumed to be unrelated, unless a program controlling their streams intentionally relates them. The expected behavior of such control program is implicit, and unknown to the network. Complex relations between multiple socket connections require complex controlling programs to run at both end-points. Connectors generalize sockets to an arbitrary number of peers, and use a declarative protocol description language to describe the expected interactions among those peers. The explicit protocol description is provided as part of configuring a connector. A control program that handles the data stream control is generated on-the-fly by an implementation of Reowolf, and is no longer the responsibility of the application. Applications thus interact with a connector, by getting and putting data, abstracting away the communication with multiple peers.

An important problem here is how programs may introduce unintended constraints, viz. a program that constrains the behavior of a connector that is not explicitly described by the configured user-defined protocol. Since Reowolf is implemented on the operating systems level, it is necessary to define what the behavior is of programs that operate on multiple connectors. Some operations may lead to a deadlock, which should be reported as a connector error. Multiple processes that access the same connector must have well-defined behavior (e.g., connectors should be thread-safe).

Synchronicity

In Reowolf, synchronous communication means there is not an *a priori* requirement on the order of the events that realize the communication. For example, if sending and receiving are synchronous, then the realization of that communication may first perform sending and then receiving (typical) or may under certain conditions optimize, in which receiving happens before sending (clairvoyant or speculative manner). Connectors can still describe asynchronous communication by imposing a requirement on the order of its realizing events using logical buffers (e.g., send happens first, receive happens later).

Synchronous communication could be implemented, e.g. using a synchronous network, in which in each round all components issue an operation. Synchronous networks can be implemented on top of asynchronous networks using a synchronizer, such as those by Awerbuch [3]. An important issue in Reowolf is to understand how to support synchronicity efficiently: under which conditions can user-defined protocols be efficiently implemented on top of asynchronous networks without the overhead of a synchronizer?

Implementation Architecture

From the OSI layering perspective, sockets directly expose applications to (an implementation of) the Transport layer; applications are responsible for session management and representation formats of transmitted data. Reowolf connectors instead expose an abstraction above the Transport layer, including the Session and Representation layers.

Session management will be handled by a Reowolf implementation: for the duration in which a connector is connected there is an on-going session. However, a Reowolf implementation is free to choose how to realize the communication protocol that underlies the session that results from combining the user-defined protocols into a composite connector: it can be a shared memory, multiple TCP connections, UDP, other network protocols, or combinations thereof.

Representation of data is low-level: it involves abstracting from network byte order and data framing. For example, three components A, B, C are connected and component A sends a megabyte of data to component B and synchronously component C sends a one byte of data to component B, then even after C has finished its own transmission it must wait for the completion of A's transmission.

Modular Validation and Verification

A program that makes use of a connector can be validated and verified with respect to the local user-defined protocol it configures. The correctness of a program has to be established under the assumption that any behavior that is valid according to the protocol description may take place, where the connector takes place of a constrained adversary. Moreover, a Reowolf implementation is correct if all true correctness properties of all programs hold even in the case where communication takes place with unknown peers. This ensures that modular development of applications is fruitful.

5   Leveraging Explicit Protocol Descriptions

We will consider four scenarios by which we exemplify how the explicit protocol descriptions of Reowolf can improve performance and detect abuse.

Shared Memory Optimization

An implementation of Reowolf that recognizes multiple peers on the same physical machine can use a different implementation to realize communication: it is not necessary to send and receive packets over the wire to communicate, as the protocol can be implemented under this circumstance by means of shared memory, possibly even without copying data [8]. Synchronization of peers can then use concurrency primitives such as mutexes and thread barriers. Research has shown that the run-time performance of compiled shared memory protocols is on par with, and sometimes even outperforms, hand-crafted concurrency code [10,11]. Connectors that involve a mix of local and remote peers can leverage both optimization strategies; local communication takes place via shared memory, while remote communication is performed by exchanging IP packets.

Informed Route Optimization

The protocol description and state of connectors are publicly visible by the network. Intermediary nodes along the routing path can use this information for route optimization. Consider three participants in a protocol: a node in Helsinki repeatedly sends the same payload to two nodes in Tokyo. With sockets there are two channels, both from Helsinki to Tokyo. Each payload has to travel twice the distance. In Reowolf, a single connector is established between the three parties. Implementations can recognize by the protocol description that data is duplicated. Instead of sending the same payload twice over the long distance, it could be sent once and forwarded in Tokyo to the other recipient. The traffic on the long distance path of the network is then halved.

Local Deviation Detection

Reowolf will inspect traffic and checks it with protocol conformity. Within a peer, conformity to the local protocol is checked before passing data to an application. Since each connector is configured with an explicit protocol description, implementations of Reowolf can locally intercept incoming network traffic that violates the configured protocol, possibly raising an exceptional network condition for the application to handle. Outgoing traffic that violates the local protocol can be dropped immediately.

From Eavesdropping to Eager Dropping

During the configuration phase of connectors, peers distribute their local protocols over the network to reach a consensus on the composed protocol. Consequently, outgoing traffic that violates the composed protocol can be dropped immediately: otherwise, it would be dropped locally at the receiving peer. Moreover, intermediate network nodes are able to snoop on the composed protocol, and use it to check packets it forwards with protocol conformity, too. Thus, all nodes cooperate in checking conformity to the protocol. Since dropping reduces superfluous network traffic, it is in the best interest of the intermediary nodes to do so. Similar to TCP session hijacking, it is a challenge when considering malicious peers too: a naïve implementation may unintentionally allow a DOS.

References

1. Arbab, F.: Abstract behavior types: a foundation model for components and their composition. Sci. Comput. Program. **55**(1–3), 3–52 (2005)

2. Arbab, F.: Proper protocol. In: Ábrahám, E., Bonsangue, M., Johnsen, E.B. (eds.) Theory and Practice of Formal Methods. LNCS, vol. 9660, pp. 65–87. Springer, Cham (2016). https://doi.org/10.1007/978-3-319-30734-3_7

3. Awerbuch, B.: Complexity of network synchronization. J. ACM **32**(4), 804–823 (1985)

4. Ciatto, G., Mariani, S., Louvel, M., Omicini, A., Zambonelli, F.: Twenty years of coordination technologies: state-of-the-art and perspectives. In: Di Marzo Serugendo, G., Loreti, M. (eds.) COORDINATION 2018. LNCS, vol. 10852, pp. 51–80. Springer, Cham (2018). https://doi.org/10.1007/978-3-319-92408-3_3

5. Claffy, K.C., Huston, G., Clark, D.: Workshop on Internet Economics (WIE2017) final report. SIGCOMM Comput. Commun. Rev. **48**(3), 42–45 (2018). https://doi.org/10.1145/3276799.3276805

6. Dharmapurikar, S., Krishnamurthy, P., Sproull, T., Lockwood, J.: Deep packet inspection using parallel Bloom filters. In: Proceedings of the 11th Symposium on High Performance Interconnects, pp. 44–51. IEEE (2003)

7. Dokter, K., Jongmans, S.S., Arbab, F., Bliudze, S.: Relating BIP and Reo. arXiv preprint arXiv:1508.04848 (2015)

8. Hergarden, M., Jongmans, S.S.: Shared memory implementations of protocol programming languages, data-race-free. In: Proceedings of the 13th Workshop on Implementation, Compilation, Optimization of Object-Oriented Languages, Programs and Systems, pp. 36–40. ACM (2018). https://doi.org/10.1145/3242947.3242952

9. Jenčik, M., Mihályi, D.: Program components & abstract behavioral types. Acta Electrotech. Inform. **12**(1), 38–43 (2012)

10. Jongmans, S.S.: Automata-theoretic protocol programming. Ph.D. thesis, Centrum Wiskunde & Informatica (CWI), Leiden University (2016)

11. Jongmans, S.-S.T.Q., Arbab, F.: Can high throughput atone for high latency in compiler-generated protocol code? In: Dastani, M., Sirjani, M. (eds.) FSEN 2015. LNCS, vol. 9392, pp. 238–258. Springer, Cham (2015). https://doi.org/10.1007/978-3-319-24644-4_17

12. Mühl, G., Fiege, L., Pietzuch, P.: Distributed Event-Based Systems. Springer, Heidelberg (2006). https://doi.org/10.1007/3-540-32653-7

13. Proença, J., Clarke, D., de Vink, E., Arbab, F.: Dreams: a framework for distributed synchronous coordination. In: Proceedings of the 27th Annual ACM Symposium on Applied Computing, pp. 1510–1515. ACM (2012). https://doi.org/10.1145/2245276.2232017

14. Tankard, C.: What the GDPR means for businesses. Netw. Secur. **2016**(6), 5–8 (2016). https://doi.org/10.1016/S1353-4858(16)30056-3. http://www.sciencedirect.com/science/article/pii/S1353485816300563

15. Yu, F., Chen, Z., Diao, Y., Lakshman, T.V., Katz, R.H.: Fast and memory-efficient regular expression matching for deep packet inspection. In: Proceedings of the 12th Symposium on Architecture for Networking and Communications Systems, pp. 93–102. IEEE (2006)

Modeling and Verifying Dynamic Architectures with FACTum Studio

Habtom Kahsay Gidey[1]([✉]), Alexander Collins[2], and Diego Marmsoler[2]

[1] Universität der Bundeswehr München, Neubiberg, Germany
habtom.gidey@unibw.de
[2] Technische Universität München, Munich, Germany

Abstract. With the emergence of ambient and adaptive computing, dynamic architectures have become increasingly important. Dynamic architectures describe an evolving state space of systems over time. In such architectures, components can appear or disappear, and connections between them can change over time. Due to the evolving state space of such architectures, verification is challenging. To address this problem, we developed FACTum Studio, a tool that combines model checking and interactive theorem proving to support the verification of dynamic architectures. To this end, a dynamic architecture is first specified in terms of component types and architecture configurations. Next, each component type is verified against asserted contracts using nuXmv. Then, the composition of the contracts is verified using Isabelle/HOL. In this paper, we discuss the tool's extended features with an example of an encrypted messaging system. It is developed with Eclipse and active on Github.

Keywords: Dynamic architectures · Model checking · Interactive theorem proving · FACTum · Eclipse/EMF · Xtext

1 Introduction

Software systems that enact self-adaptation, learning, and complex reasoning are inherently dynamic [7,9,21]. They autonomously change their structure and composition at run time [5,20]. Such systems can be specified in two steps: First, a set of component types is specified using state machines [10,18]. Next, the composition of components is specified using architectural assertions [4,13].

The dynamic nature of such systems leads to a dynamically evolving state space, which makes their verification difficult. Sometimes, we lack even upper bounds on the active number of components, and so verification requires reasoning over an unbounded state space. In FACTum [14,17], we address this problem by splitting the verification process into two steps. To verify a property for the overall architecture, we first identify suitable contracts for the component types and verify them against their implementation using model checking. In a second step, we apply interactive theorem proving (ITP) to combine these verified contracts to derive the overall system property. The restricted state space of

© Springer Nature Switzerland AG 2020
F. Arbab and S.-S. Jongmans (Eds.): FACS 2019, LNCS 12018, pp. 243–251, 2020.
https://doi.org/10.1007/978-3-030-40914-2_13

single component types enables automatic verification techniques and allows fast feedback on the satisfaction of contracts. On the other hand, the expressiveness of ITP allows us to reason about potentially unbounded numbers of components when verifying overall system properties.

In a previous work [16], we presented a preliminary version of FACTUM Studio, an Eclipse-based modeling application supporting the specification of Architectural Design Patterns and their verification in Isabelle/HOL. With this paper, we introduce an extended version, adding the following features: (i) Support for graphical modeling of component behavior using annotated state machines. (ii) Corresponding code generation for the nuXmv model checker.

2   Background

We approach development of the tool, FACTUM Studio, following the FACTUM architecture verification framework [14,17]. Figure 1 illustrates the essential parts of the implemented framework. It depicts the process of architecture specification, including the creation of theories and models for verification. It also shows where ITP and model checking tools are utilized.

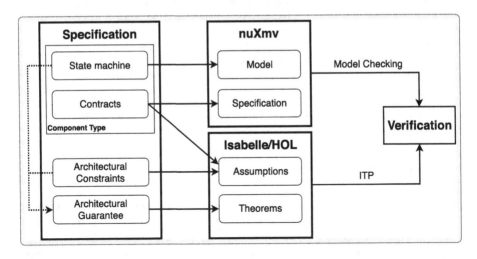

Fig. 1. Verifying dynamic architectures in FACTUM studio.

In our previous paper [16] we reported the first release of FACTUM Studio. We discussed its basic features, introducing the graphical and textual language that supports architectural pattern specifications. These core features facilitate the specification and representation of abstract data types, component types, and architectural assertions in terms of constraints and guarantees.

3   FACTum Studio 2

The second release of FACTum Studio adds the following new features: (i) textual and graphical modeling of behavior for components in terms of state machines, (ii) specification of contracts for component types in terms of LTL-formulæ, (iii) generation of models for the nuXmv model checker from component behaviors, (iv) generation of specifications for nuXmv from component type contracts, and (v) generation of additional assumptions for Isabelle locales from contracts. Moreover, the new release contains additional feature enhancements and bug fixes.

Example 1 (Secure Messaging System). Figure 2 shows an architectural diagram consisting of the key elements of the messaging protocol. The architecture represents a simple encrypted message exchange system between a sender and a receiver. Messages are encrypted with a key by the sender and decrypted with the same key on the receiving end. However, the nodes are restricted to forwarding the messages and cannot read messages in the middle. The `Encrypt` component type has an input port and output port which connects to a `Node` component type. The `Node` component type also contains ports which connect it to the `Encrypt` and `Decrypt` component types. Similarly, the `Decrypt` component type has ports connecting it to a `Node` component type. The `Node` component type represents an arbitrary number of dynamic nodes. They can join and leave the network at any time dynamically. A message moves through the active nodes until it reaches its destination. The full textual specification of the running example is provided in the project repository in GitHub [8].

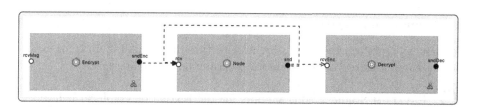

Fig. 2. Running example - secure messaging, architecture diagram.

4   Specification of Dynamic Architectures

In FACTum, specifications usually begin by describing essential architectural elements such as data types and component types. In the following, we use text boxes with a light bulb icon to highlight the new features.

4.1   Specifying Datatypes

In FACTum Studio, data types are specified using algebraic specification techniques [4, 22].

In FACTUM Studio 2 we extended the specification of data types by allowing the user to map FACTUM sorts to nuXmv data types and FACTUM operations to nuXmv operations.

Example 2 (Data Types for Secure Messaging System). Figure 3 shows a data type specification for our example. It declares a data type `Numbers` and its mapping to nuXmv data type defined in `DTMap`. The specification of data types' operations is enhanced in the new release to include several possibilities, such as ranges of numbers.

```
DTSpec {
    DT Numbers (
        Sort INT1, INT2
        Operation
                add: INT1, INT1 => INT2,
                sub: INT2, INT1 => INT1,
                key: => INT1
    DTMap {
        Sort INT1 -> 0 .. 1024,
             INT2 -> 5 .. 1029
        Operation add[x, y] -> x + y,
                  sub[x, y] -> x - y,
                  key[] -> 5
        }
    )
}
```

Fig. 3. Datatype specification

```
CType Encrypt ShortName enc {
    InputPorts {
        InputPort rcvMsg (Type: Numbers.INT1)
    }
    OutputPorts {
        OutputPort sndEnc (Type: Numbers.INT2)
    }
    Behavior {
        Variables msg1: Numbers.INT1,
                  msg2: Numbers.INT1
        States wait, encrypt
        Initial wait
        Transitions {
            wait [rcvMsg==_] -> wait
            wait -> [msg1=rcvMsg A sndEnc=Numbers.add[msg2, 5]] encrypt
            encrypt -> [msg2=rcvMsg A sndEnc=Numbers.add[msg1, 5]] wait
        }
    }
}
```

Fig. 4. Component type specification

4.2   Component Types

Component types are specified using architecture diagrams. Figure 2, for example, shows an architecture diagram for our running example.

 FACTUM Studio 2 adds support for the textual or graphical behavior specification of component types using state machines.

Example 3 (Behavior for Encrypt Component Type). In Fig. 4 we show an example of the textual description of the `Encrypt` component type. It describes the component's two ports, the input port `rcvMsg` and the output port `sndEnc`. The `Behaviour` code block describes the specification of the component type's behavior. It contains a description of states, `wait` and `encrypt`. It defines an initial state `wait` where its initial value is set to `rcvMsg`. It also defines the `wait` and `encrypt` state transitions as a start and end of the transition behavior, respectively. Figure 5 shows the behavior specified graphically. The graphical representations are also editable with text annotations.

Fig. 5. Graphical behavior specification with state machines.

4.3 Architecture Specification

In FACTUM Studio, architectural configurations are specified using architectural assertions: LTL-formulæ expressed over the architecture. In FACTUM Studio 2, the architecture specification did not change much [16].

Example 4 (Architecture Specification for Secure Messaging System). In Fig. 6, we have the architectural constraints set over sets of component instances. The formulæ **enc** and **dec** assert a property that requires a unique decryption and a unique encryption components set active at any time point. The **con** predicate specifies the connection property of components within the architecture.

```
ArchSpec {
    flex enc1 : Encrypt,
    flex enc2 : Encrypt,
    flex dec1 : Decrypt,
    flex dec2 : Decrypt,
    flex nd1 : Node,
    flex nd2 : Node

    // Only one instance of Encrypt and Decrypt
    enc: G (cAct(enc1) ∧ (∀enc2.(cAct(enc2) → eq(enc2, enc1))))
    dec: G (cAct(dec1) ∧ (∀dec2.(cAct(dec2) → eq(dec2, dec1))))

    // Specify connection
    con: ∃nd1. (∃nd2. (G (cAct(nd1) ∧ cAct(nd2) ∧ conn(enc1.sndEnc, nd1.rcv)
         ∧ conn(nd1.snd, nd2.rcv) ∧ conn(nd2.snd, dec1.rcvEnc))))
}
```

```
ArchGuarantee {
    flex enc : Encrypt,
    flex dec : Decrypt

    flex msg : Numbers.INT1

    g: G (val(enc.rcvMsg, msg) → (F(val(dec.sndDec, msg))))
}
```

Fig. 6. Architectural constraints   **Fig. 7.** Architectural guarantees

5 Verification Using Model Checking and ITP

In FACTUM, architectures are verified in two steps: First, we specify the architectural guarantees and establish suitable contracts. Contracts are then verified against described behaviors of component types using model checking. Secondly, the verified contracts are combined with the architectural specifications to verify the overall architecture using interactive theorem proving.

Example 5 (Guarantee for Secure Messaging System). In Fig. 7, the architectural guarantee **g** ensures a property that when a message **msg** is sent by the sender it will eventually be delivered to the receiver and cannot be read in between.

 FACTUM Studio 2 now supports the specification of contracts for component types using LTL-formulæ over their ports.

5.1   Verifying Component Types

As discussed in Sect. 1, FACTUM Studio automatically verifies individual implementations of component types using the model checker nuXmv. To this end, we first identify suitable contracts for each type of component. Contracts specify behaviors of component types expressed as behavior trace assertions [15] (Fig. 8).

```
Contracts {
flex msg: Numbers.INT1

    c: G (([msg=rcvMsg]) ⇒ (F([sndEnc=Numbers.add[msg, Numbers.key[]]])))
}
```

Fig. 8. Contract for the encrypt component type.

Example 6 (Contracts for Component Types). In Fig. 8, a suitable contract for the encryption component type is specified as c. The contract describes the desired property for encrypting and forwarding every message. It asserts a liveness property so that, eventually, every input is encrypted.

Then, FACTUM Studio can be used to generate nuXmv code for each component type. The corresponding algorithm is shown in Algorithm 1. It describes the mapping process followed during the model transformations.

 FACTUM Studio 2 now supports the generation of nuXmv code.

Algorithm 1. nuXmv code generation

Require: pattern
1: **for all** component types **do**
2:    **for all** states **do**
3:        add state to enum
4:    **end for**
5:    **for all** state machine variables **do**
6:        encode as variable with sort
7:    **end for**
8:    **for all** input ports and output ports **do**
9:        encode as variable with sort
10:        create noVal port as boolean
11:    **end for**
12:    encode initialization of variables
13:    encode state machines in assignment style
14:    encode transitions of noVal output ports in assignment style
15:    **for all** contracts **do**
16:        encode as LTLSpec
17:    **end for**
18: **end for**

5.2   Verifying the Architecture

As a final step, FACTum Studio generates Isabelle/HOL and nuXmv code verification. It transforms the specified architectural assertions to corresponding Isabelle/HOL assumptions and theories. Similarly, the state machines and contracts of individual components are transformed into corresponding nuXmv models and specifications, see Fig. 1. At this stage, FACTum Studio checks the state machine models of individual component types against their contracts. Next, it combines the verified contracts with the Isabelle/HOL theorems generated from the architectural guarantee. Using ITP, we can then sketch Isabelle/HOL proof interactively to finalize the verification of dynamic architectures.

 FACTum Studio 2 now also exports component contracts as assumptions for the corresponding Isabelle locale.

Example 7. The complete generated Isabelle/HOL theory and the proof for the running example are provided in the code repository [8].

6   Related Work

The first category and significant number of related works are tools such as AutoFocus, RoboChart, VerCor, and others using languages DynAlloy, Dynamic Wright [1,6,19]. These tools use automated verification with model checking or an analyzer such as the Alloy Analyzer. The tools in this category do not provide support for ITP based dynamic architecture verification.

The second category of tools supports model transformations of specifications to proof assistants. These tools are few in number and would include the previous version of FACTum Studio, Reo, and STeP [2,3,11,12,16].

However, to the best of our knowledge, the new FACTum Studio is the only tool to use a combined approach of both ITP and model checking for the verification of dynamically adapting, component-based systems.

7   Conclusion and Outlook

Verification of dynamic architecture specifications for software-intensive systems requires a combined approach to address the varying nature of challenges in the domain. In this paper, with the implementation of new features in FACTum Studio, we described how to approach and address the state-space scalability problem of dynamic architecture verification by using model checking and ITP. First, we demonstrated how individual architectural elements and their dynamic properties are specified and verified using a model checker. We then used individually checked contracts to verify assured compositional constraints using ITP.

Since ITP requires a steep learning curve, in the next versions of FACTum Studio, we plan to implement some form of automatic proof generation from specified guarantees. Moreover, we also plan to support the hierarchical specification of component types and their configurations. That can enable users to import, use, or extend already specified and verified dynamic architecture patterns.

Acknowledgments. Parts of the work on which we report in this paper were funded by the German Federal Ministry of Economics and Technology (BMWi) under grant no. 0325811A.

References

1. Aravantinos, V., Voss, S., Teufl, S., Hölzl, F., Schätz, B.: AutoFOCUS 3: tooling concepts for seamless, model-based development of embedded systems. In: CEUR Workshop Proceedings, vol. 1508, pp. 19–26. CEUR-WS.org (2015)
2. Arbab, F.: Reo: a channel-based coordination model for component composition. Math. Struct. Comput. Sci. **14**(03), 329–366 (2004). https://doi.org/10.1017/s0960129504004153
3. Baier, C., Sirjani, M., Arbab, F., Rutten, J.: Modeling component connectors in Reo by constraint automata. Sci. Comput. Program. **61**(2), 75–113 (2006)
4. Broy, M.: A model of dynamic systems. In: Bensalem, S., Lakhneck, Y., Legay, A. (eds.) ETAPS 2014. LNCS, vol. 8415, pp. 39–53. Springer, Heidelberg (2014). https://doi.org/10.1007/978-3-642-54848-2_3
5. Bruni, R., Bucchiarone, A., Gnesi, S., Melgratti, H.: Modelling dynamic software architectures using typed graph grammars. Electron. Notes Theor. Comput. Sci. **213**(1), 39–53 (2008)
6. Bucchiarone, A., Galeotti, J.P.: Dynamic software architectures verification using DynAlloy. In: Electronic Communications of the EASST, vol. 10 (2008)
7. Gerostathopoulos, I., Skoda, D., Plasil, F., Bures, T., Knauss, A.: Architectural homeostasis in self-adaptive software-intensive cyber-physical systems. In: Tekinerdogan, B., Zdun, U., Babar, A. (eds.) ECSA 2016. LNCS, vol. 9839, pp. 113–128. Springer, Cham (2016). https://doi.org/10.1007/978-3-319-48992-6_8
8. Gidey, H.K., Marmsoler, D.: FACTum Studio (2018). https://habtom.github.io/factum/
9. Koza, J.R.: Genetic programming: on the programming of computers by means of natural selection, vol. 1. MIT Press (1992)
10. Li, C., Huang, L., Chen, L., Li, X., Luo, W.: Dynamic software architectures: formal specification and verification with CSP. In: Proceedings of the Fourth Asia-Pacific Symposium on Internetware, p. 5. ACM (2012)
11. Li, Y., Sun, M.: Modeling and analysis of component connectors in Coq. In: Fiadeiro, J.L., Liu, Z., Xue, J. (eds.) FACS 2013. LNCS, vol. 8348, pp. 273–290. Springer, Cham (2014). https://doi.org/10.1007/978-3-319-07602-7_17
12. Manna, Z., Sipma, H.B.: Deductive verification of hybrid systems using step. In: Henzinger, T.A., Sastry, S. (eds.) HSCC 1998. LNCS, vol. 1386, pp. 305–318. Springer, Heidelberg (1998). https://doi.org/10.1007/3-540-64358-3_47
13. Marmsoler, D.: Towards a calculus for dynamic architectures. In: Hung, D., Kapur, D. (eds.) ICTAC 2017. LNCS, vol. 10580, pp. 79–99. Springer, Cham (2017). https://doi.org/10.1007/978-3-319-67729-3_6

14. Marmsoler, D.: A framework for interactive verification of architectural design patterns in Isabelle/HOL. In: Sun, J., Sun, M. (eds.) ICFEM 2018. LNCS, vol. 11232, pp. 251–269. Springer, Cham (2018). https://doi.org/10.1007/978-3-030-02450-5_15

15. Marmsoler, D.: Hierarchical specification and verification of architectural design patterns. In: Russo, A., Schürr, A. (eds.) FASE 2018. LNCS, vol. 10802, pp. 149–168. Springer, Cham (2018). https://doi.org/10.1007/978-3-319-89363-1_9

16. Marmsoler, D., Gidey, H.K.: FACTUM studio: a tool for the axiomatic specification and verification of architectural design patterns. In: Bae, K., Ölveczky, P.C. (eds.) FACS 2018. LNCS, vol. 11222, pp. 279–287. Springer, Cham (2018). https://doi.org/10.1007/978-3-030-02146-7_14

17. Marmsoler, D., Gidey, H.K.: Interactive verification of architectural design patterns in FACTum. Formal Aspects Comput. **31**(5), 541–610 (2019)

18. Marmsoler, D., Gleirscher, M.: Specifying properties of dynamic architectures using configuration traces. In: Sampaio, A., Wang, F. (eds.) ICTAC 2016. LNCS, vol. 9965, pp. 235–254. Springer, Cham (2016). https://doi.org/10.1007/978-3-319-46750-4_14

19. Miyazawa, A., Cavalcanti, A., Ribeiro, P., Li, W., Woodcock, J., Timmis, J.: Robochart reference manual. Technical report, University of York (2017)

20. Oquendo, F.: Dynamic software architectures: formally modelling structure and behaviour with Pi-ADL. In: 2008 The Third International Conference on Software Engineering Advances, pp. 352–359. IEEE (2008)

21. Oreizy, P., et al.: An architecture-based approach to self-adaptive software. IEEE Intell. Syst. Appl. **14**(3), 54–62 (1999)

22. Wirsing, M.: Algebraic specification. In: van Leeuwen, J. (ed.) Handbook of Theoretical Computer Science (Vol. B), pp. 675–788. MIT Press, Cambridge (1990)

Revisiting Trace Equivalences for Markov Automata

Arpit Sharma[✉]

EECS Department, Indian Institute of Science Education and Research Bhopal,
Bhopal, India
arpit@iiserb.ac.in

Abstract. Equivalences are important for system synthesis as well as system analysis. This paper defines new variants of trace equivalence for Markov automata (MAs). We perform button pushing experiments with a black box model of MA to obtain these equivalences. For every class of MA scheduler, a corresponding variant of trace equivalence is defined. We investigate the relationship among these equivalences and also prove that each variant defined in this paper is strictly coarser than the corresponding variant of trace equivalence defined originally in [12]. Next, we establish the relationship between our equivalences and bisimulation for MAs. Finally, we investigate the relationship of these equivalences with trace relations defined in the literature for some of the implied models.

Keywords: Markov · Scheduler · Equivalence · Trace · Bisimulation · Stochastic

1 Introduction

Markov automata (MAs) [8] extend probabilistic automata (PAs) [10] with stochastic aspects [9]. MAs thus support non-deterministic probabilistic branching and exponentially distributed delays in continuous time. MAs are compositional, i.e., a parallel composition operator allows one to construct a complex MA from several component MAs running in parallel. They provide a natural semantics for a variety of specifications for concurrent systems.

Behavioral equivalences are widely used for system analysis and implementation verification. For example, equivalences have been used to efficiently check if the implementation is an approximation of specification of the expected behavior. Additionally, equivalences are also used for reducing the size of stochastic models by combining equivalent states into a single state. For MAs, research has mainly concentrated on branching-time equivalences, e.g., strong bisimulation [8] and several variants of weak/stutter bisimulation [1,6,8,14]. Only recently we have studied several variants of trace equivalence for closed[1] MAs [12,13]. These trace equivalences have been defined by using button pushing experiments with

---

[1] An MA is said to be closed if it is not subject to any further synchronization.

© Springer Nature Switzerland AG 2020
F. Arbab and S.-S. Jongmans (Eds.): FACS 2019, LNCS 12018, pp. 252–260, 2020.
https://doi.org/10.1007/978-3-030-40914-2_14

stochastic trace machines. Since schedulers are used to resolve non-deterministic choices in MAs, for every class of MA scheduler, a corresponding variant of trace equivalence has been defined. Roughly speaking, two MAs $\mathcal{M}_1, \mathcal{M}_2$ are trace equivalent (w.r.t. scheduler class $\mathcal{C}$), denoted $\equiv_{\mathcal{C}}$, if for every scheduler $\mathcal{D}$ of class $\mathcal{C}$ of $\mathcal{M}_1$ there exists a scheduler $\mathcal{D}'$ of class $\mathcal{C}$ of $\mathcal{M}_2$ such that for all timed traces, i.e., (σ, θ), we have $P^{trace}_{\mathcal{M}_1, \mathcal{D}}(\sigma, \theta) = P^{trace}_{\mathcal{M}_2, \mathcal{D}'}(\sigma, \theta)$ and vice versa. Here, $P^{trace}_{\mathcal{M}_1, \mathcal{D}}(\sigma, \theta)$ denotes the probability of all timed paths that are compatible with the timed trace (σ, θ) in $\mathcal{M}_1$ under scheduler $\mathcal{D}$. An issue with these equivalences is that they are too restrictive and discriminating because they require to fix a pair of corresponding schedulers and compare the probability of paths compatible with the timed traces obtained under the effect of these schedulers. As a result, a restriction is imposed on the probability of paths compatible with every timed trace obtained by performing button pushing experiments.

This paper focuses on defining new (coarser) variants of trace equivalence for closed MAs. More specifically, we define a coarser variant of stationary deterministic trace equivalence and a coarser variant of stationary randomized trace equivalence for closed MAs. Unlike [12], we relax the definition of trace equivalence w.r.t. scheduler class $\mathcal{C}$ by comparing the probability of paths compatible with a single trace at a time. In other words, our definitions require quantification over timed traces to compare the probability of compatible paths rather than quantification over schedulers. Our work is inspired from the new definitions of trace and testing equivalences defined in [4,5] for non-deterministic and probabilistic processes. We formally prove that our equivalences are coarser than the ones defined in [12] and also study the relationship among newly defined trace equivalences. Next, we investigate the relationship between our trace equivalences and bisimulation for MAs [8]. Finally, we study the relationship between our equivalences and trace equivalences defined in the literature for Kripke structures (KSs) [3], continuous-time Markov chains (CTMCs) [15] and probabilistic automata (PAs) [11].

Organisation of the Paper. Section 2 briefly recalls the main concepts of MAs. Section 3 briefly recalls trace equivalences for closed MAs. Section 4 defines new variants of trace equivalence. Section 5 investigates their relationship with trace relations for other models. Finally, Sect. 6 concludes the paper and provides pointers for future research.

2   Preliminaries

This section presents the necessary definitions and basic concepts related to Markov automata (MA) that are needed for the understanding of the rest of this paper. Let $Distr(S)$ denotes the set of probability distribution functions over the countable set S.

Definition 1 (MA). *A* Markov automaton *(MA) is a tuple* $\mathcal{M} = (S, s_0, Act, AP, \rightarrow, \Rightarrow, L)$ *where:*

1. S is a nonempty finite set of states,   2. s_0 is the initial state,
3. Act is a finite set of actions,   4. AP is a finite set of atomic propositions,
5. $\rightarrow \subseteq S \times Act \times Distr(S)$ is the probabilistic transition relation,
6. $\Rightarrow \subseteq S \times \mathbb{R}_{\geq 0} \times S$ is the Markovian transition relation, and
7. $L: S \rightarrow 2^{AP}$ is a labeling function.

We abbreviate $(s, \alpha, \mu) \in \rightarrow$ as $s \xrightarrow{\alpha} \mu$ and similarly, $(s, \lambda, s') \in \Rightarrow$ by $s \xRightarrow{\lambda} s'$. Let $PT(s)$ and $MT(s)$ denote the set of probabilistic and Markovian transitions that leave state s. A state s is *Markovian* iff $MT(s) \neq \varnothing$ and $PT(s) = \varnothing$; it is *probabilistic* iff $MT(s) = \varnothing$ and $PT(s) \neq \varnothing$. Further, s is a *hybrid* state iff $MT(s) \neq \varnothing$ and $PT(s) \neq \varnothing$; finally s is a *deadlock* state iff $MT(s) = \varnothing$ and $PT(s) = \varnothing$. In this paper we only consider MAs that do not have any deadlock state. Let $MS \subseteq S$ and $PS \subseteq S$ denote the set of Markovian and probabilistic states in MA $\mathcal{M}$. For any Markovian state $s \in MS$, let $R(s, s') = \sum\{\lambda | s \xRightarrow{\lambda} s'\}$ be the rate to move from state s to state s'. For $C \subseteq S$, let $R(s, C) = \sum_{s' \in C} R(s, s')$ be the rate to move from state s to a set of states C. The exit rate for state s is defined by: $E(s) = \sum_{s' \in S} R(s, s')$.

It is easy to see that an MA where $MT(s) = \varnothing$ for any state s is a probabilistic automaton (PA) [10]. An MA where $PT(s) = \varnothing$ for any state s is a continuous-time Markov chain (CTMC) [2]. The semantics of MAs can thus be given in terms of the semantics of CTMCs (for Markovian transitions) and PAs (for probabilistic transitions).

The meaning of a Markovian transition $s \xRightarrow{\lambda} s'$ is that the MA moves from state s to s' within t time units with probability $1 - e^{-\lambda \cdot t}$. If s has multiple outgoing Markovian transitions to different successors, then we speak of a race between these transitions, known as the *race condition*. In this case, the probability to move from s to s' within t time units is $\frac{R(s, s')}{E(s)} \cdot (1 - e^{-E(s) \cdot t})$. In closed MAs all outgoing probabilistic transitions from every state $s \in S$ are labeled with $\tau \in Act$ (internal action). Note that in any closed MA, probabilistic transitions take precedence over Markovian transitions [8]. Intuitively, this means that in a closed MA, τ labeled transitions are not subject to interaction and thus can happen immediately[2], whereas the probability of a Markovian transition to happen immediately is zero. Accordingly, we assume that each state s has either only outgoing τ transitions or outgoing Markovian transitions. In other words, a closed MA only has probabilistic and Markovian states. We use a distinguished action $\perp \notin Act$ to indicate Markovian transitions. Let $Act_\perp = Act \cup \{\perp\}$.

Definition 2 (MA timed paths). *Let $\mathcal{M} = (S, s_0, Act, AP, \rightarrow, \Rightarrow, L)$ be an MA. An infinite path π in $\mathcal{M}$ is a sequence $s_0 \xrightarrow{\sigma_0, t_0} s_1 \xrightarrow{\sigma_1, t_1} s_2 \ldots s_{n-1} \xrightarrow{\sigma_{n-1}, t_{n-1}} s_n \ldots$ where $s_i \in S$, $\sigma_i \in Act$ or $\sigma_i = \perp$, and for each i, there exists a measure μ_i such that $(s_i, \sigma_i, \mu_i) \in \rightarrow$ with $\mu_i(s_{i+1}) > 0$. For $\sigma_i \in Act$, $s_i \xrightarrow{\sigma_i, t_i} s_{i+1}$ denotes that after residing t_i time units in s_i,*

---

[2] We restrict to models without Zenoness. Simply, this means that τ cycles are not allowed.

the MA $\mathcal{M}$ has moved via action σ_i to s_{i+1} with probability $\mu_i(s_{i+1})$. On the other hand, $s_i \xrightarrow{\perp, t_i} s_{i+1}$ denotes that after residing t_i time units in s_i, a Markovian transition led to s_{i+1} with probability $\mu_i(s_{i+1}) = P(s_i, s_{i+1})$ where $P(s_i, s_{i+1}) = \frac{R(s_i, s_{i+1})}{E(s_i)}$. A finite path π is a finite prefix of an infinite path. The length of an infinite path π, denoted $|\pi|$ is ∞; the length of a finite path π with $n+1$ states is n.

Let $Paths^{\mathcal{M}} = Paths_{fin}^{\mathcal{M}} \cup Paths_{\omega}^{\mathcal{M}}$ denotes the set of all paths in $\mathcal{M}$ that start in s_0, where $Paths_{fin}^{\mathcal{M}} = \bigcup_{n \in \mathbb{N}} Paths_n^{\mathcal{M}}$ is the set of all finite paths in $\mathcal{M}$ and $Paths_n^{\mathcal{M}}$ denotes the set of all finite paths of length n that start in s_0. Let $Paths_{\omega}^{\mathcal{M}}$ be the set of all infinite paths in $\mathcal{M}$ that start in s_0. Trace of an infinite path $\pi = s_0 \xrightarrow{\sigma_0, t_0} s_1 \xrightarrow{\sigma_1, t_1} s_2 \ldots s_{n-1} \xrightarrow{\sigma_{n-1}, t_{n-1}} s_n \ldots$ denoted $Trace(\pi)$ is given as $L(s_0)\sigma_0 L(s_1)\sigma_1 \ldots L(s_{n-1}) \sigma_{n-1} L(s_n) \ldots$. Trace of a finite path π can be defined in an analogous manner. Let $|\varsigma|$ denotes the number of actions in a finite trace ς. Non-determinism in an MA is resolved by a scheduler. Schedulers are also known as adversaries or policies. More formally, schedulers are defined as follows:

Definition 3 (Scheduler). *A scheduler $\mathcal{D}$ for MA $\mathcal{M}$ is*

- *stationary deterministic (SD) if $\mathcal{D}: S \to Act_{\perp}$ such that $\mathcal{D}(s) \in Act(s)$*
- *stationary randomized (SR) if $\mathcal{D}: S \to Distr(Act_{\perp})$ such that $\mathcal{D}(s)(\alpha) > 0$ implies $\alpha \in Act(s)$*

Let $Sched(\mathcal{M})$ denotes the set of all schedulers of $\mathcal{M}$. Let $Sched_{\mathcal{C}}(\mathcal{M})$ denotes the set of all schedulers of class $\mathcal{C}$. Let $Paths_{\mathcal{D}}^{\mathcal{M}}$ denotes the set of all infinite paths of $\mathcal{M}$ under $\mathcal{D} \in Sched(\mathcal{M})$ that start in s_0.

3 Trace Equivalences

In [12], several variants of trace equivalence for closed MAs have been defined. These equivalences were obtained by performing push-button experiments with a stochastic trace machine $\mathcal{M}$. The machine is equipped with an action display, a state label display, a timer and a reset button. Consider a run of the machine (under scheduler $\mathcal{D}$ of class $\mathcal{C}$) which always starts from the initial state. The state label shows the label of the current state and action display shows the last action[3] that has been executed. Note that the action display remains unchanged until the next action is executed by the machine. The timer display shows the absolute time. The reset button is used to restart the machine for another run starting from the initial state. An external observer records the sequence of state labels, actions and time checks where each time check is recorded at an arbitrary time instant between the occurrence of two successive actions. The observer can press the reset button to stop the current run. Once the reset button is pressed, the action display will be empty and the state label display shows the set of

---

[3] For Markovian states, action display shows the distinguished action $\perp$.

atomic propositions that are true in the initial state. The machine then starts for another run and the observer again records the sequence of actions, state labels and time checks. Note that the machine needs to be executed for infinitely many runs to complete the whole experiment. It is assumed that the observer can distinguish between two successive actions that are equal. An outcome of this machine is $(\sigma, \theta) = (<L(s_0)\sigma_0 L(s_1)\sigma_1 \ldots L(s_{n-1})\sigma_{n-1}L(s_n)>, <t'_0, t'_1, \ldots, t'_n>)$, where $\sigma_0, \ldots, \sigma_{n-1} \in \{\tau, \bot\}$. This outcome can be interpreted as follows: for $0 \leq m < n$, action σ_m of the machine is performed in the time interval $(y_m, y_{m+1}]$ where $y_m = \Sigma_{i=0}^m t'_i$.

Definition 4 (Compatibility). *Let* $(\sigma, \theta) = (<L(s_0)\sigma_0 L(s_1)\sigma_1 \ldots L(s_{n-1})$ $\sigma_{n-1}L(s_n)>, <t'_0, t'_1, \ldots, t'_n>)$ *be an outcome of* $\mathcal{M}$ *under* $\mathcal{D} \in Sched(\mathcal{M})$, *then a path* $\pi = s_0 \xrightarrow{\sigma_0, t_0} s_1 \xrightarrow{\sigma_1, t_1} s_2 \ldots s_{n-1} \xrightarrow{\sigma_{n-1}, t_{n-1}} s_n \ldots \in Paths_\mathcal{D}^\mathcal{M}$ *is said to be compatible with* (σ, θ), *denoted* $\pi \triangleright (\sigma, \theta)$, *if the following holds:*

$$Trace(\pi[0 \ldots n]) = \sigma \quad and \quad \Sigma_{j=0}^i t_j \in (y_i, y_{i+1}] \; for \; 0 \leq i < n$$

where $y_i = \Sigma_{j=0}^i t'_j$.

The probability of all the paths compatible with an outcome is defined as follows:

Definition 5. *Let* (σ, θ) *be an outcome of trace machine* $\mathcal{M}$ *under* $\mathcal{D} \in$ $Sched(\mathcal{M})$. *Then the probability of all the paths compatible with* (σ, θ) *is defined as follows:*

$$P_{\mathcal{M}, \mathcal{D}}^{trace}(\sigma, \theta) = Pr_\mathcal{D}(\{\pi \in Paths_\mathcal{D}^\mathcal{M} | \pi \triangleright (\sigma, \theta)\})$$

Informally, $P_{\mathcal{M}, \mathcal{D}}^{trace}$ is a function that gives the probability to observe (σ, θ) in machine $\mathcal{M}$ under scheduler $\mathcal{D}$.

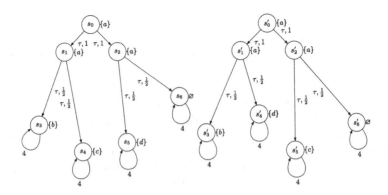

Fig. 1. MAs $\mathcal{M}$ (left) and $\mathcal{M}'$ (right)

Definition 6 (Set of observations). *Let $P^{trace}_{\mathcal{M},\mathcal{D}}$ be an observation of machine $\mathcal{M}$ under $\mathcal{D} \in Sched(\mathcal{M})$. Then the set of observations for scheduler class $\mathcal{C}$, denoted $O_\mathcal{C}(\mathcal{M})$, is defined as follows:*

$$O_\mathcal{C}(\mathcal{M}) = \{P^{trace}_{\mathcal{M},\mathcal{D}} | \mathcal{D} \in Sched_\mathcal{C}(\mathcal{M})\}$$

Informally, $O_\mathcal{C}(\mathcal{M})$ denote a set of functions where each function assigns a probability value to every possible outcome of the trace machine, i.e., (σ, θ).

Definition 7 (Trace equivalence). *Two MAs $\mathcal{M}_1$, $\mathcal{M}_2$ are trace equivalent w.r.t. scheduler class $\mathcal{C}$ denoted $\mathcal{M}_1 \equiv_\mathcal{C} \mathcal{M}_2$ iff $O_\mathcal{C}(\mathcal{M}_1) = O_\mathcal{C}(\mathcal{M}_2)$.*

This definition says that for every $\mathcal{D} \in Sched_\mathcal{C}(\mathcal{M}_1)$ there exists a scheduler $\mathcal{D}' \in Sched_\mathcal{C}(\mathcal{M}_2)$ such that for all outcomes (σ, θ) we have $P^{trace}_{\mathcal{M}_1,\mathcal{D}}(\sigma, \theta) = P^{trace}_{\mathcal{M}_2,\mathcal{D}'}(\sigma, \theta)$ and vice versa. In [12], connections among these equivalences have been investigated, i.e., $\equiv_{SD} \implies \equiv_{SR}$ and $\equiv_{SR} \not\implies \equiv_{SD}$.

4   Coarser Trace Equivalences

This section defines coarser variants of trace equivalence for MAs. We obtain these equivalences by relaxing the definition of traces defined in Sect. 3. Our new definitions require quantification over timed traces to compare the probability of compatible paths.

Definition 8. *Two MAs $\mathcal{M}_1$, $\mathcal{M}_2$ are trace equivalent w.r.t. scheduler class $\mathcal{C}$ denoted $\mathcal{M}_1 \equiv_{\mathcal{C},new} \mathcal{M}_2$ iff for all (σ, θ) it holds that:*

1. *For each $\mathcal{D} \in Sched_\mathcal{C}(\mathcal{M}_1)$ there exists $\mathcal{D}' \in Sched_\mathcal{C}(\mathcal{M}_2)$ such that: $P^{trace}_{\mathcal{M}_1,\mathcal{D}}(\sigma, \theta) = P^{trace}_{\mathcal{M}_2,\mathcal{D}'}(\sigma, \theta)$*
2. *For each $\mathcal{D}' \in Sched_\mathcal{C}(\mathcal{M}_2)$ there exists $\mathcal{D} \in Sched_\mathcal{C}(\mathcal{M}_1)$ such that: $P^{trace}_{\mathcal{M}_2,\mathcal{D}'}(\sigma, \theta) = P^{trace}_{\mathcal{M}_1,\mathcal{D}}(\sigma, \theta)$*

Next, we investigate the connections between trace equivalence ($\equiv_\mathcal{C}$) and new trace equivalence ($\equiv_{\mathcal{C},new}$).

Theorem 1. *The following holds:*

1. *$\equiv_{SD} \implies \equiv_{SD,new}$, $\equiv_{SD,new} \not\implies \equiv_{SD}$, $\equiv_{SD} \implies \equiv_{SR,new}$ and $\equiv_{SR,new} \not\implies \equiv_{SD}$*
2. *$\equiv_{SR} \implies \equiv_{SR,new}$, $\equiv_{SR,new} \not\implies \equiv_{SR}$, $\equiv_{SR} \not\implies \equiv_{SD,new}$ and $\equiv_{SD,new} \not\implies \equiv_{SR}$*
3. *$\equiv_{SD,new} \implies \equiv_{SR,new}$ and $\equiv_{SR,new} \not\implies \equiv_{SD,new}$*

Example 1. Consider the two MAs $\mathcal{M}$ and $\mathcal{M}'$ shown in Fig. 1. These two MAs are $\equiv_{SD,new}$ and $\equiv_{SR,new}$ but they are $\not\equiv_{SD}$ and $\not\equiv_{SR}$.

Example 2. Consider the two MAs $\mathcal{M}$ and $\mathcal{M}'$ shown in Fig. 2. These two MAs are $\equiv_{SR}$ and $\equiv_{SR,new}$ but they are $\not\equiv_{SD,new}$.

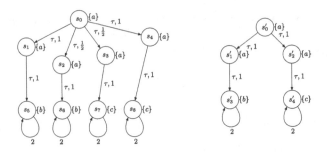

Fig. 2. MAs $\mathcal{M}$ (left) and $\mathcal{M}'$ (right)

Next, we investigate the relationship of bisimulation to new variants of trace equivalence defined in this paper. Informally, two states are bisimilar if they are able to mimic each other's behavior step-wise. Due to space limit we do not provide the definition of bisimulation (denoted $\sim$), but refer the interested reader to [8,12]. In [12], it has been proved that $\sim \not\Rightarrow \equiv_{SD}$, $\equiv_{SD} \not\Rightarrow \sim$, $\sim \Rightarrow \equiv_{SR}$ and $\equiv_{SR} \not\Rightarrow \sim$.

Theorem 2. *The following holds:*

1. $\sim \not\Rightarrow \equiv_{SD,new}$ *and* $\equiv_{SD,new} \not\Rightarrow \sim$ *2.* $\sim \Rightarrow \equiv_{SR,new}$ *and* $\equiv_{SR,new}$
$\not\Rightarrow \sim$

Example 3. Consider the two MAs $\mathcal{M}$ and $\mathcal{M}'$ shown in Fig. 2. These two MAs are $\equiv_{SR,new}$ but they are $\not\sim$.

5   Connections with Other Models

MAs subsume many different concurrent models [7,8]. In the following we will prove that the trace equivalences defined in this paper are backward compatible with the trace equivalences defined in the literature for some of these implied models. An MA where $PT(s) = \varnothing$ for all $s \in S$ (fully stochastic) is a CTMC. Similarly, an MA where $MT(s) = \varnothing$ for all $s \in S$ (non-deterministic and probabilistic) is a PA. Alternatively, an MA where $MT(s) = \varnothing$ for all $s \in S$ and all transitions are Dirac[4] (fully non-deterministic) is a Kripke structure (KS). Let $\equiv_{AT}$ denotes trace equivalence for CTMCs [15]. Let $\equiv_{FN}$ denotes trace equivalence for KSs [3]. Similarly, let $\sim_{PTr,dis}$ denotes SD trace equivalence and $\sim_{PTr,dis}^{ct}$ denotes SR trace equivalence for PAs [11].

Theorem 3. *Let $\mathcal{M}_1$, $\mathcal{M}_2$ be two fully stochastic MAs. Then the following holds:*

– $\mathcal{M}_1 \equiv_{SD} \mathcal{M}_2 \Leftrightarrow \mathcal{M}_1 \equiv_{SR} \mathcal{M}_2$

---

[4] We call a distribution Dirac, if there exists only one successor for each action.

- $\mathcal{M}_1 \equiv_{\mathcal{SR}} \mathcal{M}_2 \Leftrightarrow \mathcal{M}_1 \equiv_{\mathcal{SD},new} \mathcal{M}_2$
- $\mathcal{M}_1 \equiv_{\mathcal{SD},new} \mathcal{M}_2 \Leftrightarrow \mathcal{M}_1 \equiv_{\mathcal{SR},new} \mathcal{M}_2$
- $\mathcal{M}_1 \equiv_{\mathcal{SR},new} \mathcal{M}_2 \implies \mathcal{M}_1 \equiv_{AT} \mathcal{M}_2$
- $\mathcal{M}_1 \equiv_{AT} \mathcal{M}_2 \nRightarrow \mathcal{M}_1 \equiv_{\mathcal{SD}} \mathcal{M}_2$

Theorem 4. *Let $\mathcal{M}_1$, $\mathcal{M}_2$ be two fully non-deterministic MAs. Then the following holds:*

- $\mathcal{M}_1 \equiv_{\mathcal{SD}} \mathcal{M}_2 \Leftrightarrow \mathcal{M}_1 \equiv_{\mathcal{SR}} \mathcal{M}_2$
- $\mathcal{M}_1 \equiv_{\mathcal{SR}} \mathcal{M}_2 \Leftrightarrow \mathcal{M}_1 \equiv_{\mathcal{SD},new} \mathcal{M}_2$
- $\mathcal{M}_1 \equiv_{\mathcal{SD},new} \mathcal{M}_2 \Leftrightarrow \mathcal{M}_1 \equiv_{\mathcal{SR},new} \mathcal{M}_2$
- $\mathcal{M}_1 \equiv_{\mathcal{SR},new} \mathcal{M}_2 \Leftrightarrow \mathcal{M}_1 \equiv_{FN} \mathcal{M}_2$

Theorem 5. *Let $\mathcal{M}_1$, $\mathcal{M}_2$ be two non-deterministic and probabilistic MAs. Then the following holds:*

- $\mathcal{M}_1 \sim_{PTr,dis} \mathcal{M}_2 \nRightarrow \mathcal{M}_1 \equiv_{\mathcal{SD},new} \mathcal{M}_2$
- $\mathcal{M}_1 \equiv_{\mathcal{SD},new} \mathcal{M}_2 \implies \mathcal{M}_1 \sim_{PTr,dis} \mathcal{M}_2$
- $\mathcal{M}_1 \sim_{PTr,dis}^{ct} \mathcal{M}_2 \nRightarrow \mathcal{M}_1 \equiv_{\mathcal{SR},new} \mathcal{M}_2$
- $\mathcal{M}_1 \equiv_{\mathcal{SR},new} \mathcal{M}_2 \implies \mathcal{M}_1 \sim_{PTr,dis}^{ct} \mathcal{M}_2$

6   Conclusions

We have defined new variants of trace equivalence for closed MAs. We have proved that our variants of trace equivalence are strictly coarser that the ones defined in [12]. We have investigated the relationship among these new equivalences and have also compared them with bisimulation for MAs. Finally, we have investigated the relationship of our equivalences with the trace equivalences defined in literature for implied models. Sketching the complete linear-time branching-time spectrum of equivalences for MAs is left for future work.

References

1. Aldini, A., Bernardo, M.: Expected-delay-summing weak bisimilarity for Markov automata. In: QAPL. EPTCS, vol. 194, pp. 1–15 (2015)
2. Baier, C., Haverkort, B.R., Hermanns, H., Katoen, J.P.: Model-checking algorithms for continuous-time Markov chains. IEEE Trans. Software Eng. **29**(6), 524–541 (2003)
3. Baier, C., Katoen, J.P.: Principles of Model Checking. MIT Press, Cambridge (2008)
4. Bernardo, M., De Nicola, R., Loreti, M.: Revisiting trace and testing equivalences for nondeterministic and probabilistic processes. In: Birkedal, L. (ed.) FoSSaCS 2012. LNCS, vol. 7213, pp. 195–209. Springer, Heidelberg (2012). https://doi.org/10.1007/978-3-642-28729-9_13
5. Bernardo, M., De Nicola, R., Loreti, M.: Revisiting trace and testing equivalences for nondeterministic and probabilistic processes. In: LMCS, vol. 10, no. 1 (2014)

6. Deng, Y., Hennessy, M.: On the semantics of Markov automata. Inf. Comput. **222**, 139–168 (2013)
7. Eisentraut, C., Hermanns, H., Zhang, L.: Concurrency and composition in a stochastic world. In: Gastin, P., Laroussinie, F. (eds.) CONCUR 2010. LNCS, vol. 6269, pp. 21–39. Springer, Heidelberg (2010). https://doi.org/10.1007/978-3-642-15375-4_3
8. Eisentraut, C., Hermanns, H., Zhang, L.: On probabilistic automata in continuous time. In: LICS, pp. 342–351 (2010)
9. Hermanns, H.: Interactive Markov Chains: And the Quest for Quantified Quality. LNCS, vol. 2428. Springer, Heidelberg (2002). https://doi.org/10.1007/3-540-45804-2
10. Segala, R.: A compositional trace-based semantics for probabilistic automata. In: Lee, I., Smolka, S.A. (eds.) CONCUR 1995. LNCS, vol. 962, pp. 234–248. Springer, Heidelberg (1995). https://doi.org/10.1007/3-540-60218-6_17
11. Segala, R.: Modelling and verification of randomized distributed real time systems. Ph.D. thesis, MIT (1995)
12. Sharma, A.: Trace relations and logical preservation for Markov automata. In: Jansen, D.N., Prabhakar, P. (eds.) FORMATS 2018. LNCS, vol. 11022, pp. 162–178. Springer, Cham (2018). https://doi.org/10.1007/978-3-030-00151-3_10
13. Sharma, A.: Stuttering for Markov automata. In: TASE. IEEE Computer Society (2019)
14. Song, L., Zhang, L., Godskesen, J.C.: Late weak bisimulation for Markov automata. CoRR abs/1202.4116 (2012)
15. Wolf, V., Baier, C., Majster-Cederbaum, M.E.: Trace machines for observing continuous-time Markov chains. ENTCS **153**(2), 259–277 (2006)

Author Index

Printed in the United States
By Bookmasters